Springer Series in Biophysics 11

Boris Martinac

Editor

Sensing with Ion Channels

 Springer

Professor Boris Martinac
Foundation Chair of Biophysics School
  of Biomedical Sciences
University of Queensland
Brisbane QLD 4072
Australia
b.martinac@uq.edu.au

ISBN: 978-3-540-72683-8   e-ISBN: 978-3-540-72739-2

Springer Series in Biophysics ISSN: 0932-2353

Library of Congress Control Number: 2007932178

Printed on acid-free paper

9 8 7 6 5 4 3 2 1

springer.com

Preface

Life as we know it would not exist without the ability of living organisms to sense the surrounding environment and respond to changes within it. All living cells are able to detect and translate environmental stimuli into biologically meaningful signals. Without mechanisms to receive sensations of touch, hearing, sight, taste, smell or pain, the outside world would cease to exist for any living being that has to rely on these mechanisms for its survival. The importance of sensory input for the existence of life thus seems obvious, justifying the effort made to understand its molecular origins.

Living cells are surrounded by a plasma membrane that forms a boundary between the cell interior and the external physical world. As a consequence, the cellular plasma membrane presents a major target for environmental stimuli acting upon a living cell. The membrane contains protein molecules that confer various functions on it. Many such membrane protein molecules are ion channels, which function as molecular sensors of physical and chemical stimuli and convert these stimuli into biological signals vital for the existence of every living organism, be it microbe, plant, animal or human being. As molecular transducers of mechanical, electrical, chemical, thermal or electromagnetic (light) stimuli, ion channels contribute to changes in electrical, chemical or osmotic activity within cells by gating between the two basic conformations in which they exist – open and closed. By opening and closing, ion channels regulate the transport of ions (in some cases also other solutes), which can thus enter or exit living cells and affect their activity.

The last two decades have been exceptionally exciting for research in the field of ion channels. Much progress has resulted from using a multidisciplinary approach to elucidate the structure and function of ion channels and their role in various aspects of cell physiology, including sensory physiology. Molecular biology and genetics have provided the primary structures of a very large number of ion channel proteins and have helped identify their contribution to various cellular functions. The patch clamp technique has provided the means to study the functional properties of single ion channels with unprecedented precision. X-ray and electron crystallography have provided structural snapshots of a number of ion channel molecules at near atomic resolution, whereas magnetic resonance spectroscopy and fluorescence spectroscopy have provided means to access the dynamics of these molecules. Using the structural and functional information obtained by these experimental techniques, computer-assisted molecular modelling has brought ion channels to life by visualizing the molecular events that shape their function. In a nutshell, the multidisciplinary approach to the study of ion channels has yielded an unprecedented wealth of new knowledge that, in the not too distant future, can be expected to lead to a thorough understanding of the molecular mechanisms underlying sensory transduction in living cells.

The aim of this volume is to illustrate the broad spectrum of sensory transduction mechanisms found in diverse types of living cells, and to provide a comprehensive account of the molecular events on which these mechanisms are based. The chapters in this volume focus on ion channels as key molecules enabling biological cells to sense and process the physical and chemical stimuli they are exposed to in their environment. The first two chapters (Chap. 1 by C. Kung, X.-L. Zhou, Z.-W. Su, W.J. Heynes, S.H. Loukin and Y. Saimi; and Chap. 2 by P. Blount, I. Isla and Y. Li) describe mechanosensitive channels designed to detect osmotic forces acting upon cell membranes of eukaryotic and prokaryotic microbes. Chapter 3 (by T. Furuichi, T. Kawano, H. Tatsumi and M. Sokabe) illustrates the variety of ion channels that play a role in the response of plants to environmental stimuli. In Chapter 4, F. Lang, E. Gulbins, I. Szabo, A. Vereninov and Stephan Huber summarize how sensing of cellular volume contributes to the regulation of cell proliferation and apoptosis. The following four chapters (Chap. 5 by W. Liedtke; Chap. 6 by K. Talavera, T. Voets and B. Nilius; Chap. 7 by O. Hamill and R. Maroto; and Chap. 8 by A. Patel, P. Delmas and Eric Honoré) focus on different members of the superfamily of TRP (transient receptor potential) ion channels, which have recently emerged as key players in the physiology of sensory transduction in animals and humans. Chapter 9 (by P. Barry, W. Qu and A. Moorhouse) concentrates on the biophysical aspects of cyclic nucleotide gated (CNG) channels and includes a brief overview of the physiological function of CNG channels in both olfaction and phototransduction. Chapter 10 (by A. L. Brown, D. Ramot and M. Goodman) summarizes what is known about the ion channels that mediate sensation in the roundworm *Caenorhabditis elegans*, which has served as an excellent model organism in many areas of biological research. In Chapter 11, S. Kellenberger describes our current understanding of the physiological functions, and the mechanisms, of ion permeation, gating and regulation of epithelial sodium and acid-sensing ion channels. The focus of the final two chapters (Chap. 12 by C. Kennedy and Chap. 13 by T. Yasuda and D. Adams) is on ion channels that play a role in sensation of pain.

In summary, this volume provides a comprehensive overview of the progress that has been made towards understanding the molecular basis of a great variety of sensory transduction mechanisms found in living cells. Its main purpose is to serve as a reference to ion channel specialists and as a source of new information to non specialists who wish to learn about the structural and functional diversity of ion channels and their role in sensory physiology.

Brisbane, March 2007   Boris Martinac

Contents

Contributors

David J. Adams
School of Biomedical Sciences, The University of Queensland, Brisbane,
Queensland 4072 Australia, dadams@uq.edu.au

Peter H. Barry
Department of Physiology and Pharmacology, School of Medical Sciences, The
University of New South Wales, UNSW Sydney 2052, Australia,
P.Barry@unsw.edu.au

Paul Blount
Department of Physiology, University of Texas-Southwestern Medical Center,
5323 Harry Hines Blvd., Dallas, TX 75390-9040, USA,
Paul.Blount@utsouthwestern.edu

Austin L. Brown
Program in Biophysics, Stanford University School of Medicine, B-111 Beckman
Center, 279 Campus Drive, Stanford, CA 94305-5345, USA

Patrick Delmas
Faculté de Médecine, IFR Jean Roche, Laboratoire de Neurophysiologie
Cellulaire, CNRS-UMR 6150, Bd Pierre Dramard, 13916 Marseille Cedex 20,
France

Takuya Furuichi
Graduate School of Medicine, Nagoya University, 65 Tsurumai, Nagoya
466-8550, Japan

Miriam B. Goodman
Department of Molecular and Cellular Physiology, Stanford University School of
Medicine, B-111 Beckman Center, 279 Campus Drive, Stanford, CA 94305-5345,
USA, mbgoodman@stanford.edu

Erich Gulbins
Department of Physiology, University of Essen, Essen, Germany

Owen P. Hamill
Department of Neuroscience and Cell Biology, University of Texas Medical Branch, Galveston, TX 77555, USA, ohamill@utmb.edu

W. John Haynes
Laboratory of Molecular Biology, University of Wisconsin, Madison, WI 53706, USA

Honoré, Eric
Institut de Pharmacologie Moléculaire et Cellulaire, CNRS-UMR 6097, 660 Route des Lucioles, 06560 Valbonne, France, honore@ipmc.cnrs.fr

Stephan M. Huber
Department of Physiology, University of Tübingen, Gmelinstraae 5, 72076 Tübingen, Germany

Irene Iscla
Department of Physiology, University of Texas-Southwestern Medical Center, 5323 Harry Hines Blvd., Dallas, TX 75390-9040, USA

Tomonori Kawano
Graduate School of Environmental Engineering, the University of Kitakyushu, 1-1 Hibikino, Wakamatsuku, Kitakyushu 808-0135, Japan

Stephan Kellenberger
Department of Pharmacology and Toxicology, University of Lausanne, Rue du Bugnon 27, 1005 Lausanne, Switzerland, Stephan.Kellenberger@unil.ch

Charles Kennedy
Strathclyde Institute of Pharmacy and Biomedical Sciences, University of Strathclyde, John Arbuthnott Building, 27 Taylor Street, Glasgow G4 ONR, UK, c.kennedy@strath.ac.uk

Ching Kung
Laboratory of Molecular Biology and Department of Genetics, University of Wisconsin, Madison, WI 53706, USA, ckung@wisc.edu

Florian Lang
Department of Physiology, University of Tübingen, Gmelinstrabe 5, 72076 Tübingen, Germany, florian.lang@uni-tuebingen.de

Yuezhou Li
Department of Physiology, University of Texas-Southwestern Medical Center,
5323 Harry Hines Blvd., Dallas, TX 75390-9040, USA

Wolfgang Liedtke
Center for Translational Neuroscience, Duke University, Durham, NC 27710,
USA, wolfgang@neuro.duke.edu

Sephan H. Loukin
Laboratory of Molecular Biology, University of Wisconsin, Madison, WI 53706,
USA

Andrew J. Moorhouse
Department of Physiology and Pharmacology, School of Medical Sciences, The
University of New South Wales, UNSW Sydney 2052, Australia

Rosario Maroto
Department of Neuroscience and Cell Biology, University of Texas Medical
Branch, Galveston, TX 77555, USA

Bernd Nilius
Laboratorium voor Fysiologie, Campus Gasthuisberg, KULeuven, 3000 Leuven,
Belgium

Amanda Patel
Institut de Pharmacologie Moléculaire et Cellulaire, CNRS-UMR 6097,
660 Route des Lucioles, 06560 Valbonne, France

Wei Qu
Department of Physiology and Pharmacology, School of Medical Sciences,
The University of New South Wales, UNSW Sydney 2052, Australia

Daniel Ramot
Program in Neuroscience, Stanford University School of Medicine, B-111
Beckman Center, 279 Campus Drive, Stanford, CA 94305-5345, USA

Yoshiro Saimi
Laboratory of Molecular Biology, University of Wisconsin, Madison, WI 53706,
USA

Masahiro Sokabe
Graduate School of Medicine, Nagoya University, 65 Tsurumai, Nagoya 466-8550,
Japan

Zhen-Wei Su
Laboratory of Molecular Biology, University of Wisconsin, Madison, WI 53706,
USA

Ildiko Szabo
Department of Biology, University of Padua, Italy

Karel Talavera
Laboratorium voor Fysiologie, Campus Gasthuisberg, KULeuven, 3000 Leuven,
Belgium, Karel.talavera@med.kuleuven.be

Hitoshi Tatsumi
Department of Physiology, Graduate School of Medicine, Nagoya University,
65 Tsurumai, Nagoya 466-8550, Japan, tatsumi@med.nagoya-u.ac.jp

Alexey Vereninov
Institute of Cytology, Russian Academy of Sciences, St. Petersburg,
Russia

Thomas Voets
Laboratorium voor Fysiologie, Campus Gasthuisberg, KULeuven, 3000 Leuven,
Belgium

Takahiro Yasuda
School of Biomedical Sciences, The University of Queensland, Brisbane,
Queensland 4072 Australia

Xin-Liang Zhou
Laboratory of Molecular Biology, University of Wisconsin, Madison, WI 53706,
USA

List of Abbreviations

AA	arachidonic acid
ABA	abscisic acid
AChR	acetylcholine receptor
ADH	anti-diuretic hormone
ADPKD	autosomal dominate polycystic kidney disease
AID	α_1 subunit interaction domain
AMFE	anomalous mole fraction effects
AP	Action potential
2-APB	2-aminoethoxydiphenyl borate
AQP-5	aquaporin 5
ARC	AA-activated ROC
ASDN	aldosterone-sensitive distal nephron
ASIC	acid-sensing ion channel
ASO	antisense oligonucleotides
ATP	adenosine 5'-triphosphate
BEL	bromoenol lactone
BL	blue light
Ca^{2+}	calcium ion
$[Ca^{2+}]_c$	cytosolic free Ca^{2+} concentration
CaM	calmodulin
cAMP	3', 5'-cyclic monophosphate
CC	conformational coupling
CDPK	calcium-dependent protein kinase
CFA	complete Freund's adjuvant
CFTR	cystic fibrosis transmembrane resistance
cGMP	3', 5'-cyclic monophosphate
CGRP	calcitonin gene-related peptide
CHO	Chinese hamster ovary
CI	Ca^{2+} inactivation

CICR	Ca^{2+}-induced Ca^{2+} release
CIF	Ca^{2+} influx factor
CNBD	cyclic nucleotide-binding domain
CNG	Cyclic nucleotide-gated
CNGC	cyclic nucleotide-gated cation channel
CNS	central nervous system
CSK	cytoskeleton
CuP	copper phenanthroline
DAG	diacylglycerol
dAVP	desmopressin (1-desamino-8-D-arginine vasopressin)
DEG	degenerin
DEG/ENaC	degenerin/epithelial Na^+ channel
DHPR	dihydropyridine receptors
DMD	Duchenne muscular dystrophy
DPI	diphenyleneiodonium
DRG	dorsal root ganglia
E–C	excitation–contraction
ECM	extracellular matrix
5′6′-EET	5′, 6′-epoxyeicosatrienoic acid
EIPA	ethylisopropylamiloride
ENaC	epithelial Na^+ channel
EPR	electron paramagnetic resonance
EPSP	excitatory postsynaptic potentials
ER	endoplasmic reticulum
FRAP	fluoride-resistant acid phosphatase
FSGS	focal segmental glomerulosclerosis
GABA	gamma -aminobutyric acid
GBM	glomerular basement membrane
GFP	green fluorescent protein
GHK	Goldman-Hodgkin-Katz
GpBp	G$\beta\gamma$ protein-binding pocket
GPC	glycerophosphorylcholine
GPCR	G-protein coupled receptor
GPS	G protein coupled receptor proteolytic site
H^+	proton
HEK	human embryonic kidney
HEK293T	human embryonic kidney cell line 293, transformed by large-T antigen
20-HETE	20-hydroxyeicosatetraenoic acid
HO$^{\cdot}$	hydroxyl radical
HP	holding potential

HR	hypersensitive response
HVA	high-voltage activated
i.v.	intravenous
IAV	inactive (drosophila melanogaster mutant line of flies)
I_{CRAC}	Ca^{2+} release-activated currents
iGluR	ionotropic glutamate receptor
iGluRs	ionotropic glutamate receptors
$InsP_3$	inositol 1,4,5-trisphosphate
IP_3R	inositol 1,4,5-trisphosphate receptor
$iPLA_2$	Ca^{2+}-independent phospholipase A2
I–V	current–voltage
LPL	lysophospholipid
L–R	left–right
LVA	low-voltage activated
MeT	mechano-electrical transduction
MOC	mechano-operated channel
MS	mechanosensitive
MSCC	mechanosensitive nonselective cation channel
MscCa	mechanosensitive Ca^{2+}-permeable cation channel
MT	Mechanotransduction
MTS	methanethiosulfonate
MTX	maitotoxin
NAN	nanchung (drosophila melanogaster mutant line of flies; *Korean*: deaf)
NH_4^+	ammonium ion
NHERF	Na^+/H^+ exchange regulatory factor
NMDA	N-methyl-D-aspartate
NO·	nitric oxide
NO_{3^-}	nitrate
NompC	no mechano-receptor potential mutant C
OAG	1-oleoyl-2-acetylglycerol
OCR-2	C. elegans mutant OSM-9 like, capsaicin-receptor related gene 2
ORL1	opioid receptor-like 1
OSM-9	Caenorhabditis elegans mutant with an osmotic avoidance phenotype, mutant line number 9
OSN	olfactory sensory neuron
OVLT	organum vasculosum laminae terminalis
PAF	platelet activating factor
PAR-2	proteinase-activated-receptor-2
PC	polycystin

4-α-PDD	4 alpha-phorbol 12,13- didecanoate
PDE	phosphodiesterase
PDZ	PSD-95/disc large protein/zona occludens 1
PIP_2	phosphatidylinositol 4,5-bisphosphate
PKC	protein kinase C
PKD1	polycystin 1
PKD2	polycystin kidney disease 2; polycystin 2
PLA_2	phospholipase A2
PLC	phospholipase C; periciliary liquid layer
PNS	peripheral nervous system
PP2	4-amino-5-(4-chlorophenyl)-7-(t-butyl) pyrazolo[3,4,d] pyrimidine
PUFA	Polyunsaturated fatty acid
ROC	receptor-operated channel
ROS	reactive oxygen species
RVD	regulatory cell volume decrease
RVD	regulatory volume decrease
RVI	regulatory cell volume increase
RyR1	ryanodine receptors
s.c.	subcutaneous
SA	salicylic acid
SDSL	site-directed spin labeling
SG model	Sukharev and Guy model
sGC	soluble guanylate cyclase
siRNA	small interfering RNA
SOC	store-operated channel
SR	sarcoplasmic reticulum
STIM	stromal interaction molecule
TEA	tetraethylammonium
TM	transmembrane
TMA	tetramethylammonium
TMD	Transmembrane domain
TPC1	two pore channel 1
TRP	transient-receptor-potential
TRPA	transient receptor potential ankyrin
TRPC	canonical transient receptor potential
TRPV	transient receptor potential vanilloid
VDCC	voltage-dependent Ca^{2+}-permeable channel
VGCCs	Voltage-gated calcium channels
VNO	vomeronasal organ
VOCC	voltage-operated Ca^{2+} channels
VP	variation potential

PLC	periciliary liquid layer
CNS	central nervous system
PNS	peripheral nervous system
DRG	dorsal root ganglion
DEG	degenerin

L

ATP	adenosine 5′-triphosphate
DRG	dorsal root ganglia
CGRP	calcitonin gene-related peptide
FRAP	fluoride-resistant acid phosphatase
ASO	antisense oligonucleotides
CFA	complete Freund's adjuvant
siRNA	small interfering RNA
CNS	central nervous system
s.c.	subcutaneous
i.v.	intravenous
VOCC	voltage-operated Ca^{2+} channels

M

VGCCs	Voltage-gated calcium channels
HVA	high-voltage activated
LVA	low-voltage activated
DRG	dorsal root ganglion
AID	α_1 subunit interaction domain
GpBp	G$\beta\gamma$ protein-binding pocket
ER	endoplasmic reticulum
I–V	current–voltage
CI	Ca^{2+} inactivation
HP	holding potential
CGRP	calcitonin-gene-related peptide
EPSP	excitatory postsynaptic potentials
NMDA	N-methyl-D-aspartate
HEK	human embryonic kidney
GABA	gamma -aminobutyric acid
ORL1	opioid receptor-like 1

Chapter 1
Microbial Senses and Ion Channels

Ching Kung(✉), Xin-Liang Zhou, Zhen-Wei Su, W. John Haynes, Sephan H. Loukin, and Yoshiro Saimi

Abstract The complexity of animals and plants is due largely to cellular arrangement. The structures and activities of macromolecules had, however, evolved in early microbes long before the appearance of this complexity. Among such molecules are those that sense light, heat, force, water, and ligands. Though historically and didactically associated with the nervous system, ion channels also have deep evolutionary roots. For example, force sensing with channels, which likely began as water sensing through membrane stretch generated by osmotic pressure, must be ancient and is universal in extant species. Extant microbial species, such as the model bacterium *Escherichia coli* and yeast *Saccharomyces cerevisiae*, are equipped with stretch-activated channels. The ion channel proteins MscL and MscS show clearly that these bacterial channels receive stretch forces from the lipid bilayer. TRPY1, the mechanosensitive channel in yeast, is being

305 R.M. Bock Laboratories, 1525 Linden Drive, University of Wisconsin Madison, WI 53706, USA, ckung@wisc.edu

B. Martinac (ed.), *Sensing with Ion Channels. Springer Series in Biophysics 11*
© 2008 Springer-Verlag Berlin Heidelberg

developed towards a similar basic understanding of channels of the TRP (transient-receptor-potential) superfamily. TRPY1 resides in the vacuolar membrane and releases Ca^{2+} from the vacuole to the cytoplasm upon hyperosmotic shock. Unlike in most TRP preparations from animals, the mechanosensitivity of TRPY1 can be examined directly under patch clamp in either whole-vacuole mode or excised patch mode. The combination of direct biophysical examination in vitro with powerful microbial genetics in vivo should complement the study of mechanosensations of complex animals and plants.

1.1 The Microbial World

The game of 20 questions teaches children that the world consists of animals, vegetables, and minerals. Children become adults who continue to ignore the bulk of the biological world – the microbes. The origin of this ignorance is deeper than childhood indoctrination: we came from animals that deal with predators, prey, parents, progeny, peers, and possessions, all about our own size. Although science has revealed invisible microbes, to most people they are but parasites, pathogens, and pests. Even for scientists, the hard-wired animal-plant-mineral illusion, like the geocentric illusion of sunrise and sunset, is hard to dispel in daily life.

1.1.1 Microbial Dominance

Our ignorance and bias notwithstanding, microbes reign supreme on this planet in diversity, in number, and in mass (Woese 1994). This is true in the past, the present, and, in all likelihood, also in the future. Earth was formed ~4.6×10^9 years ago. Life began ~3.9×10^9 years ago at high temperatures, with liquid water, and under an anoxic reducing atmosphere. The appearance of ancestral cyanobacteria (~2.9×10^9 years) gradually built up O_2 in the atmosphere, predating the appearance of modern eukaryotes (~1.5×10^9 years). Thus a large part of our planetary history (from ~3.9×10^9 years ago to 1.5×10^9 years ago) saw forms that we would now classified as bacteria, archaea, and the microbial ancestors of eukaryotes. There were no animals or plants, let alone the human species, which is only 10^5 years old. The microbial way of life has continued to be successful to the present. Even a casual look at the current tree of life reveals that the greatest diversities are among microbes (Woese 2000) (Fig. 1.1). Contrary to the notion of evolution being a progression, with new forms replacing old, plants and animals are add-ons pasted onto the microbial diversity. Today, besides the niches occupied by animals and plants, microbes continue to thrive deep underground, in arctic waters, in hydrothermal vents, in the "Dead Sea". They are also present on, and in, just about every animal and plant. [There are more that 500 kinds of bacteria in our oral cavity! (Becker et al. 2002)]. As to the future, this planet will not become sterile with yet

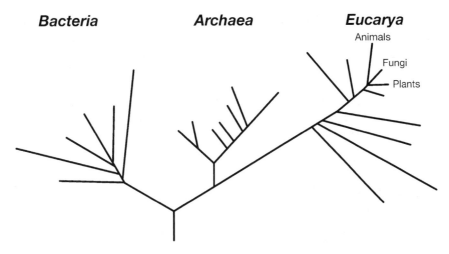

Fig. 1.1 The tree of life showing the relatedness of the three branches: *Bacteria*, *Archaea*, and *Eucarya*. Note that even among eukaryotes, multicellular macrobes are the minority (see Woese 2000)

another "wave of mass extinction", natural or man-made. Given the variety of ways in which microbes extract energy and the variety of niches they currently occupy, there is little doubt that microbes will survive such "disasters" and carry on for billions of years to come.

Another common misconception is that eukaryotes mean animals and plants. Though the visible animals and plants loom large in our mind, they are in fact a small part of the eukaryotic diversity (Embley and Martin 2006). Currently, taxonomists divide *Eukarya* into six clusters (Adl et al. 2005), one of which comprises both animals and fungi. The nondescript term "protists" in the common currency of scientific discourse in fact comprises the greatest variations Nature has devised for eukaryotes. The animalcentric, if not anthropocentric, view of physiology often overlooks this true diversity. For example, description and classification of transient-receptor-potential (TRP) channels usually deal only with those in mammals, with those of the fly and the worm thrown in as honorary guests. However, in reality, TRP channel genes are found in fungal genomes as well as those of ciliates, flagellates, slime molds, Trypanosome, *Leishmania*, etc., indicating an early origin (see Fig. 1.2 and Sect. 1.5.3 below).

1.1.2 Molecular Mechanisms Invented and Conserved in Microbes

We commonly speak of higher and lower organisms. Our self-appointed "higher" status can be defended only on the grounds of complexity. Complexity in biology has to do with the arrangement of cells and does not correlate with the plurality of

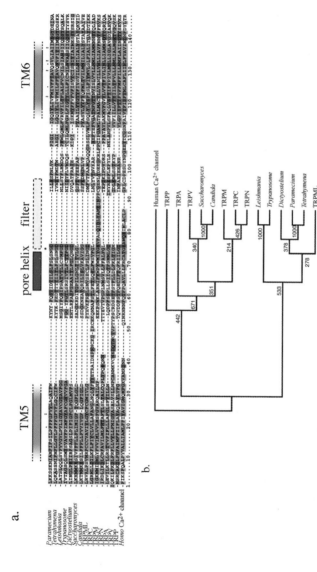

Fig. 1.2 An alignment and unrooted cladogram of the major family members of TRP (transient-receptor-potential) channels. **a** A Clustal W (Gonnet 250) alignment made using the program Clustal X. Several representative TRP genes found in various protists were aligned with a member of each major family of TRP channel along with a calcium channel (conserved in multicellular organisms) for comparison. The protist channels were found by BLAST searching genomic sequences currently available for each organism listed. The majority of these protist sequences were predicted by automated annotation procedures at the respective sequencing centers. The *Leishmania* sequence was recently described (Chenik et al. 2005). The color coding represents both a frequency of conservation and the chemical relatedness of residue side chains. **b** A single unrooted bootstrapped cladogram constructed using the neighbor joining method (Saitou and Nei 1987) drawn from the Clustal W (Gonnet 250) alignment (with all gapped sequence removed) shown above made using the program Clustal X. From a comparison of 1,000 possible trees, the numbers represent the number of trees in which the branches shown were present. The calcium channel was selected as the outgroup for the purpose of drawing this tree

molecular components. While the chimp and the mouse have the same number of genes as we do, the unicellular *Paramecium* actually has nearly twice that number (Aury et al. 2006)!

Not only do animals not have more types of molecules, their molecules are no more intricate than those of microbes. Students of modern biology are familiar with bacterial cytochromes, rhodopsins, ribosomes, Kreb's cycle, oxidative phosphorylation, photosynthesis, etc. Even the cytoskeleton, which is often cited as a hallmark of eukaryotes, is not exclusive. Distant homologs of actin, tubulin, etc., have been found to function in prokaryotes (Shih and Rothfield 2006). Ion channels – the focus of this volume – are clearly ancient and are almost ubiquitous among free-living microbes (see Sect. 1.3).

Of course, the universality of the structures and functions of different kinds of DNAs, RNAs, proteins, and lipids reflects their early evolutionary origin. Honed by selection among early cellular forms, they now continue to serve in all three branches of life. It therefore seems obvious that, if one is to study the basic physical and chemical working of these molecules, it makes little difference whether they are taken from a bacterium, a worm, or a human.

Unlike humans or worms, microbes can be cultivated in small spaces, in short time frames, and therefore with little expense. The cultures can be clonal, therefore having cells of the same genotype and phenotype. The streamlined genomes of most microbes obviate the need to deal with introns and other intervening sequences. Their smaller genomes make genetic and genomic exploration much simpler. The molecular tools collected through the last 50 years of the molecular biology revolution have made *Escherichia coli* and *Saccharomyces cerevisiae* convenient in vivo laboratories and factories. Moreover, there is little expressed ethical concern and therefore little risk of animal-rights objections when sacrificing billions of microbes. It is therefore no accident that much of the molecular insights into the workings of DNA, RNA, enzymes, and now ion channels, have come from investigation of microbial materials.

1.2 Microbial Senses

A large part of the bacteriology literature deals with microbial responses to various "environmental stresses", i.e., changes in temperature, hydration, pH, carbon or nitrogen source, and energy source. From this literature, such basic concepts as end-product feedback inhibition of enzymes, promoter regulation by transcription factors, and the concept of second messengers, such as cAMP, have been derived. Here, the sensors are often the enzymes and the promoters. In these cases, the routes from sensing to response do not involve ion channels, and are not further reviewed here.

Many bacteria and archaea are also capable of chemotaxis – the active seeking of attractants and avoiding of repellents. Here, the receptors are trimers of dimeric binding subunits that straddle the plasma membrane, and the binding signal is transmitted to flagella through a phosphorylation relay (Baker et al. 2006). While active locomotion brings animal behavior to mind, prokaryotic chemotaxis, as it is currently understood, does not employ ion channels. The roles, if any, of ion channels in the sensing and response to environmental changes differ in different microbes. In ciliates such as *Paramecium* and *Tetrahymena*, channels are known to generate receptor potentials and action potentials like those in the excitable membranes of multicellular animals (see Sect. 1.3.1). In yeast, a 36-pS channel of unknown composition in the plasma membrane responds to osmotic downshock, and a vacuolar TRP channel detects and responds to osmotic upshock (see Sect. 1.5.4). Prokaryotic mechanosensitive channels detect and respond to osmotic downshock (see Sect. 1.4, and Chap. 2 by Blount et al., this volume). In the great majority of the cases, however, the physiological roles of microbial ion channels are simply unknown (see Sect. 1.3.2).

1.3 Microbial Channels

1.3.1 The Study of Microbial Ion Channels

The study of ion channels originated from that of nerve and muscle. The excitable membranes of these tissues support action potentials best studied by electric means. Even today, optical methods of registering action potentials remain difficult and cumbersome. Much of our earlier understanding of bioelectrics followed advances in technology, e.g., from extracellular to intracellular recording, and from current clamp to voltage clamp. Throughout advances in this field, the electrode-to-cell size ratio limits the quality and the quantity of the information we extract from a preparation. This is why the giant squid axon and barnacle muscle were once popular and why the electrophysiology of *Caenorhbditis elegans* remains difficult, despite great advances in its neurogenetics. Given this historic limitation, it is no surprise that electrophysiology of microbes is underdeveloped.

The first microbe penetrated with a microelectrode was *Paramecium*. A paramecium may be a unicell, but it is a large one. Depending on the species, a paramecium is some 100–300 µm in length – visible to the naked eye, albeit as a speck. Under a low power microscope, it is obvious that paramecia transiently stop or back up briefly by reversing the direction of their ciliary beat. Ciliary reversals (avoidance reactions) occur spontaneously and can also be induced by a variety of stimuli. Intracellular recordings showed that each such avoidance reactions is correlated with an action potential (Eckert 1972). Voltage-clamp experiments were used to sort out the Ca^{2+}, Na^+, and K^+ currents that constitute various action potentials. Paramecium is also a model organism for genetics studies; *Paramecium tetraurelia*, in particular, has a convenient system for transmission genetic analyses. Accordingly, it was possible to isolate behavioral mutants, whose avoidance reactions to various stimuli were missing or altered, and to sort out the corresponding electrophysiological changes using intracellular electrodes (Saimi and Kung 1987).

With the advent of molecular genetics, the genes of some of the corresponding mutations have been cloned (Saimi and Kung 2002). The entire genome of *P. tetraurelia* has recently been sequenced and many genes corresponding to Ca^{2+}-, Na^+-, and K^+-specific channels as well as other cationic or anionic channels can be recognized (Aury et al. 2006).

The study of the two major microbial experimental models, *E. coli* and *S. cerevisiae*, as well as other microbes, has advanced largely without direct measurement of transmembrane voltage and current. Chemosmotic hypotheses notwithstanding, the bulk of contemporary microbiology literature is about genes and their expression. However, vertebrate physiology has finally met up with microbiology at two technical fronts. First, the advent of patch clamp technology (Hamill et al. 1981) removed the constraint of having to work electrically only with large cells. This made possible the direct examination of ion conductances in the membranes of microbes. For example, in the mid-1980s, our laboratory made a patch-clamp foray into the membranes of *E. coli* (Martinac et al. 1987; see also Chap. 2 by Blount et al., this volume) and *S. cerevisiae* (Gustin et al. 1986), resulting in the discovery of the activities of MscL, MscS, and TOK1 (Zhou et al. 1995). Unitary conductance have since also been recorded from *Dictyostelium* (slime mold), *Chlamydomonas* (a green flagellate), *Paramecium* (a ciliate), *Neurospora* (bread mold), *Uromyces* (a parasitic bean rust fungus), *Schizosaccharomyces* (fission yeast), *Streptomyces* (Gram-negative bacterium), *Bacillus* (Gram-positive bacterium) and *Haloferax* (an archaeon) (Saimi et al. 1999; Palmer et al. 2004). Second, the large and rapidly increasing genome sequence information from microbes has revealed genes similar to those encoding animal ion channels. This information explosion allows broad-scale comparisons. For example, the sequences of 270 bacterial and archaeal genomes were analyzed for the presence and classification of their K^+ channels (Kuo et al. 2005a). Various prokaryotic ion channels recognized through their genes have recently been brought into the limelight, because crystal structures of these channels at atomic resolution have yielded unprecedented insights into the workings of this class of channel protein. Such works, culminating in the 2003 Nobel Prize to Rod MacKinnon, need not be reviewed here.

1.3.2 The Lack of Functional Understanding of Microbial Channels

The gulf between microbiology and neurobiology remains deep. Despite the contribution of the crystallized prokaryotic channels to our profound understanding of channel structures and mechanisms, and the widespread presence of channel genes in microbial genomes, most microbiologists are unaware of and remain unconcerned with ion channels. At the same time, most neurobiologists use bacteria as tools and factories, but are unconcerned with the role of channels in the physiology of the bacteria themselves. Therefore, there is currently a vacuum in the understanding of what most microbial channels do for the microbes themselves (Kung and Blount 2004; Kuo et al. 2005a). An exception is the finding that the anion channel/exchanger functions as an electrical shunt to maintain internal electroneutrality so as to sustain a virtual proton pump in the face of an extremely acidic environment encountered by $E.$ $coli$ (Iyer et al. 2002). As to the roles of cation-specific channels in prokaryotes, little is known about their biological functions (Kung and Blount 2004; Kuo et al. 2005a). Experiments with wild-type and gain-of-function mutants of Kch, the K^+ channel of $E.$ $coli$, indicate that it likely serves to regulate membrane potential rather than uptake of bulk K^+ (Kuo et al. 2003), but the circumstances in which this regulation becomes necessary have not been defined. As briefly reviewed above, the functions of the Ca^{2+} and K^+ channels of ciliates are understood using concepts developed in animal biology. However, the roles of most ion channels in fungi and protists remain to be elucidated. These roles should be an intellectually rich field to explore, since fungi and protists are highly diverse and occupy widely different niches. The structures and biophysical properties of their channels exhibit variations not seen in plants and animals. For example, yeast has an eight-transmembrane helices (8-TM), two pore-domain K^+ channel subunit of an $S_1 S_2 S_3 S_4 S_5 P_1 S_6 S_7 P_2 S_8$ arrangement (Ketchum et al. 1995; Zhou et al. 1995), and the $Paramecium$ genome reveals a 12-TM subunit with an $S_1 S_2 S_3 S_4 S_5 P_1 S_6 S_7 S_8 S_9 S_{10} S_{11} P_2 S_{12}$ arrangement (Kuo 2005a, 2005b). The yeast K^+ channel is particularly interesting; it is gated by the total K^+ electrochemical gradient instead of voltage and this gate has been thoroughly dissected by genetic and biophysical means (Loukin and Saimi 2002). In the two key experimental models – $E.$ $coli$ and $S.$ $cerevisiae$ – gain-of-activity mutants have been isolated, the expression of which hampers growth (Ou et al. 1998; Loukin et al. 1997). However, the loss of K^+-channel activities, as in the case of knock-out mutants, led to no discernable phenotype in the laboratory in repeated extensive searches (W.J. Haynes, S.H. Loukin, Y. Saimi and C. Kung, unpublished results). The current inability to discern a laboratory phenotype cannot be taken as evidence of the frivolity of Nature, however. The widespread presence of channel genes in the streamlined genomes of these microbes testifies to the selective advantages in the wild of having these channels. It seems likely that these channels function under conditions not yet simulated in the laboratory, or that these functions cannot easily be converted into difference in growth rates or survival that are commonly monitored as laboratory microbiological phenotype. Although the physical and chemical principles of channel activities should be universal, the biological roles of microbial ion channels will likely be broader than the roles played by such channels in the nervous system, although the latter are best known and well taught.

1.4 Prokaryotic Mechanosensitive Channels

The first patch-clamp survey of the $E.$ $coli$ membrane revealed large conductance channels that can be activated by pipette suction (Martinac et al. 1987). Fortuitously, these channels can be extracted and reconstituted into lipid bilayers while retaining their ion conductance and mechanosensitivity (Sukharev et al. 1997). Using patch clamp to follow the channel activity in various column fractions, the protein of one such channel, MscL, was isolated and its gene cloned (Sukharev et al. 1994). A reasonable hypothesis is that MscL serves as an emergency valve to release solute upon osmotic downshock. After cloning of a second channel (MscS) by Booth and co-workers, the hypothesis was proven by showing that the *mscL* Δ *mscS* Δ double mutant lyses upon medium dilution (Levina et al. 1999). The crystal structures of both MscL and MscS have been solved (Chang et al. 1998; Bass et al. 2002). Building on genetic, biochemical, biophysical and structural understanding, these bacterial channels are currently important models with which to analyze mechanosensitivity at the molecular level. See Chapter 2 by Blount et al. (this volume) for a detailed review and the current status of MscL and MscS research.

1.5   Mechanosensitive Channels of Unicellular Eukaryotes

Mechanical impact at the anterior of a paramecium elicits a cation-based receptor potential, which can, in turn, elicit a Ca^{2+}-based action potential; impact at its posterior elicits a K^+-based hyperpolarization (Eckert 1972). However, the genes and proteins behind these receptor potentials are yet to be identified. While physical impact can be important to larger unicells, the most fundamental force that all cells have to contend with is osmotic force. Thus, when the 36-pS mechanosensitive conductance on the plasma membrane of budding yeast was discovered, it was assumed to function in osmotic defense (Gustin et al. 1988), much like MscL and MscS function in *E. coli*. Unfortunately, the molecular identity of this conductance remains obscure to date. The proposal that Mid1 was that channel protein (Kanzaki et al. 1999) was not substantiated, since *mid1*Δ yeast retains 36-pS mechanosensitive conductance (X.-L. Zhou, C.P. Palmer, and C. Kung, unpublished results).

TRPY1 (YVC1) and its immediate relatives are the only bona fide mechanosensitive channels in eukaryotic microbes whose genes, proteins, and macroscopic and unitary conductance are known. Because this channel is found in budding yeast, one can bring to bear on its analysis molecular genetic tools not yet available to animal channel research. Research in TRPY1 will therefore likely complement that on the animal TRP channels. A more extensive description on this channel is therefore provided below; a recap of animal TRP research is also given for contrast (Saimi et al. 2007).

1.5.1   A Brief History of TRP Channel Studies

The bulk of current research, mostly on mammalian TRPs, is the derivative of some ten different genetic prospecting adventures. With no known sequence targets to start with, these projects independently arrived at different TRP channels. The term TRP – transient receptor potential – describes the electroretinographic phenotype of a near-blind mutant *Drosophila* isolated in the Pak laboratory in 1975 (Minke et al. 1975); the corresponding gene was cloned by Montell et al. in 1985 in the Rubin laboratory (Montell et al. 1985). This TRP channel is the founding member of TRPC, (C for canonical) and is the crux of phototransduction in insects, although how it is activated in vivo remains unclear (Minke and Parnas 2006). In 1997, TRP channels were "rediscovered" in two different contexts. In one, Julius and co-workers used expression cloning to search for and find a heat/pain receptor, by using a heat surrogate (the pepper essence capsaicin) as a probe. This vanilloid receptor, VR1, turned out to have clear homology to *Drosophila* TRP and is now called TRPV1 (Caterina et al. 1997). In the other, the Bargmann Laboratory isolated and analyzed mutant worms (*C. elegans*) defective in their avoidance of 4 M fructose. Position cloning led to identification of a gene, *osm-9*, which is homologous to TRPV1 (Colbert et al. 1997). More recently, expression cloning using a surrogate of cold (menthol) revealed its receptor, now a TRPM (McKemy et al. 2002). In the fly, mutations that cause defects in balance and touch response led to *NOMPC*, now a TRPN (Walker et al. 2000); those insensitive to pain to *PAINLESS*, now a TRPA (Tracey et al. 2003). In the worm, mutations causing a defect in the males' ability to locate the vulvas of hermaphrodites were traced to LOV-1, a homolog of PKD1, which forms channels by associating with PKD2, now TRPP (Barr and Sternberg 1999). Cloning genes of heritable diseases is the medical equivalent of phenotype-to-gene forward genetics. Thus polycystic kidney disease was traced to PKD1 and PKD2, now TRPPs (Hughes et al. 1995; Mochizuki et al. 1996). Likewise, mucolipisosis type IV was traced to MCOLN1, now TRPML (Sun et al. 2000). Thus, each founding member of the TRP subfamily, TRPC, TRPV, TRPN, TRPP, TRPM, or TRPML, was independently discovered by forward genetics. [TRPA was found almost simultaneously by the identification of *PAINLESS* (Tracey et al. 2003) and through candidate sequence homology (Story et al. 2003)]. The convergence of these multiple original studies onto the same superfamily of ion channels endorses the view that TRPs are central to many aspects of sensory biology.

Once these molecular targets are found, their sequence homologs can be recognized and used in further research. Commonly, mammalian homologs are heterologously expressed in oocytes or cultured cells and examined biophysically or biochemically. Knock-out mice are also generated to examine possible phenotypes. These studies are generically referred to as "reverse genetics", and constitute the bulk of current research in this field, as reviewed in the chapters by Hamill and Moroto (Chap. 7), Liedtke (Chap. 5), and Talavera et al. (Chap. 6) in this volume.

1.5.2 *Mechanosensitivity of TRP Channels*

Some members of each subfamily of animal TRP channels (TRPC, TRPV, TRPN, TRPP, TRPM, or TRPML) have been associated with mechanosensitivity. The evidence for this association varies greatly from case to case. At the organism level, evidence comes from mutant behavioral phenotypes such as deafness (Gong et al. 2004; Kim et al. 2003), touch-blind (Walker et al. 2000), osmotactic failure (Colbert et al. 1997), drinking behavior (Liedtke and Friedman 2003), bladder malfunction

(Birder et al. 2002), etc. At the cellular and tissue level, circumstantial evidence includes the presence of the TRP proteins or their mRNAs being in the expected places or at the expected developmental timepoint (Corey et al. 2004). At the molecular level, evidence most commonly comes from experimentation through heterologous expression. Here, mechanosensitivity is commonly indicated by osmotic-downshock-induced entry of Ca^{2+} (monitored with a dye) into cultured cells expressing a foreign TRP transgene (Kim et al. 2003; Gong et al. 2004; Liedtke et al. 2000; Strotmann et al. 2000).

Direct electrophysiological evidence for mechanosensitivities of animal TRP channels is rare. Under a current clamp, TRPV1 has been found to correlate with hypotonically induced spikes in isolated magnocellular neurosecretory cells (Naeini et al. 2006). Under patch clamp, in a cell-attached mode, heterologously expressed TRPV4 [previously OTRPC4 (Strotmann et al. 2000) and VR-OAC (Liedtke et al. 2000)] and TRPV2 (Muraki et al. 2003) were shown to be activated by hypotonicity. A ~30 pS-stretch-activated conductance native to HeLa cells can apparently be abolished with a small interfering RNA targeted against TRPM7 (Numata et al. 2006). A TRPC1-rich detergent-solubilized fraction of frog-oocyte membrane is found to correlate with unitary conductances that are activated by direct suction exerted on the bilayer patch. The same study showed that human *TRPC1*, expressed in oocytes, correlates with a tenfold increase in stretch-activated current (Maroto et al. 2005).

Part of the difficulty in the analysis of animal TRP channels is that they are often located in specialized cells and strategically located even within those cells. For example, the transducing channels for hearing are located near the tips of stereocilia of vertebrate hair cells (Corey et al. 2004) or the sensory cilia of insect chordotonal organ (Kim et al. 2003; Gong et al. 2004). TRPP is located in the primary cilia of renal epithelial cells (Nauli et al. 2003), TRPML in intracellular endosomes and lysosomes (Di Palma et al. 2002). Others are in the compound eyes, taste buds, and Merkel cells, Meissner corpuscle etc. These locations are currently nearly inaccessible to the patch clamp pipette. Since these TRP channels cannot be studied in situ, they are expressed heterologously in arenas such as oocytes or culture cells and examined therein. Results from heterologous experiments may include artifacts such as contributions (or the lack of such contribution) from host subunits or host enzyme modifications. Indirect experimentation and circumstantial evidence can be misleading. The association of TRPA1 with mechano-transduction conductance in mammalian hair cells (Corey et al. 2004) and questions surrounding this association is a case in point (Bautista et al. 2006; Kwan et al. 2006). By contrast, yeast TRPY1 is among the few channels that can be directly patch-clamped and examined for mechanosensitivity in its natural location (the vacuolar membrane; see below).

1.5.3 Distribution of TRPs and their Unknown Origins

The classification of TRPs iterated in most review papers is by primary-sequence comparison and not by biophysical characteristics or biological function (see below). Primary sequence cannot predict confidently any tertiary or quaternary structures of proteins using current bioinformatics. Thus, without a crystal structure, the commonly cited model of a TRP as a tetramer with a funnel-like center fitted with a filter is assumed partly by analogy to the known crystal structure of K^+ channels (Doyle et al. 1998; Jiang et al. 2002), based on the belief that these cation channels are distantly related (Yu and Catterall 2004) and thus should have similar structures. However, primary sequence can predict secondary structures and general topology with some confidence. The sequence of *TRP* genes predicts products with six transmembrane α helices (TMs) with extensive N- and C-terminal domains in the cytoplasm. These cytoplasmic domains contain recognizable regions with proposed (e.g. ankyrin repeats, calmodulin-binding sites, etc.) or unknown ("TRP box", "TRPM homology" etc.) functions that sort the members found into seven subfamilies (TRPC, TRPV, TRPA, TRPN, TRPM, TRPP, and TRPML). The resemblances between these subfamilies are limited. Most similarities are found in the sequence from the predicted TM5 to slightly beyond the C-terminus of the predicted TM6, a region that comprises the presumed filter and gate (Fig. 1.2).

Using the above key sequence (TM5 through TM6) as the criterion, searches in the existing databases recognize TRP-channel genes without ambiguity in the genomes of *Paramecium* and *Tetrahymena* (both ciliates), *Dictyostelium* (cellular slime mold), *Trypanosoma* (an agent of African sleeping sickness) and *Leishmania* (leishmainasis; W.J. Haynes, unpublished results; Saimi et al. 2007). Fragments of similarities can also be found in the genomes of *Chlamydomonas* (a green flagellate), *Plasmodium* (malaria), and *Thalassiosira* (diatom) etc., although full-length TRP-channel genes have not been recognized or assembled from these genomes due to technical difficulties. To date, no experimental work has been reported on these putative TRP homologs in protists.

The same search criterion revealed a TRP-channel gene in the genome of the budding yeast *Saccharomyces cerevisiae*, which has been experimentally studied at length (see below; Palmer et al. 2001; Zhou et al. 2003). By the criterion of sequence similarity to TM5-6, the yeast channel TRPY1 is more similar to animal TRPVs than animal TRPA, TRPC, TRPM, TRPP, TRPN, TRPML are to animal TRPVs. TRPY has homologs in some 30 different fungal genomes. An additional channel gene in the fission yeast *Schizosaccharomyces pombe* has similarity to that of *Drosophila* TRPP (Palmer et al. 2005).

The putative TRP channels in fungal and protist genomes usually do not bear the cytoplasmic features (ankyrin, "TRP box" etc.) used to distinguish the animal TRP subtypes. This makes it difficult to fit these channels into the official, but animal-centric, classification system (Montell et al. 2002; Clapham et al. 2003). BLAST searches using the vertebrate TM5-to-TM6 sequences have protist TRPs aligned with a greater bit score to TRPML, while fungal sequences align with other TRP subtypes (W.J. Haynes, unpublished). The cladogram (Fig. 1.2b) drawn from global alignment (Fig. 1.2a) also shows this same tendency for clustering with different TRP subtypes. Whether this clustering is evolutionarily meaningful cannot be asserted at the moment since available microbial genome sequences are limited and include many reduced genomes of parasites.

The sequence criteria commonly used have so far failed to identify TRP candidates among bacteria and archaea. The deep relationship among the three domains of life (*Bacteria, Archaea,* and *Eukarya*) remains unclear at present (Embley and Martin 2006). The commingled gene pool of primordial cell communities (Embley and Martin 2006) presumably encoded the first detectors of force and heat, from which the first TRPs were derived. In any event, TRP channels are not likely to appear de novo in their present forms without any evolutionary predecessors, which may or may not have left discernable footprints.

1.5.4 TRP Channel of Budding Yeast

Long before the identification of the *TRPY1* gene, Wada et al. (1987) first described a ~300-pS conductance observed with a planar lipid bilayer, into which a vacuolar-membrane fraction of yeast had been reconstituted. Others have observed a similar conductance by patch-clamping the vacuolar membrane after releasing the vacuoles from yeast spheroplasts (Batiza et al. 1996; Minorsky et al. 1989; Saimi et al. 1992) (Fig. 1.3a). This conductance rectifies inwardly, i.e., from the vacuole into the cytoplasm (Fig. 1.3b). It is cation selective, $PNa^+ = PK^+ \gg PCl^-$ and also passes divalent cations, $PCa2^+ \approx PBa^{2+} > PMg^{2+}$ (X.-L. Zhou, unpublished result), and it passes the physiologically important Ca^{2+}, even when it is the sole cation (Palmer et al. 2001). Vacuolar Ca^{2+} (mM) or low pH (< 5, vacuolar or cytoplasmic) inhibits its activity. More importantly, cytoplasmic Ca^{2+} (μM) enhances its activity, allowing a positive-feedback loop in the process of Ca^{2+}-induced Ca^{2+} release (CICR; see below) (Zhou et al. 2003).

The genome of *S. cerevisiae* was completely sequenced in 1996 – the first among eukaryotes (Goffeau et al. 1996). A search in this genome revealed an open reading frame that corresponds to known TRP-channel amino-acid sequences. By using a combination of gene deletion, re-expression, and direct patch clamping on the yeast vacuolar membrane, Palmer et al. (2001) found the full-length gene to be necessary for the above cation conductance (Fig. 1.3c). This observation was confirmed by others (Bihler et al. 2005) and this gene product was first named Yvc1, for yeast vacuolar channel (Palmer et al. 2001; Zhou et al. 2003), and later assigned as TRPY1 (Zhou et al. 2005) since it is a TRP homolog. This identification of TRPY1 took this vacuolar channel beyond biophysical description into the realm of cell and molecular biology.

The first finding on the cell-biological significance of TRPY1 was made in the Cyert laboratory (Denis and Cyert 2002). Using transgene-produced aequorin as a Ca^{2+} reporter in luminometry, Denise and Cyert found that osmotic upshock induces a transient rise in cytoplasmic free Ca^{2+}, and that this pulse of Ca^{2+} is missing in the *TRPY1*-deleted strain, leading to the conclusion that the TRPY1 channel is the

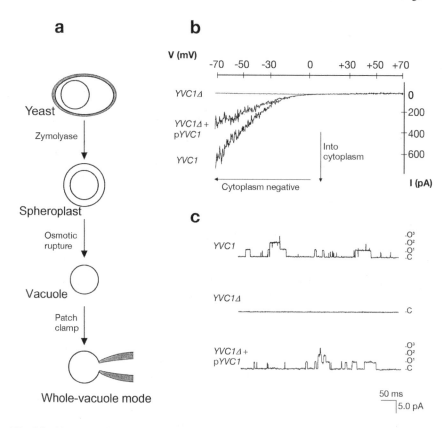

Fig. 1.3 Absence and restoration of the yeast vacuolar conductance are correlated with the deletion and expression of the *TRPY1* gene, formerly *YVC1*. **a** Diagrammatic representation of the method of vacuole preparation and patch-clamping. **b** Whole vacuole macroscopic currents upon applying a voltage ramp from +70 mV to −70 mV (bath voltage, cytoplasmic side) from each of the three strains: *YVC* wild type, *YVC1* Δ knockout, *YVC1* Δ + *pYCV1* re-expression from a plasmid. **c** Sample traces from whole vacuoles held at +10 mV. C closed level; O_1, O_2, O_3, open levels (from Palmer et al. 2001)

conduit for this Ca^{2+} passage (Denis and Cyert 2002). This work implicates TRPY1 as the link between the osmotic stimulus and the Ca^{2+}-release response (Fig. 1.4).

The observation by Zhou et al. (2003) that this channel is in fact mechanosensitive connects the osmotic-upshock-induced Ca^{2+} response in vivo and the TRPY1-channel activities in vitro. Whether examined in the whole-vacuole mode or in the excised cytoplasmic-side-out mode, application of pressure on the order of a few microNewtons/meter through the patch-clamp pipette activates the TRPY1 unitary conductances. Such conductances are always observed regardless of whether the *TRPY1* gene resides in the chromosome or on a plasmid, but are never observed in cells from a *TRPY1*-knockout strain (Fig. 1.5). Technical ease demands that a positive pressure be applied, inflating the vacuole. Suction (negative pressure) through the pipette usually break the gigaseal or confound the recording modes. Nonetheless,

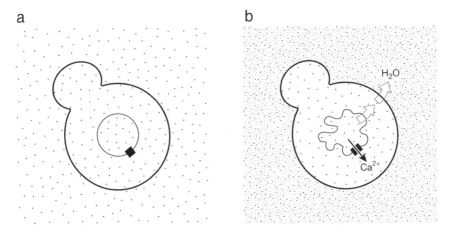

Fig. 1.4 A cell-biological model of TRPY1 (Yvc1p) function. **a** The channel in the vacuolar membrane is closed when the cell is in an osmotic steady state. **b** A sudden increase in external osmolarity creates a disequilibrium. The evacuation of water (*open arrows*) causes the vacuole to shrink, deforming the vacuolar membrane. Local membrane stretch force generated by the deformation opens the TRPY1 channel, releasing Ca^{2+} into the cytoplasm (see Denis and Cyert 2002)

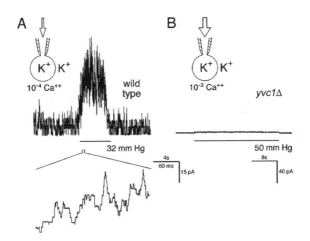

Fig. 1.5 The mechanosensitivity of TRPY1 and an extended cell-biological model. Channel activities recorded in whole-vacuole mode. Pressure pulse (in mm Hg) activates unitary conductances in the wild-type vacuole, but not in the knockout mutant. The TPRY1 is also activated by Ca^{2+} (Palmer et al. 2001). The comparison shown here illustrates that even at a higher concentration of Ca^{2+}, no mechanosensitive conductance can be elicited from the knockout

direct application of osmoticum to the bath, which visibly shrinks the vacuole on the pipette, can be shown to activate *TPRY1* (Zhou et al. 2003). This observation echoes the upshock-induced Ca^{2+} response in vivo. It also shows that the stretch force generated by membrane deformation through inflation or shrinkage can activate this channel. The cell-biological and the biophysical observations together led to a model that explains mechanistically how an osmotic upshock causes a rise in cytoplasmic free Ca^{2+}, a presumed defensive response (Fig. 1.4).

1.5.5 Other Fungal TRP Homologs

The TRPY1 channel of the budding yeast *S. cerevisiae* has homologs in some 30 different genomes of fungi spanning the two major fungal divisions: ascomycetes (molds, yeasts, truffles, lichens, etc.) and basidiomyces (smuts, mushrooms, etc.) (Zhou et al. 2005). A more distant homolog in *Schizosaccharomyces pombe* (fission yeast), with some similarity to TRPP, appears essential and relates to cell-wall synthesis (Palmer et al. 2005).

TRPY2, from *Kluyveromyces lactis*, and TRPY3, from the infectious yeast *Candida albicans*, have been studied recently. They were examined by expressing their corresponding genes, borne on a plasmid, in *S. cerevisiae* cells from which the native *TRPY1* has been deleted. Patch-clamp examination of these fungal TRP channels in this heterologous setting showed that their unitary conductance, ion selectivity, rectification, and Ca^{2+} sensitivity are similar to those of TRPY1. Most importantly, their mechanosensitivity is preserved in such a setting. This was demonstrated by following the hyperosmotically induced Ca^{2+} release from the vacuole to the cytoplasm in vivo (Fig. 1.6a) as well as by following the channel's response to direct pressure applied through the patch-clamp pipette (Fig. 1.6b). Thus, mechanosensitivity of TRPY2 and TRPY3 does not require their native membranes.

1.5.6 The Submolecular Basis of TRP Mechanosensitivity – a Crucial Question

As reviewed above, TRP channels and their genes are found in diverse eukaryotes, including many microbes (Fig. 1.2). In the few cases where microbial TRPs have been examined, namely several fungal channels, these channels are mechanosensitive (Figs. 1.5, 1.6). The universal presence of mechanosensitive channels should not be surprising since water is crucial to all life forms and life-threatening sudden de- or over-hydration can be detected as changes in the osmotic force exerted on the membrane (Kung 2005). The presumed primordial origin of such a water-sensing devise argues for a basic and evolutionarily preserved molecular mechanism. Can this mechanism be divined from the primary sequence of TRPs?

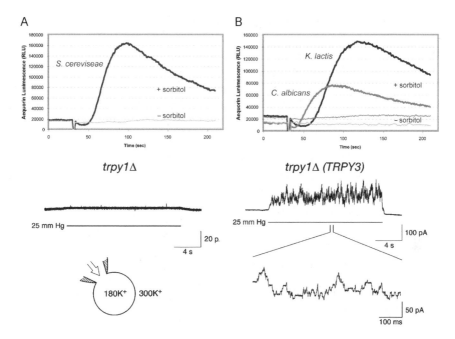

Fig. 1.6 TRPY2 from *Kluyveromyces lactis* and TRPY3 from *Candida albicans* function as mechanosensitive channels when heterologously expressed in the vacuoles of *Saccharomyces cerevisiae. Top* Ca^{2+} responses to osmotic upshock from live cells, registered as luminescence from trangenically expressed aequorin. Addition of 3 M sorbitol leads to robust responses from the host (*TPRY1, left*) as well as the guests (*TPRY2* and *TRPY3* in *trpy1* Δ cells, *right*). *Bottom TRPY3* is mechanosensitive in the vacuole of *S. cerevisiae* with *TRPY1* deleted. In whole vacuole mode, no currents can be evoked with pressure from the *trpy1* Δ vacuole (*left*) but can clearly be evoked from one expressing the *TRPY3* gene (*right*). Adapted from Zhou et al. 2005

Unlike in K$^+$ channels, where the voltage sensors can be easily recognized from the sequence, TRP-channel sequences are not useful in predicting the biophysical properties or the physiological functions of TRPs. The various cytoplasmic domains (ankyrin, calmodulin-binding domain, "TRP box", etc.) that are used to divide TRPs into subfamilies do not correspond to any gating principles. Mechanosensitivity, variously evidenced, has been reported in all eight subfamilies: TRPC (Chen and Barritt 2003; Maroto et al. 2005; Strotmann et al. 2000), TRPV (Birder et al. 2002; Gong et al. 2004; Liedtke and Friedman 2003; Liedtke et al. 2003; Mizuno et al. 2003; Suzuki et al. 2003), TRPA(Corey et al. 2004; Walker et al. 2000), TRPP (Nauli et al. 2003), TRPN (Li et al. 2006; Sidi et al. 2003; Walker et al. 2000), TRPM (Grimm et al. 2003), TRPML (Di Palma et al. 2002) and now TRPY (Zhou et al. 2003). Furthermore, where investigated thoroughly, TRPs are polymodal. For example, TRPV1, the original vanilloid receptor, is activated by heat (Caterina et al. 1997), acidic pH and (Tominaga et al. 1998), and inhibited by PIP$_2$ (Prescott and Julius 2003). It is also activated nonphysiologically by irritants evolved in plants (peppers, etc.) and by toxins in spiders (Siemens et al. 2006). At the same time, there is evidence that TRPV1 (Naeini et al. 2006), or its spicing variants (Birder et al. 2002), are also used as osmoreceptors. Thus, at least some TRPs may integrate several stimuli (e.g., stretch, heat, ligands) into a single Ca^{2+} flux into the receptor cell and one meaning in the central nervous system (e.g., pain). In addition, the same TRP channels may serve different physiological functions in separate tissues in organisms that differentiate tissues. Labels such as thermoTRPs or mechanoTRPs refer to the history of their discovery and context for their continued investigation and not necessarily to their true or only biological roles.

Since only the sequence from TM5 to just beyond TM6 is conserved significantly among all TRPs (Fig. 1.2), the central mechanism for force-to-flux transduction should therefore lie within this region, which is largely buried in the membrane. By analogy to K^+ channels, this region covers the filter and the gate. For want of a crystal structure of a TRP channel, the precise location and the 3-D arrangement of the gate is currently unknown. How stretch force, from the lipid bilayer (Kung 2005) or cytoskeleton (Sukharev and Corey 2004), is transmitted to this gate is the crux of mechanosensitivity. This question should be one of the foci of future research. Evidence for the crucial importance of the lipid bilayer in activating animal mechanosensitive channels has recently been summarized by Kung (2005). It is argued that, even if forces are transmitted to these channels through matrix or cytoskeletal proteins, the molecular displacement at the channel-lipid interface may be the ultimate energetic cause of mechanosensitive channel activation (Kung 2005).

1.6   Conclusion

There are two views of biology today. In one, we strive to understand all life forms and hope to find principles that apply to all or most or many of them. Here, we experiment on whatever organisms that happen to offer experimental advantages. In the other, we strive to understand human beings. When we cannot address our questions directly with humans, we use animals that are as closely related as possible. Even if one's goal is the betterment of mankind in its narrow sense, understanding basic principles remain crucial. Unlike those in chemistry or physics, many principles in biology remain to be understood. Mechanosensation is a case in point.

If all TRPs have the same origin, and the conserved TM5–TM6 region holds the secret, then any TRP channel should be equally appealing as a subject for investigation. Given our innate interest in humans, as well as the possible medical benefits, it seems natural to gravitate towards studying human TRPs or their mammalian equivalents. Why then study microbial TRPs? As stated above, microbes offer tremendous experimental advantages in understanding molecular mechanisms, as evidenced from the last 50 years' revolution in biology. DNA replication, transcription, and translation were largely solved through the study of bacteria and their phages. Central metabolism was understood through the study of mitochondria and yeast. Massive bacterial cultures have led to the crystal structures of K^+ channels and have provided great insights into ion filtration and gating mechanisms. The study of bacterial MscL and MscS clearly showed that these mechanosensitive channels receive their gating force from the lipid bilayer (Sukharev et al. 1997; Kung 2005). Crystallography, genetics, spectroscopy, and molecular-dynamic simulations have made MscL and MscS the concrete models for investigating mechanosensitivity at the molecular and submolecular level. The key question of how force is transmitted to the TRP-channel gate requires that we define the location and the 3-D arrangement of the gate, and where and how it is connected to other structures that receive the stretch force. The yeast system, with its demonstrated prowess of genetic manipulation, is being brought to bear on this question. While understanding TRPY will unlikely bring clinical or commercial harvest, efforts in TRPY research should complement the bulk of animal TRP channel research towards a deeper molecular understanding of mechanosensation in general.

Note added to the proof For recent findings on the yeast TRP channel, see Zhou *et al.* (2007).

Acknowledgments Supported by NIH GM054867 (Y.S.), GM047856 (C.K.) and the Vilas Trust of the University of Wisconsin – Madison.

References

Adl SM, Simpson AG, Farmer MA, Andersen RA, Anderson OR, Barta JR, Bowser SS, Brugerolle G, Fensome RA, Fredericq S, James TY, Karpov S, Kugrens P, Krug J, Lane CE, Lewis LA, Lodge J, Lynn DH, Mann DG, McCourt RM, Mendoza L, Moestrup O, Mozley-Standridge SE, Nerad TA, Shearer CA, Smirnov AV, Spiegel FW, Taylor MF (2005) The new higher level classification of eukaryotes with emphasis on the taxonomy of protists. J Eukaryot Microbiol 52:399–451

Aury JM, Jaillon O, Duret L, Noel B, Jubin C, Porcel BM, Segurens B, Daubin V, Anthouard V, Aiach N, Arnaiz O, Billaut A, Beisson J, Blanc I, Bouhouche K, Camara F, Duharcourt S, Guigo R, Gogendeau D, Katinka M, Keller AM, Kissmehl R, Klotz C, Koll F, Le Mouel A, Lepere G, Malinsky S, Nowacki M, Nowak JK, Plattner H, Poulain J, Ruiz F, Serrano V, Zagulski M, Dessen P, Betermier M, Weissenbach J, Scarpelli C, Schachter V, Sperling L, Meyer E, Cohen J, Wincker P (2006) Global trends of whole-genome duplications revealed by the ciliate *Paramecium tetraurelia*. Nature 444:171–178

Baker MD, Wolanin PM, Stock JB (2006) Signal transduction in bacterial chemotaxis. Bioessays 28:9–22

Barr MM, Sternberg PW (1999) A polycystic kidney-disease gene homologue required for male mating behaviour in *C. elegans*. Nature 401:386–389

Bass RB, Strop P, Barclay M, Rees DC (2002) Crystal structure of *Escherichia coli* MscS, a voltage-modulated and mechanosensitive channel. Science 298:1582–1587

Batiza AF, Schulz T, Masson PH (1996) Yeast respond to hypotonic shock with a calcium pulse. J Biol Chem 271:23357–23362

Bautista DM, Jordt SE, Nikai T, Tsuruda PR, Read AJ, Poblete J, Yamoah EN, Basbaum AI, Julius D (2006) TRPA1 mediates the inflammatory actions of environmental irritants and proalgesic agents. Cell 124:1269–1282

Becker MR, Paster BJ, Leys EJ, Moeschberger ML, Kenyon SG, Galvin JL, Boches SK, Dewhirst FE, Griffen AL (2002) Molecular analysis of bacterial species associated with childhood caries. J Clin Microbiol 40:1001–1009

Bihler H, Eing C, Hebeisen S, Roller A, Czempinski K, Bertl A (2005) TPK1 is a vacuolar ion channel different from the slow-vacuolar cation channel. Plant Physiol 139:417–424

Birder LA, Nakamura Y, Kiss S, Nealen ML, Barrick S, Kanai AJ, Wang E, Ruiz G, De Groat WC, Apodaca G, Watkins S, Caterina MJ (2002) Altered urinary bladder function in mice lacking the vanilloid receptor TRPV1. Nat Neurosci 5:856–860

Caterina MJ, Schumacher MA, Tominaga M, Rosen TA, Levine JD, Julius D (1997) The capsaicin receptor: a heat-activated ion channel in the pain pathway. Nature 389:816–824

Chang G, Spencer RH, Lee AT, Barclay MT, Rees DC (1998) Structure of the MscL homolog from *Mycobacterium tuberculosis*: a gated mechanosensitive ion channel. Science 282:2220–2226

Chen J, Barritt GJ (2003) Evidence that TRPC1 (transient receptor potential canonical 1) forms a Ca(2+)-permeable channel linked to the regulation of cell volume in liver cells obtained using small interfering RNA targeted against TRPC1. Biochem J 373:327–336

Chenik M, Douagi F, Achour YB, Khalef NB, Ouakad M, Louzir H, Dellagi K (2005) Characterization of two different mucolipin-like genes from *Leishmania major*. Parasitol Res 98:5–13

Clapham DE, Montell C, Schultz G, Julius D (2003) International Union of Pharmacology. XLIII. Compendium of voltage-gated ion channels: transient receptor potential channels. Pharmacol Rev 55:591–596

Colbert HA, Smith TL, Bargmann CI (1997) OSM-9, a novel protein with structural similarity to channels, is required for olfaction, mechanosensation, and olfactory adaptation in *Caenorhabditis elegans*. J Neurosci 17:8259–8269

Corey DP, Garcia-Anoveros J, Holt JR, Kwan KY, Lin S-Y, Volrath MA, Amalfitano A, Cheung EL-M, Derfler BH, Duggan A, Geleoc GS, Gray PA, Hoffman MP, Rehm KL, Tamasauskas D, Zhang DS (2004) TRPA1 is a candidate for the mechanosensitive transduction channel of vertebrate hair cells. Nature 432:723–730

Denis V, Cyert MS (2002) Internal Ca(2+) release in yeast is triggered by hypertonic shock and mediated by a TRP channel homologue. J Cell Biol 156:29–34

Di Palma F, Belyantseva IA, Kim HJ, Vogt TF, Kachar B, Noben-Trauth K (2002) Mutations in Mcoln3 associated with deafness and pigmentation defects in varitint-waddler (Va) mice. Proc Natl Acad Sci USA 99:14994–14999

Doyle DA, Morais Cabral J, Pfuetzner RA, Kuo A, Gulbis JM, Cohen SL, Chait BT, MacKinnon R (1998) The structure of the potassium channel: molecular basis of K+ conduction and selectivity. Science 280:69–77

Eckert R (1972) Bioelectric control of ciliary activity. Science 176:473–481

Embley TM, Martin W (2006) Eukaryotic evolution, changes and challenges. Nature 440:623–630

Goffeau A, Barrell BG, Bussey H, Davis RW, Dujon B, Feldmann H, Galibert F, Hoheisel JD, Jacq C, Johnston M, Louis EJ, Mewes HW, Murakami Y, Philippsen P, Tettelin H, Oliver SG (1996) Life with 6000 genes. Science 274:546

Gong S, Son W, Chung YD, Kim J, Shin DW, McClung CA, Lee Y, Lee HW, Chang DJ, Kaang B-K, Cho H, Oh U, Hirsh J, Kernan MJ, Kim C (2004) Two interdependent TRPV channel subunits, Inactive and Nanchung, mediate hearing in Drosophila. J Neurosci 24:9059–9066

Grimm C, Kraft R, Sauerbruch S, Schultz G, Harteneck C (2003) Molecular and functional characterization of the melastatin-related cation channel TRPM3. J Biol Chem 278:21493–21501

Gustin MC, Martinac B, Saimi Y, Culbertson MR, Kung C (1986) Ion channels in yeast. Science 233:1195–1197

Gustin MC, Zhou XL, Martinac B, Kung C (1988) A mechanosensitive ion channel in the yeast plasma membrane. Science 242:762–765

Hamill OP, Marty A, Neher E, Sakmann B, Sigworth FJ (1981) Improved patch-clamp techniques for high-resolution current recording from cells and cell-free membrane patches. Pfluegers Arch 391:85–100

Hughes J, Ward CJ, Peral B, Aspinwall R, Clark K, San Millan JL, Gamble V, Harris PC (1995) The polycystic kidney disease 1 (PKD1) gene encodes a novel protein with multiple cell recognition domains. Nat Genet 10:151–160

Iyer R, Iverson TM, Accardi A, Miller C (2002) A biological role for prokaryotic ClC chloride channels. Nature 419:715–718

Jiang Y, Lee A, Chen J, Cadene M, Chait BT, MacKinnon R (2002) The open pore conformation of potassium channels. Nature 417:523–526

Kanzaki M, Nagasawa M, Kojima I, Sato C, Naruse K, Sokabe M, Iida H (1999) Molecular identification of a eukaryotic, stretch-activated nonselective cation channel. Science 285:882–886

Ketchum KA, Joiner WJ, Sellers AJ, Kaczmarek LK, Goldstein SA (1995) A new family of outwardly rectifying potassium channel proteins with two pore domains in tandem. Nature 376:690–695

Kim J, Chung YD, Park DY, Choi S, Shin DW, Soh H, Lee HW, Son W, Yim J, Park CS, Kernan MJ, Kim C (2003) A TRPV family ion channel required for hearing in Drosophila. Nature 424:81–84

Kung C (2005) A possible unifying principle for mechanosensation. Nature 436:647–654

Kung C, Blount P (2004) Channels in microbes: so many holes to fill. Mol Microbiol 53:373–380

Kuo MM, Saimi Y, Kung C (2003) Gain-of-function mutations indicate that *Escherichia coli* Kch forms a functional K^+ conduit in vivo. EMBO J 22:4049–4058

Kuo MM, Haynes WJ, Loukin SH, Kung C, Saimi Y (2005a) Prokaryotic K(+) channels: from crystal structures to diversity. FEMS Microbiol Rev 29:961–985

Kuo MM-C, Kung C, Saimi Y (2005b) K^+ channels: a survey and a case study of Kch of *Escherichia coli*. In: Kubalski A, Martinac B (eds) Bacterial ion channels and their eukaryotic homologs. ASM, Washington DC, pp 1–20

Kwan KY, Allchorne AJ, Vollrath MA, Christensen AP, Zhang DS, Woolf CJ, Corey DP (2006) TRPA1 contributes to cold, mechanical, and chemical nociception but is not essential for hair-cell transduction. Neuron 50:277–289

Levina N, Totemeyer S, Stokes NR, Louis P, Jones MA, Booth IR (1999) Protection of *Escherichia coli* cells against extreme turgor by activation of MscS and MscL mechanosensitive channels: identification of genes required for MscS activity. EMBO J 18:1730–1737

Li W, Feng Z, Sternberg PW, Xu CZS (2006) A *C. elegans* stretch receptor neuron revealed by a mechanosensitive TRP channel homologue. Nature 440:684–687

Liedtke W, Friedman JM (2003) Abnormal osmotic regulation in trpv4$^{-/-}$ mice. Proc Natl Acad Sci USA 100:13698–13703

Liedtke W, Choe Y, Marti-Renom MA, Bell AM, Denis CS, Sali A, Hudspeth AJ, Friedman JM, Heller S (2000) Vanilloid receptor-related osmotically activated channel (VR-OAC), a candidate vertebrate osmoreceptor. Cell 103:525–535

Liedtke W, Tobin DM, Bargmann CI, Friedman JM (2003) Mammalian TRPV4 (VROAC) directs behavioral responses to osmotic and mechanical stimuli in *Caenorhabditis elegans*. Proc Natl Acad Sci USA 100[Suppl 2]:14531–14536

Loukin SH, Saimi Y (2002) Carboxyl tail prevents yeast K(+) channel closure: proposal of an integrated model of TOK1 gating. Biophys J 82:781–792

Loukin SH, Vaillant B, Zhou XL, Spalding EP, Kung C, Saimi Y (1997) Random mutagenesis reveals a region important for gating of the yeast K^+ channel Ykc1. EMBO J 16:4817–4825

Maroto R, Raso A, Wood TG, Kurosky A, Martinac B, Hamill OP (2005) The role of TRPC1 in forming the mechanosensitive cation channel in frog oocytes. Nat Cell Biol 7:179–185

Martinac B, Buechner M, Delcour AH, Adler J, Kung C (1987) Pressure-sensitive ion channel in *Escherichia coli*. Proc Natl Acad Sci USA 84:2297–2301

McKemy DD, Neuhausser WM, Julius D (2002) Identification of a cold receptor reveals a general role for TRP channels in thermosensation. Nature 416:52–58

Minke B, Parnas M (2006) Insight on TRP channels from in vivo studies in Drosophila. Annu Rev Physiol 68:649–684

Minke B, Wu C-F, Pak WL (1975) Induction of photoreceptor voltage noise in the dark in Drosophila mutant. Nature 258:84–87

Minorsky P, Zhou XL, Culbertson M, Kung C (1989) A patch-clamp analysis of a cation-current in the vacuolar membrane of the yeast Saccharomyces. Plant Physiol 89[Suppl]:882

Mizuno A, Matsumoto N, Imai M, Suzuki M (2003) Impaired osmotic sensation in mice lacking TRPV4. Am J Physiol Cell Physiol 285:C96–101

Mochizuki T, Wu G, Hayashi T, Xenophontos SL, Veldhuisen B, Saris JJ, Reynolds DM, Cai Y, Gabow PA, Pierides A, Kimberling WJ, Breuning MH, Deltas CC, Peters DJ, Somlo S (1996) PKD2, a gene for polycystic kidney disease that encodes an integral membrane protein. Science 272:1339–1342

Montell C, Jones K, Hafen E, Rubin G (1985) Rescue of the Drosophila phototransduction mutation trp by germline transformation. Science 230:1040–1043

Montell C, Birnbaumer L, Flockerzi V, Bindels RJ, Bruford EA, Caterina MJ, Clapham DE, Harteneck C, Heller S, Julius D, Kojima I, Mori Y, Penner R, Prawitt D, Scharenberg AM, Schultz G, Shimizu N, Zhu MX (2002) A unified nomenclature for the superfamily of TRP cation channels. Mol Cell 9:229–231

Muraki K, Iwata Y, Katanosaka Y, Ito T, Ohya S, Shigekawa M, Imaizumi Y (2003) TRPV2 is a component of osmotically sensitive cation channels in murine aortic myocytes. Circulation Res 93:829–838

Naeini RS, Witty MF, Seguela P, Bourque CW (2006) An N-terminal variant of Trpv1 channel is required for osmosensory transduction. Nat Neurosci 9:93–98

Nauli SM, Alenghat FJ, Luo Y, Williams E, Vassilev P, Li X, Elia AE, Lu W, Brown EM, Quinn SJ, Ingber DE, Zhou J (2003) Polycystins 1 and 2 mediate mechanosensation in the primary cilium of kidney cells. Nat Genet 33:129–137

Numata T, Shimizu T, Okada Y (2006) TRPM7 is a stretch- and swelling-activated cation channel involved in volume regulation in human epithelial cells. Am J Physiol Cell Physiol 292: C460–C467

Ou X, Blount P, Hoffman RJ, Kung C (1998) One face of a transmembrane helix is crucial in mechanosensitive channel gating. Proc Natl Acad Sci USA 95:11471–11475

Palmer CP, Zhou XL, Lin J, Loukin SH, Kung C, Saimi Y (2001) A TRP homolog in Saccharomyces cerevisiae forms an intracellular Ca(2+)-permeable channel in the yeast vacuolar membrane. Proc Natl Acad Sci USA 98:7801–7805

Palmer CP, Batiza A, Zhou X-L, Loukin SH, Saimi Y, Kung C (2004) Ion channels of microbes. In: Fairweather I (ed) Cell signalling in prokaryotes and lower metazoa. Kluwer, Dordrecht, pp 325–345

Palmer CP, Aydar E, Djamgoz MB (2005) A microbial TRP-like polycystic-kidney disease-related ion channel gene. Biochem J 387:211–219

Prescott ED, Julius D (2003) A modular PIP2 binding site as a determinant of Capsaicin receptor sensitivity. Science 300:1284–1288

Saimi Y, Kung C (1987) Behavioral genetics of Paramecium. Annu Rev Genet 21:47–65

Saimi Y, Kung C (2002) Calmodulin as an ion channel subunit. Annu Rev Physiol 64:289–311

Saimi Y, Loukin SH, Zhou XL, Martinac B, Kung C (1999) Ion channels in microbes. Methods Enzymol 294:507–524

Saimi Y, Martinac B, Delcour AH, Minorsky PV, Gustin MC, Culbertson MR, Adler J, Kung C (1992) Patch clamp studies of microbial ion channels. Methods Enzymol 207:681–691

Saimi Y, Zhou X-L, Loukin SH, Haynes WJ, Kung C (2007) Microbial TREP channels and their mechanosensitivity. In: Hamill O (ed) Mechanosensitive ion channels. Elsevier, Amsterdam (in press)

Saitou M, Nei M (1987) The neighbor-joining method: a new method for reconstructing phylogenetic trees. Mol Biol Evol 4:406–425

Shih YL, Rothfield L (2006) The bacterial cytoskeleton. Microbiol Mol Biol Rev 70:729–754

Sidi S, Friedrich RW, Nicolson T (2003) NompC TRP channel required for vertebrate sensory hair cell mechanotransduction. Science 301:96–99

Siemens J, Zhou S, Piskorowski R, Nikai T, Lumpkin EA, Basbaum AI, King D, Julius D (2006) Spider toxins activate the capsaicin receptor to produce inflammatory pain. Nature 444:208–212

Story GM, Peier AM, Reeve AJ, Eid SR, Mosbacher J, Hricik TR, Earley TJ, Hergarden AC, Andersson DA, Hwang SW, McIntyre P, Jegla T, Bevan S, Patapoutian A (2003) ANKTM1, a

TRP-like channel expressed in nociceptive neurons, is activated by cold temperatures. Cell 112:819–829

Strotmann R, Harteneck C, Nunnenmacher K, Schultz G, Plant TD (2000) OTRPC4, A nonselective cation channel that confers sensitivity to extracellular osmolarity. Nat Cell Biol 2:695–702

Sukharev S, Corey DP (2004) Mechanosensitive channels: multiplicity of families and gating paradigms. Sci STKE 219:re4

Sukharev SI, Blount P, Martinac B, Blattner FR, Kung C (1994) A large-conductance mechanosensitive channel in *E. coli* encoded by mscL alone. Nature 368:265–268

Sukharev SI, Blount P, Martinac B, Kung C (1997) Mechanosensitive channels of *Escherichia coli*: the MscL gene, protein, and activities. Annu Rev Physiol 59:633–657

Sun M, Goldin E, Stahl S, Falardeau JL, Kennedy JC, Acierno JS Jr, Bove C, Kaneski CR, Nagle J, Bromley MC, Colman M, Schiffmann R, Slaugenhaupt SA (2000) Mucolipidosis type IV is caused by mutations in a gene encoding a novel transient receptor potential channel. Hum Mol Genet 9:2471–2478

Suzuki M, Mizuno A, Kodaira K, Imai M (2003) Impaired pressure sensation in mice lacking TRPV4. J Biol Chem 278:22664–22668

Tominaga M, Caterina MJ, Malmberg AB, Rosen TA, Gilbert H, Skinner K, Raumann BE, Basbaum AI, Julius D (1998) The cloned capsaicin receptor integrates multiple pain-producing stimuli. Neuron 21:531–543

Tracey WD Jr, Wilson RI, Laurent G, Benzer S (2003) *painless*, a Drosophila gene essential for nociception. Cell 113:261–273

Wada Y, Ohsumi Y, Tanifuji M, Kasai M, Anraku Y (1987) Vacuolar ion channel of the yeast, *Saccharomyces cerevisiae*. J Biol Chem 262:17260–17263

Walker RG, Willingham AT, Zuker CS (2000) A Drosophila mechanosensory transduction channel. Science 287:2229–2234

Woese CR (1994) There must be a prokaryote somewhere: microbiology's search for itself. Microbiol Rev 58:1–9

Woese CR (2000) Interpreting the universal phylogenetic tree. Proc Natl Acad Sci USA 97:8392–8396

Yu FH, Catterall WA (2004) The VGL-chanome: a protein superfamily specialized for electrical signaling and ionic homeostasis. Sci STKE 253:re15

Zhou XL, Vaillant B, Loukin SH, Kung C, Saimi Y (1995) YKC1 encodes the depolarization-activated K^+ channel in the plasma membrane of yeast. FEBS Lett 373:170–176

Zhou XL, Batiza AF, Loukin SH, Palmer CP, Kung C, Saimi Y (2003) The transient receptor potential channel on the yeast vacuole is mechanosensitive. Proc Natl Acad Sci USA 100:7105–7110

Zhou X-L, Loukin SH, Coria R, Kung C, Saimi Y (2005) Heterologously expressed fungal transient receptor potential channels retain mechanosensitivity in vitro and osmotic response in vivo. Eur Biophys J 34:413–422

Zhou X-L, Su Z-W, Anishkin A, Haynes WJ, Friske EM, Loukin SH, Kung C, and Saimi Y (2007) Yeast screens show aromatic residues at the end of the sixth helix anchor TRP-channel gate. Proc Natl Acad Sci USA (in press)

Chapter 2
Mechanosensitive Channels and Sensing Osmotic Stimuli in Bacteria

Paul Blount(✉), Irene Iscla, and Yuezhou Li

Abstract Microbes are directly exposed to the elements, and one of the most acute insults they can experience is a rapid change in osmotic environment. The forces that are generated by even small changes in the osmolarity are massive. In response to increases in the osmolarity of the medium, also called osmotic upshock, the cell transports and synthesizes cytoplasmic osmoprotectants, which thus help to maintain its turgor. A subsequent severe osmotic downshock is more than an inconvenience, it is life threatening. The cell swells, its cell-wall becomes compromised, and without immediate action the organism would lyse. Large-conductance mechanosensitive channels within the cytoplasmic membrane prevent this needless death by serving as biological emergency release valves. Two such bacterial mechanosensitive channel families have been extensively studied: MscL and MscS. For each, a crystal structure of a family member has been obtained, and detailed models for structural changes that occur upon gating have been postulated. As we learn more of the molecular mechanisms by which these channels sense and respond to membrane tension, we discover similarities not only between these two relatively distant families, but also potentially with the more complex mechanosensory systems of eukaryotic organisms.

Department of Physiology U. T. Southwestern Med. Ctr, 5323 Harry Hines Blvd, (For FedEX: 6001 Forest Park) Dallas, TX 75390-9040, Paul.Blount@utsouthwestern.edu

2.1   Osmotic Regulation of Bacteria

Approximately 40 years ago, Britten and McClure studied the bacterial "amino acid pool" of *Escherichia coli* (Britten and McClure 1962). The question was simple: did amino acid concentrations within the cell change upon changing environmental conditions? What they found for proline was truly astounding. If cells are grown in a high osmolarity media containing any proline, then this amino acid accumulates to very high levels in the cell. The experimenters then subsequently challenged the cells by diluting them into a low osmotic environment; this rapid and substantial environmental shift is often called osmotic downshock. Upon examination, it was discovered that very little proline remained associated with the cells. A trivial explanation would be that the cells had lysed. However, viability studies demonstrated that although virtually all of the proline was released into the medium, cell viability remained extremely high. This simple observation led to decades of research defining the systems and molecules involved in the osmo-regulatory process. A simple overview is shown in Fig. 2.1.

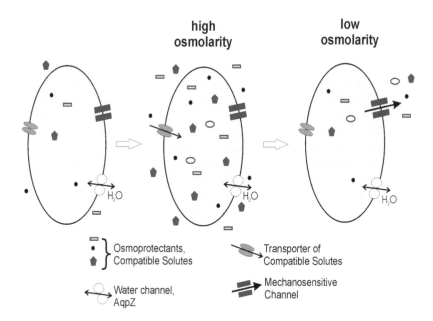

Fig. 2.1 Cellular responses to changes in the osmotic environment. When bacterial cells are exposed to a high osmolarity medium, water fluxes out of the cell through water channels, AqpZ (depicted in the lower right side of the cell in this schematic), and a series of adjustments occur in order to maintain cell turgor. Specifically, cells actively accumulate solutes such as K^+, betaine, and proline through transporters (shown on the left of the cell), as well as synthesis of osmolytes like trehalose and glutamate. Upon exposure to a medium with low osmolarity (osmotic downshock), the influx of water increases first turgor, then membrane tension, thus activating the mechanosensitive channels that act as emergency release valves by allowing a rapid efflux of solutes (upper right of the cells)

2.1.1 Maintaining Cell Turgor with Compatible Solutes

Several experimental approaches have been utilized in an attempt to determine the turgor within *E. coli* cells. In the first and most classic approach, turgor is estimated by determining the threshold osmotic upshock required for plasmolysis, the separation of the cytoplasmic membrane from the cell wall (Knaysi 1951; Cayley et al. 1992). However, this technique can overestimate the forces if, as we now know, the cell wall is elastic. Another approach is to calculate the turgor by determining the concentrations of the osmotically active solutes (Cayley et al. 1991, 1992). Finally, the osmolarity of cell lysates can be measured (Mitchell and Moyle 1956). While each of these approaches has its own flaws, all estimates have predicted that the cytoplasm of the cell is under great pressure, somewhere between 3 and 10 atmospheres (Knaysi 1951; Munro et al. 1991; Cayley et al. 1992). Estimates by measuring concentrations of cytoplasmic solutes (Epstein and Schultz 1965; Cayley et al. 1991, 1992) suggest that the pressure is largely independent of external osmolarity during steady state growth.

An increase in cell volume, as occurs normally during cell growth, must be compensated by the accumulation of cytoplasmic solutes. In addition, to maintain cell turgor in a high osmotic environment, additional solutes are required; indeed, some solutes achieve molar concentrations. However, some molecules and ions can compromise the integrity and functionality of many enzymes. Hence, only molecules that can be accumulated to these high levels without deleterious effects on the cell are accumulated; these are referred to as 'compatible' solutes. Not only are these compounds tolerated by the cell; in some instances compatible solutes have been shown to stabilize protein structure (Arakawa and Timasheff 1985). These solutes are either pumped into the cytoplasm or synthesized. Many pumps, including those for proline, potassium and quaternary ammonium compounds such as glycine betaine, proline betaine, γ-butyrobetaine, and carnitine, are expressed on the cell surface and sense changes in osmolarity by various mechanisms (Wood 1999, 2006; Poolman et al. 2002). One of these mechanisms appears to include changes in protein–anionic lipid interactions as a result of high concentrations of cytoplasmic cations in cells undergoing plasmolysis (Poolman et al. 2002). Many of these compounds are associated with decay of plants and meat and are found in environments in which *E. coli* thrive, including the intestinal tract.

Once osmolytes accumulate within the cell, turgor is restored. However, if the cell is challenged with an osmotic downshock, the cell will swell, thus running the risk of lysing. Osmotic forces can truly be massive. A gradient of 250 mM of a solute with two components, such as NaCl, across a membrane translates to greater than 11 atmospheres of pressure! To make things worse for the cell, its cytoplasmic membrane contains water channels – aquaporins – that under normal conditions may help preserve turgor by facilitating water entrance in rapidly growing cells (Calamita et al. 1998), but under osmotic stresses would only accentuate the problem. As mentioned above, Britten and McClure found that *E. coli* survived such an insult, but lost essentially all of their proline pool

(Britten and McClure 1962). Subsequent studies demonstrated that not just proline was jettisoned from the cell, but all of the compatible solutes (Tsapis and Kepes 1977; Schleyer et al. 1993); even a handful of not-so-small enzymes including thioredoxin, elongation factor Tu and DnaK (Ajouz et al. 1998; Berrier et al. 2000) (discussed more fully in Sect. 2.2, below) have been shown to leave the cell. In the earlier studies, the conduit through which this efflux took place was unclear; it seemed possible that transporters ran backwards, perhaps uncoupled to metabolic energy, or that a number of stretch-activated channels may exist that would show some specificity for each of the solutes studied (Schleyer et al. 1993). It was not until native *E. coli* membranes were studied using patch clamp that a clearer picture emerged.

2.1.2   Measuring Mechanosensitive Channel Activities in Native Membranes

Bacterial cells are too small to be directly accessible to patch clamp. However, a technique was developed to generate giant spheroplasts; briefly, septation is inhibited and the membrane of several cells is collapsed together into a sphere by compromising the cell wall. Electrophysiological approaches seemed feasible because such a preparation had been used to measure the bacterial voltage potential (Felle et al. 1980). These giant cells are on the order of 4–10 μm and are large enough to be subjected to patch clamp. Hence, in all studies in which native bacterial membranes have been investigated using this approach, giant spheroplasts were utilized. The most obvious channel activities observed in such native bacterial membranes are mechanosensitive channels. In the original report, only a single activity was reported (Martinac et al. 1987); however, we now know that there are at least four mechanosensitive channel activities in *E. coli*: MscL (mechanosensitive channel of large conductance), MscS (smaller), MscK (K^+-regulated) and MscM (mini). MscL truly is of large conductance, being over 3 nanosiemens (nS), which is approximately 100-fold greater than that of most eukaryotic channels. MscS and MscK are both approximately 1 nS, while MscM is a little less than 300 picosiemens (pS). MscL and MscS are the most prevalent and are easily distinguished from each other (Sukharev et al. 1993). An additional study describing the new activity, MscM, also noted that the conductances of the channels was proportional to their membrane stretch threshold: Upon mild membrane tension, the smaller MscM opens, greater stimulus opens MscS-like activities, and additional stimulus is required to open MscL. This observation led to the hypothesis that the channels were tailored in sensitivity and conductance so that the response of the size and amount of solutes released from the cell would be proportional to the amount of downshock (Berrier et al. 1996). Early studies did not distinguish between MscS and MscK activities because of their similar conductance and activation by membrane stretch (MscK is only slightly more sensitive than MscS). Now, because their molecular identities

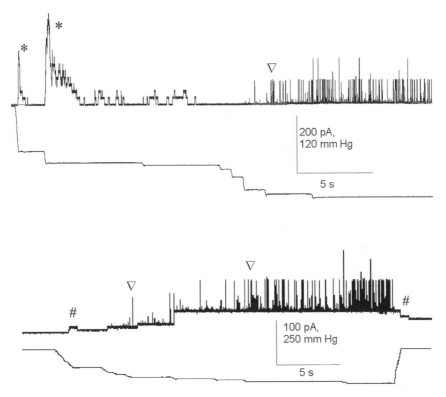

Fig. 2.2 Typical current traces of mechanosensitive channels in *Escherichia coli*. Recordings were generated from patches derived from *E. coli* giant spheroplasts at −20 mV. For each panel, activities from native membranes are shown in the top trace, while the bottom trace shows the negative pressure applied to the patch. *Upper trace* Typical trace containing MscS (*) and MscL (∇) activities. *Lower trace* Strain MJF451 (*ΔmscS*) was utilized, and shows one of the minority of traces that also shows MscK (#) activity

have been revealed and null-mutants generated, we know that MscK is less prevalent and has less of a tendency to desensitize, relative to MscS. Figure 2.2 shows typical traces showing the primary activities discussed in this chapter: MscS, MscK and MscL.

Mechanosensitive channel activities were found in other bacterial species, including Gram positive organisms (Zoratti and Petronilli 1988; Zoratti et al. 1990; Berrier et al. 1992); thus, the data indicated that these activities reflected channels with a conserved function, not an artifact unique to the *E. coli* organism. Thus, it was hypothesized that these large-conducting mechanosensitive channels in the bacterial envelope were the conduits through which solutes are released from the cell upon osmotic downshock. However, the presence of an inner and outer membrane and the cell wall of *E. coli* complicated the interpretation.

2.1.3   Getting Solutes Out of the Cytoplasm:
Cell Wall, Turgor and Elasticity

One of the intriguing aspects of the finding that solutes were jettisoned from *E. coli* upon osmotic downshock was the complexity that the molecules must transverse three barriers. Not only did the cells have a cytoplasmic membrane, but they also possessed a cell wall and an outer membrane. The outer membrane is largely thought to be a molecular sieve, selectively allowing nutrients and other beneficial molecules to pass. It is unlikely that this membrane normally absorbs any of the tension resulting from cellular turgor, and it is unclear what, if any, structural changes this membrane undergoes upon osmotic downshock. In contrast, the cell wall has many of the properties to absorb such forces.

The cell wall is made of peptidoglycan, which consists of oligosaccharide chains cross-linked by peptides (Glauner et al. 1988); this forms a rather imperfect web around the cell. As described in Sect. 2.1.1, the cytoplasm of *E. coli* is thought to be under 3–10 atmospheres of pressure; presumably, it is the cell wall that absorbs much of the resulting tension within the cellular envelope. However, the cell wall cannot be thought of as a sheet of armor since several studies using independent approaches suggest that this structure has highly elastic properties (Koch and Woeste 1992; Doyle and Marquis 1994; Yao et al. 1999). In addition, it undergoes significant remodeling with approximately 50% recycled per generation, and there is significantly decreased cross-linking in rapidly growing cells. Hence, the cell wall is not an impenetrable boundary; it seems likely that osmotic forces, which may lead to the doubling or tripling of turgor, may compromise the cell wall, exposing the inner membrane to tensions that would normally lead to its rupture; it is at these times that mechanosensitive channels play their important physiological role.

We now have strong phenotypic evidence that bacterial mechanosensitive channels truly play the physiological role of a biological "emergency release valve". When a double null mutant of *mscL* and *mscS* is challenged with osmotic downshock, viability decreases and the cells appear to lyse (Levina et al. 1999). This phenotype is not observed with single null mutant strains, indicating that MscL and MscS are redundant in function. The amount of osmotic downshock required to gate the mechanosensitive channels in vivo, and to lyse cells deficient in them, is approximately 200–400 mOsmol, which translates to a potential increase of about 4.5 to over 9 atmospheres of pressure within the cell. On the other hand, in patch clamp channel activities are routinely seen at about 0.1 to 0.5 atmospheres above ambient. Evidently, the cell wall can protect the cells from the first few additional atmospheres of pressure.

Both the MscL and MscS channels appear to be constitutively expressed, but are up-regulated upon entry into stationary phase or during adaptation to osmotic stress (Stokes et al. 2003). This regulation is due to the sigma factor RpoS. In *rpoS* mutant cells, the expression of MscL and MscS is lower, but rapidly growing cells still survive osmotic shock, suggesting that the basal expression level is sufficient. This is consistent with the finding that 'leak' or uninduced levels of channels from expression plasmids, estimated to be less than ten channels per cell, suppresses the osmotic-lysis phenotype of the MscL/MscS double null mutant. Presumably not many channels are necessary because the channel pore is large and the cell is small. On the other hand, RpoS mutants that have recently entered stationary phase at high osmolarity become acutely sensitive to osmotic downshock, lysing to an even greater extent than the MscL/MscS double mutant. These latter data suggest a complex relationship between mechanosensitive channels and changes in RpoS-induced cell structure, such as morphological shifts observed in cellular shape and size (Popham and Young 2003), as bacteria enter stationary phase.

To summarize, it is now clear that bacterial mechanosensitive channels do play an important role in sensing and adapting the cell to acute osmotic downshock. But why are there so many activities? Is the multitude of channels all from the same family? Do they share common molecular mechanisms for sensing forces dependent upon the osmotic environment? Unfortunately, little is known of MscM except its existence. On the other hand, MscL and MscS have demonstrated themselves to be extremely tractable and, as described in the following sections, these channels are now revealing the answers to many of these questions and are giving us a first glimpse of how channels can sense and respond to mechanical forces.

2.2 MscL

The first of the bacterial mechanosensitive channels to be cloned and sequenced was MscL from *E. coli* (Eco-MscL) (Sukharev et al. 1993). It was also the first channel from any organism that was definitively demonstrated to directly produce a mechanosensitive channel activity. To date, it remains the best-characterized mechanosensitive channel from any species. As described below, researchers have the advantage of a plethora of mutated channels (Ou et al. 1998; Maurer and Dougherty 2003; Levin and Blount 2004), a crystal structure, and detailed models for how the channel senses and responds to mechanical forces. While many of the issues concerning the structure and gating models are briefly outlined in this section, a more comprehensive review of these issues can be found in Blount et al. (2007a).

Early studies demonstrated that MscL and MscS could be solubilized with mild detergents, reconstituted into membranes and yet remain functional (Sukharev et al. 1993). It was this observation that allowed the biochemical enrichment of the MscL activity and, ultimately, the identification of the protein responsible for this activity and the gene that encoded it (Sukharev et al. 1994). The *mscL* gene predicted a small protein of only 136 amino acids consistent with two α helical transmembrane domains (TMDs). Subsequent studies supported this prediction and suggested that the N- and C-termini were cytoplasmic (Blount et al. 1996a). This study and others also predicted that the channel was a homohexamer (Blount et al. 1996a; Saint et al. 1998), but we now know that the data were misleading and the channel is in reality

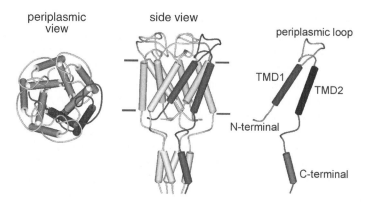

Fig. 2.3 Schematic representation of the crystal structure of MscL from *Mycobacterium tuberculosis* in its closed or mostly closed state. The periplasmic and side views of the molecule show the pentameric structure of the channel. The different domains of a single subunit are shown for clarity on the right

a homopentamer (Chang et al. 1998; Sukharev et al. 1999a). Site-directed mutagenesis confirmed that channel activity could be modified by structural changes to the protein (Blount et al. 1996b, 1997), and a random mutagenesis study implicated the cytoplasmic half of the first transmembrane domain (TMD1) as a mutagenic 'hot spot', implying its importance in mechanosensitive channel function (Ou et al. 1998; Maurer and Dougherty 2003).

Many of the predictions were confirmed and others resolved when a crystal structure of a homologue from *Mycobacterium tuberculosis* (Tb-MscL) was obtained (Chang et al. 1998). A structural model derived from X-ray crystallography of this channel is presented in Fig. 2.3. In this model, the complex is definitively shown to be a homopentamer. The cytoplasmic half of TMD1 appeared to be the central pore or constriction site of the channel, thus providing an explanation of why this was a mutagenic hot spot. TMD2 appears to face the lipids, there is a periplasmic loop between the two TMDs – thought to provide a torsional spring component to the channel (Blount et al. 1996b; Ajouz et al. 2000; Park et al. 2004; Tsai et al. 2005) – and there is a bundle of helixes at the C-terminal end of the protein. Because only a small opening was observed at the pore of the channel (~4 Å), it appeared to be in a closed, or nearly closed, conformation. Many of the predictions of the relative locations of specific residues in the closed structure have been supported by electron paramagnetic resonance (EPR) spectroscopy in combination with site-directed spin labeling (SDSL) (Perozo et al. 2001). However, some evidence suggests a slightly altered model around the constriction point of the pore for the "fully closed" *E. coli* MscL channel (Bartlett et al. 2004, 2006; Iscla et al. 2004; Levin and Blount 2004; Li et al. 2004). Several experiments using relatively independent approaches performed with the *E. coli* MscL channel now suggest that the structure does not represent the fully closed state found in membranes.

The highly conserved MscL channel is almost ubiquitous within the bacterial kingdom, and many homologues of the *E. coli* MscL channel have been studied and shown to encode mechanosensitive channel activity (Moe et al. 1998). There is at least one marine bacterial species, *Vibrio alginolyticus*, that does not appear to contain a MscL and is sensitive to osmotic downshift; interestingly, expression of the *E. coli* MscL in trans allows this cell to survive such a challenge (Nakamaru et al. 1999). When the Tb-MscL structure was published, there was no functional data on this homologue. Subsequent examination of Tb-MscL activity demonstrated that it was indeed functional, but not in a normal physiological range, at least when expressed in *E. coli*; the Tb-MscL channel needed far more energy to gate than Eco-MscL, and did not rescue the osmotic-lysis phenotype of the MscS/MscL *E. coli* double mutant (Moe et al. 2000). On the other hand, experiments in which analogous residues were similarly mutated suggested that the Eco- and Tb-MscL functioned by similar mechanisms (Moe et al. 2000). It is important to note that, because of the marked functional differences, one must be cautious when trying to correlate functional studies performed with the Eco-MscL with aspects of the Tb-MscL structure.

A current model for the gating of MscL by Sukharev and Guy (SG model), based on *M. tuberculosis* MscL crystal structure and crosslinking experiments, is shown in Fig. 2.4. Upon channel opening, the TMDs are thought to tilt, and the channel opens like the iris of a camera, with residues from TMD1 forming the lumen of the pore. A second model of MscL gating, based on experimental data from EPR studies, is

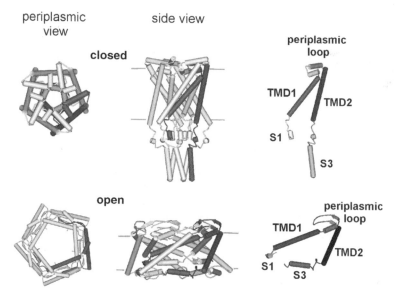

Fig. 2.4 One set of proposed models for the closed and open structures of *E. coli* MscL. Closed and fully open states of the channel are shown from periplasmic and side views; *horizontal gray lines* in the middle panel indicate the approximate position of the cell membrane. The spatial disposition of the different domains of the molecule in closed and open states in a single subunit is shown on the right

in agreement with these gross features of the SG model. However, many other details of channel gating are still in debate. The SG model predicts precisely which residues line the pore (Sukharev et al. 2001b), but subsequent studies have cast doubt on many of these predictions (Perozo et al. 2002a; Bartlett et al. 2004, 2006; Li et al. 2004). Another feature of the SG model for MscL gating is the prediction that the separation of the TMD1s is independent of ion permeation, and a stable "closed-expanded" state of the channel can exist. By this model, the extreme N-terminal end of the protein, not resolved in the crystal structure and referred to as S1, functions as a second gate (Sukharev et al. 2001a, 2001b). An alternative view is that it is the separation of the TMD1s, and transient exposure of non-hydrophilic residues to an aqueous environment, i.e., the lumen of the channel, that is the primary energy barrier to channel gating (Chang et al. 1998; Blount and Moe 1999); in essence, the channel tears open before the membrane rips. This latter hypothesis has the advantage that it would explain both the increased sensitivity and shortened open dwell times experimentally observed in channels with mutations to more hydrophilic residues in the pore. Hence, although the general features of the open structure seem likely, many details have yet to be resolved.

The positioning of the C-termini in the closed and open state is of physiological importance because it may determine what can transverse the channel. In the original paper describing the Tb-MscL crystal structure, it was noted that "it is possible that the cytoplasmic helix is stabilized in this structure by the low pH of the crystallization conditions." Indeed, an alternative structure has been proposed (Anishkin et al. 2003). The model predicts, however, that the alternative helix-bundle is extremely stable and remains intact even upon channel gating. By this model, the C-terminal acts as a sieve. On the other hand, experimental molecular sieving experiments (Cruickshank et al. 1997), and a more recent study using labeled compounds (van den Bogaart et al. 2007), have suggested that compounds greater than 30 Å, or proteins on the order of 6.5 kDa, transverse the MscL pore; these data demonstrate that the C-terminal bundle does not play much of a sieving role. Indeed, as stated above, some studies suggest that even larger proteins, such as thioredoxin, the elongation factor Tu, and the heat shock protein DnaK are released by MscL upon osmotic downshock (Ajouz et al. 1998; Berrier et al. 2000), although it seems unlikely that these proteins flux as globular proteins. Other studies dispute the finding that such large proteins transit through MscL (Vazquez-Laslop et al. 2001; van den Bogaart et al. 2007). There may be a resolution: one study suggested that the discrepancy among other previous studies is simply due to a variation of the protocol; some large proteins do indeed flux through MscL, but only under certain experimental conditions (Ewis and Lu 2005). Clearly, more research is needed in this area.

2.3   MscS

In the very first experiments in which patch clamp was applied to native *E. coli* membranes it is likely that it is the MscS channel whose activity was first observed and characterized, given the conductance and channel kinetics reported (Martinac et al. 1987, 1990). This also makes sense because MscS is prevalent in native membranes and is more sensitive to tension than MscL. As with MscL, the gene corresponding to MscS activity has been cloned, and a crystal structure as well as proposed models for gating exist. However, the molecular identity of the channel was not found until several years after that of MscL, and thus the MscS field is not yet as well developed.

The MscS and MscL families appear to be quite distinct from each other. While MscL is strongly conserved, even amongst very diverse bacterial species, MscS shows much more variation; even a single organism often contains numerous homologues. On the other hand, a distant evolutionary origin between the family of MscS and MscL mechanosensitive channels has been proposed (Kloda and Martinac 2001, 2002; Martinac 2004). This is based on the apparent sequence homology of a few channels from archaea with MscL that are clearly related to MscS in structure; these chimera-like channels could potentially be 'missing links' between the MscL and MscS families.

The *mscS* gene was discovered, somewhat serendipitously, by classical microbial genetics. An *E. coli* mutant strain, generated by UV mutagenesis, that showed impaired growth in the presence of both high K^+ and betaine or proline was isolated (McLaggan et al. 2002). The lesion was identified as a missense mutation within a gene called *kefA* (also called *aefA*). The authors realized that the phenotype had some of the characteristics of what one may anticipate from a dysfunctional bacterial mechanosensitive channel. Hence, even though the *kefA*-null mutant still showed what appeared to be normal activities in native *E. coli* membranes, the authors pursued and investigated homologues of this gene found in the *E. coli* genome. This persistence led to identification of a gene, *yggB*, which correlated well with MscS activity (Levina et al. 1999). A closer investigation of null strains led to the discovery that *kefA* did encode a channel activity that was masked by MscS; the former activity was observed only in high K^+ buffer, which suggested its current name, MscK (for K^+ regulated) (Li et al. 2002), while the same publication suggested YggB be renamed MscS; these names have since been in general use.

Many of the early studies characterizing the voltage-dependent nature (Martinac et al. 1987) and inactivation properties (Koprowski and Kubalski 1998) of "mechanosensitive channels of *E. coli*" were probably characterizing a combination of MscS and MscK. The MscS and MscK activities are of similar conductance and both are more sensitive to membrane stretch than is the MscL channel. Hence, a more recent study using a *mscK* null mutant strain of *E. coli* allowed for the analysis of MscS in isolation (Akitake et al. 2005; Sotomayor et al. 2006). These latter works serve as the current definitive studies on the properties of MscS activity. The findings include the observation that MscS exhibits essentially voltage-independent activation by tension, but strong voltage-dependent inactivation under depolarizing conditions. In addition, the channel appears to respond preferentially to acute stimuli but inactivates prior to opening if the stimulus is applied slowly over the course of several seconds.

The structure of *E. coli* MscS was solved to 3.9 Å resolution by X-ray crystallography (Bass et al. 2002) where residues 27–280, of the total 287, were resolved

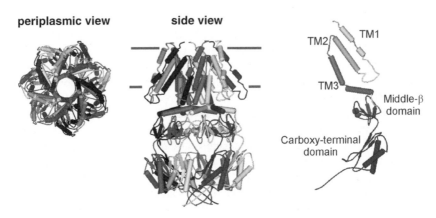

periplasmic view **side view**

TM2 TM1

TM3

Middle-β domain

Carboxy-terminal domain

Fig. 2.5 The mechanosensitive channel of small conductance from *E. coli* as solved in the crystal structure. The heptameric structure of *E. coli* MscS channel is shown from periplasmic and side views. The *horizontal gray lines* in the latter represent the approximate position of the cell membrane. The different domains are shown in a single subunit on the right for clarity

(Fig. 2.5). Three helical TMDs were found at the N-terminal region of the protein. The channel, however, appeared to be a homoheptamer (Bass et al. 2002), rather than the hexamer predicted from crosslinking experiments (Sukharev 2002). TMD3, which is rich in glycine and alanine, appears to form the constriction point or pore of the channel. A glycine at position 113 induces a turn within the α-helical domain, realigning the helix along the presumed membrane/cytoplasmic interface. Distal to this structure is a region that is relatively high in β-sheet character, which appears to form a cytoplasmic "cage" with seven pores, or portals, at the subunit interfaces. Finally, all of the subunits interact to terminate in a short β-barrel-like 'crown' at the extreme C-terminal end. This cage-like structure has been proposed to act as a molecular sieve that determines the size, and perhaps ionic preference, of the pore. The sizes of the pores in the crystal structure are as follows: the pore formed by TMD3 is ~11 Å, the portals in the cytoplasmic cage are ~14 Å, and the extreme C-terminal β-barrel is ~8 Å. Given the size of the potential pores, especially the transmembrane constriction point, it was presumed that the channel might be in an open state. However, molecular dynamic simulations now suggest that a hydrophobic barrier may deter permeation (Anishkin and Sukharev 2004). These data have led to the speculation that the channel is in an inactivated or desensitized state, which may indeed be a low-energy state, especially for a channel no longer under the constraints of the lateral pressures of the biological membrane. Further investigation will be required to determine if the structure is open, partially open or inactivated.

The constriction defining the pore of the MscS channel is composed of a region of TMD3 in which glycine and alanine residues appear to be tightly packed, with the closest association between glycine 108 and alanine 106. Assuming the channel is open or partially open, a model for the closure of MscS has been derived from mutagenesis experiments in which each of the glycines in this region has been substituted with serine (Edwards et al. 2005). By this model, the TMD3s become more vertical or normal to the membrane plane by sliding along each other and rotating. The resulting change in the packing of the small amino acids, glycine and alanine, in this region would then lead to a channel with a smaller closed pore. The leucines at positions 105 and 109 would come into closer proximity, thus forming a tighter constriction point and more efficient hydrophobic barrier leading to an unambiguous non-conducting state.

The importance of the cytoplasmic cage-like structure is emphasized by the observation that deletions at the C-terminal end of the protein are poorly tolerated, often yielding channels that are not expressed well in the membrane (Schumann et al. 2004). One study found that Ni^{2+} binding to MscS poly-histidine tagged at the C-terminus inhibited gating, suggesting movement in this region (Koprowski and Kubalski 2003). Another study tried to determine the proximity of specific residues within the cage-like structure of the closed MscS channel by disulfide trapping and crosslinking studies, and concluded that, in the closed conformation, this structure assumes a much more compact structure than that observed in the solved crystal structure (Miller et al. 2003). This hypothesis has been referred to as the "Chinese-lantern" model by analogy with lanterns whose intensity is adjusted by either collapsing or expanding the lamp (Edwards et al. 2004). Taken to the extreme, it seems possible that the cytoplasmic portholes not only serve as a molecular sieve, but that they also constrict enough to inhibit or retard permeation when the channel is in the closed state; this may be a sort of second gate.

Overall, the closed structure of the MscS channel may be more compact than that observed in the crystal structure. One emerging model for the gating of MscS is shown schematically in Fig. 2.6. Similar to MscL, in the open structure the

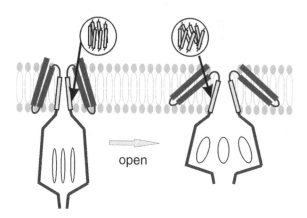

Fig. 2.6 A schematic of the "Chinese Lantern" model describing the structural rearrangements thought to occur upon gating of the MscS channel. The three transmembrane domains (TMDs), depicted as *gray cylinders*, are within the bilayer. Nine pores exist: one in the bilayer formed by TM3, seven at the interface of the subunits within the cytoplasmic "cage", and one, a β-barrel, at the extreme cytoplasmic end. This model maintains that each of these pores expand upon opening of the channel. *Circled insets* Details of the TMD3 domains (*arrow*) emphasizing the predicted change in tilt of these domains in upon gating

orientation of the TMDs forming the pore may adjust, becoming more tilted relative to the normal. But the structural rearrangements in MscS gating are thought not to be limited to the TMDs. The potential collapse of the cytoplasmic cage domain may define, at least in part, the potential permeation barriers in the closed structure. The TMD1 and TMD2 domains may also move upon gating, flapping out like wings upon stimulation (Bass et al. 2002). Consistent with this hypothesis, placing a negative charge at position 40, which is just below the lipid-aqueous interface on the periplasmic side, should increase the probability of the extension of the these 'wings', and indeed does effect a gain-of-function phenotype (Okada et al. 2002); on the other hand, data derived from a recent asparagine scan was interpreted to suggest that this movement is trivial or non-existent (Nomura et al. 2006). While Fig. 2.6 and the discussion above describe some of the current views, one must appreciate that less is known of MscS than MscL, and the models for its gating are still in flux; for a more complete discussion of the evidence for and against aspects of gating models for MscS, see Blount et al. (2005).

2.4   How Do Bacterial Mechanosensitive Channels Sense Osmolarity?

There are a number of ways in which a sensor can sense osmolarity (see Poolman et al. 2002 for a review of potential sensory information for both hyper- and hypo-osmosensors). For mechanosensitive channels, this has become an issue of whether cytoskeletal and extracellular tethers are required for many of the sensors. There are several lines of evidence that such tethers play little role in MscL and MscS gating. The strongest evidence is that both channels can be solubilized, purified, reconstituted into synthetic membranes and yet remain functional; no additional proteins are required (Häse et al. 1995; Blount et al. 1996a; Okada et al. 2002; Sukharev 2002).

Although it was clear from reconstitution experiments that only the channel and a lipid bilayer are required for MscL or MscS channel activity, there were still two possible mechanisms: do the channels sense pressure across the membrane or tension within it? A clue was given by a previous experiment performed in the yeast *Saccharomyces cereviseae*. The authors used data from whole-cell patch clamp experiments of yeast of varying size, and thus different radius of curvature, and applied Laplace's law (tension in a membrane equals the pressure across it times the radius of curvature divided by 2) to calculate tension. When the probability of channel opening was plotted versus the positive pressure in the electrode required to gate the channel, yeast of three different sizes fit three independent Boltzmann curves. However, when the radius of curvature was measured, the tension in the membrane could be calculated, and the probability of channel opening (P_o) could then be plotted against tension in the membrane rather than pressure across it. When this was done, all of the data points from the three experiments merged to fit a single curve (Gustin et al. 1988). Hence, in yeast it appeared that the channel gated in response to the tension in the membrane, not the pressure across it.

Assuming that MscL and MscS also gated in response to tension within the membrane, and utilizing special imaging equipment to determine the radius of curvature in the patch, the spatial and energetic parameters of channel gating were calculated by using the Boltzmann plots of P_o versus tension. For MscL, such an analysis has been performed (Sukharev et al. 1999b), updated (Chiang et al. 2004), and utilized for some of the mutated channels (Anishkin et al. 2005). The current derived values for opening the wild type channel suggest that it takes approximately 7–13 dynes/cm (mN/m) of tension in the membrane to achieve a 50% probability of channel gating, that the energy required to gate the channel at this level (ΔE) is $51 \pm 13\,kT$, and that the change in area (ΔA) is $20 \pm 5\,nm^2$. The latter parameter is consistent with current predictions for the approximate pore size of the open channel. A similar study for MscS estimated the energy, area, and gating charge for the closed-to-open transition of MscS to be $24 kT$, $18 nm^2$, and $+0.8$, respectively (Akitake et al. 2005). All of these data have been consistent with the hypothesis that these channels sense tension in the membrane; a more recent study formally demonstrated that MscL does indeed sense membrane tension (Moe and Blount 2005).

Amphipaths, which can intercalate into the membrane asymmetrically, provide additional evidence that the MscL and MscS channels sense membrane tension. One of the early studies demonstrated that bacterial mechanosensitive channels were modulated by amphipaths (Martinac et al. 1990); although at the time it was not known that there were multiple activities in the *E. coli* membrane, the sensitivity and conductance were consistent with MscS activity. A subsequent and more detailed study with MscL demonstrated that the activity of this channel is also modulated by amphipaths and could be gated by lysophospholipids (Perozo et al. 2002b). These findings are consistent with the hypothesis that these channels sense physical changes in the membrane.

Other stimuli have been proposed to play a role in bacterial mechanosensitive channel gating. For example, decreasing the thickness of the membrane by reconstituting into lipids with shorter chain lengths appears to make MscL more sensitive to membrane tension (Perozo et al. 2002b). This suggests that membrane thinning in response to stretch may play a modulatory role, but does not directly gate the channel. It is as yet unclear if membrane thinning affects MscS gating. Adding curvature to the membrane may be a way of adding tension (Perozo et al. 2002b; Meyer et al. 2006), but again curvature itself does not appear to be the major stimulus for MscL gating (Moe and Blount 2005). Finally, studies have implied that direct interactions may occur between lipid headgroups and MscL (Elmore and Dougherty 2001; Yoshimura et al. 2004; Powl et al. 2005) and MscS (Nomura et al. 2006); however, at least for MscL, one study tested lipids with various headgroups and found that neither negatively charged lipid headgroups nor the major endogenous headgroups expressed in *E. coli* appear to favorably affect channel gating (Moe and Blount 2005). Hence, it appears that the primary stimulus is a change in physical properties of the membrane, most likely the lateral pressure profile within the membrane and a differential tension between the bilayer leaflets; consistent with this theory, a study of MscS under high hydrostatic pressure suggested that it is lateral compression of the bilayer that is intimately involved in the expansion of the channel area as the channel opens (Macdonald and Martinac 2005; see (Blount et al. 2007b) for a more detailed description of current theories of stimuli for various mechanosensitive channels).

2.5   Perspective: Other Mechanosensitive Channels from Bacteria and Other Organisms

From physiological studies it is clear that the MscS and MscL channels are major players in the cell's ability to rapidly adapt to osmotic downshocks. But are these the only players? Unfortunately, little is known of MscM, so its role cannot be evaluated. As mentioned earlier, MscK is a homologue of MscS that is found in *E. coli* (Levina et al. 1999). This channel activity has not yet been functionally reconstituted into membranes, but has been shown in native membranes to be regulated by ion concentration; perhaps the most physiologically relevant is K^+ (Li et al. 2002). In addition, in spheroplasts MscK activity is seen in only one of five patches, even though the amount of membrane in the patch is greater than that in an entire bacterium. Finally, the MscK protein is predicted to be significantly larger than MscS, with eight additional TMDs and a large periplasmic region (McLaggan et al. 2002). Hence, it seems possible that MscK may have tethers to periplasmic or cell wall components that play a role in gating – only spheroplast patches that maintained these functional tethers would contain channel activity. At least three additional homologues of MscS in *E. coli* are predicted from the genomic sequence; so far, no activity has been measured from them. One hypothesis is that since MscK functions in patch clamp only under very specific environmental conditions, this may also be true in vivo (although conditions for MscK to protect the cell from osmotic downshock have not been found); thus, perhaps we have yet to define the proper conditions to measure these other putative channels (Li et al. 2002). Homologues of MscS have also been found in plants, where one study has demonstrated that they play a role in regulating plastid shape and size (Haswell and Meyerowitz 2006). These may also be modified mechanosensitive channels with specially designed properties for highly specific functions.

Although MscL and MscS are quite distinct mechanosensitive channels, there appear to be several similarities in how they sense and respond to mechanical forces. They both appear to sense changes in the physical properties of the membrane that occur when the membrane is under tension, and the α-helices that form the pore, by current models, tilt relative to the membrane. But the question remains: are other channels similar in these molecular mechanisms? There is accumulating evidence that there may be some shared molecular mechanisms (Blount et al. 2007b). Several mammalian channels appear to be gated by amphipaths (e.g. see Patel et al. 2001) and one can be solubilized and functionally reconstituted (Maroto et al. 2005). Hence, bacterial channels may serve as a paradigm for how mechanosensitive channels can sense and respond to membrane tension.

Acknowledgments The authors are supported by Grant I-1420 of the Welch Foundation, Grant FA9550-05-1-0073 of the Air Force Office of Scientific Review, Grant 0655012Y of the American Heart Association – Texas Affiliate, and Grant GM61028 from the National Institutes of Health.

References

Ajouz B, Berrier C, Garrigues A, Besnard M, Ghazi A (1998) Release of thioredoxin via the mechanosensitive channel MscL during osmotic downshock of *Escherichia coli* cells. J Biol Chem 273:26670–26674

Ajouz B, Berrier C, Besnard M, Martinac B, Ghazi A (2000) Contributions of the different extramembranous domains of the mechanosensitive ion channel MscL to its response to membrane tension. J Biol Chem 275:1015–1022

Akitake B, Anishkin A, Sukharev S (2005) The "dashpot" mechanism of stretch-dependent gating in MscS. J Gen Physiol 125:143–154

Anishkin A, Sukharev S (2004) Water dynamics and dewetting transitions in the small mechanosensitive channel MscS. Biophys J 86:2883–2895

Anishkin A, Gendel V, Sharifi NA, Chiang CS, Shirinian L, Guy HR, Sukharev S (2003) On the conformation of the COOH-terminal domain of the large mechanosensitive channel MscL. J Gen Physiol 121:227–244

Anishkin A, Chiang CS, Sukharev S (2005) Gain-of-function mutations reveal expanded intermediate states and a sequential action of two gates in MscL. J Gen Physiol 125:155–170

Arakawa T, Timasheff SN (1985) The stabilization of proteins by osmolytes. Biophys J 47:411–414

Bartlett JL, Levin G, Blount P (2004) An in vivo assay identifies changes in residue accessibility on mechanosensitive channel gating. Proc Natl Acad Sci USA 101:10161–10165

Bartlett JL, Li Y, Blount P (2006) Mechanosensitive channel gating transitions resolved by functional changes upon pore modification. Biophys J 91:3684–3691

Bass RB, Strop P, Barclay M, Rees DC (2002) Crystal structure of *Escherichia coli* MscS, a voltage-modulated and mechanosensitive channel. Science 298:1582–1587

Berrier C, Coulombe A, Szabo I, Zoratti M, Ghazi A (1992) Gadolinium ion inhibits loss of metabolites induced by osmotic shock and large stretch-activated channels in bacteria. Eur J Biochem 206:559–565

Berrier C, Besnard M, Ajouz B, Coulombe A, Ghazi A (1996) Multiple mechanosensitive ion channels from *Escherichia coli*, activated at different thresholds of applied pressure. J Membr Biol 151:175–187

Berrier C, Garrigues A, Richarme G, Ghazi A (2000) Elongation factor Tu and DnaK are transferred from the cytoplasm to the periplasm of *Escherichia coli* during osmotic downshock presumably via the mechanosensitive channel mscL. J Bacteriol 182:248–251

Blount P, Moe P (1999) Bacterial mechanosensitive channels: integrating physiology, structure and function. Trends Microbiol 7:420–424

Blount P, Sukharev SI, Moe PC, Schroeder MJ, Guy HR, Kung C (1996a) Membrane topology and multimeric structure of a mechanosensitive channel protein of *Escherichia coli*. EMBO J 15:4798–4805

Blount P, Sukharev SI, Schroeder MJ, Nagle SK, Kung C (1996b) Single residue substitutions that change the gating properties of a mechanosensitive channel in *Escherichia coli*. Proc Natl Acad Sci USA 93:11652–11657

Blount P, Schroeder MJ, Kung C (1997) Mutations in a bacterial mechanosensitive channel change the cellular response to osmotic stress. J Biol Chem 272:32150–32157

Blount P, Iscla I, Li Y, Moe PC (2005) The bacterial mechanosensitive channel MscS and its extended family. In: Kubalski A, Martinac B (eds) Bacterial channels and their eukaryotic homologues. ASM, Washington, DC

Blount P, Iscla I, Moe PC, Li Y (2007a). MscL: The bacterial mechanosensitive channel of large conductance. In: Mechanosensitive Ion Channels (a volume in the Current Topics in Membranes series). O. P. Hamill (ed) St. Louis, MO, Elsevier Press 58:202–233

Blount P, Li Y, Moe PC, Iscla I (2007b). Mechanosensitive channels gated by membrane tension: Bacteria and beyond. In: Mechanosensitive ion channels (a volume in the Mechanosensitivity in Cells and Tissues, Moscow Academia series). A. Kamkin and I. Kiseleva (ed) New York, Springer Press 70–100

Britten RJ, McClure FT (1962) The amino acid pool in *Escherichia coli*. Bacteriol Rev 26:292–335

Calamita G, Kempf B, Bonhivers M, Bishai WR, Bremer E, Agre P (1998) Regulation of the *Escherichia coli* water channel gene aqpZ. Proc Natl Acad Sci USA 95:3627–3631

Cayley S, Lewis BA, Guttman HJ, Record MT Jr (1991) Characterization of the cytoplasm of *Escherichia coli* K-12 as a function of external osmolarity. Implications for protein–DNA interactions in vivo. J Mol Biol 222:281–300

Cayley S, Lewis BA, Record MT Jr (1992) Origins of the osmoprotective properties of betaine and proline in Escherichia coli K-12. J Bacteriol 174:1586–1595

Chang G, Spencer RH, Lee AT, Barclay MT, Rees DC (1998) Structure of the MscL homolog from *Mycobacterium tuberculosis*: a gated mechanosensitive ion channel. Science 282:2220–2226

Chiang CS, Anishkin A, Sukharev S (2004) Gating of the large mechanosensitive channel in situ: estimation of the spatial scale of the transition from channel population responses. Biophys J 86:2846–286

Cruickshank CC, Minchin RF, Le Dain AC, Martinac B (1997) Estimation of the pore size of the large–conductance mechanosensitive ion channel of *Escherichia coli*. Biophys J 73:1925–1931

Doyle RJ, Marquis RE (1994) Elastic, flexible peptidoglycan and bacterial cell wall properties. Trends Microbiol 2:57–60

Edwards MD, Booth IR, Miller S (2004) Gating the bacterial mechanosensitive channels: MscS a new paradigm? Curr Opin Microbiol 7:163–167

Edwards MD, Li Y, Kim S, Miller S, Bartlett W, Black S, Dennison S, Iscla I, Blount P, Bowie JU, Booth IR (2005) Pivotal role of the glycine-rich TM3 helix in gating the MscS mechanosensitive channel. Nat Struct Mol Biol 12:113–119

Elmore DE, Dougherty DA (2001) Molecular dynamics simulations of wild-type and mutant forms of the *Mycobacterium tuberculosis* MscL channel. Biophys J 81:1345–1359

Epstein W, Schultz SJ (1965) Cation transport in *Escherichia coli*. J Gen Physiol 49:221–234

Ewis HE, Lu CD (2005) Osmotic shock: a mechanosensitive channel blocker can prevent release of cytoplasmic but not periplasmic proteins. FEMS Microbiol Lett 253:295–301

Felle H, Porter JS, Slayman CL, Kaback HR (1980) Quantitative measurements of membrane potential in *Escherichia coli*. Biochemistry 19:3585–3590

Glauner B, Holtje JV, Schwarz U (1988) The composition of the murein of *Escherichia coli*. J Biol Chem 263:10088–10095

Gustin MC, Zhou XL, Martinac B, Kung C (1988) A mechanosensitive ion channel in the yeast plasma membrane. Science 242:762–765

Häse CC, Le Dain AC, Martinac B (1995) Purification and functional reconstitution of the recombinant large mechanosensitive ion channel (MscL) of *Escherichia coli*. J Biol Chem 270:18329–18334

Haswell ES, Meyerowitz EM (2006) MscS-like proteins control plastid size and shape in *Arabidopsis thaliana*. Curr Biol 16:1–11

Iscla I, Levin G, Wray R, Reynolds R, Blount P (2004) Defining the physical gate of a mechanosensitive channel, MscL, by engineering metal-binding sites. Biophys J 87:3172–3180

Kloda A, Martinac B (2001) Molecular identification of a mechanosensitive channel in archaea. Biophys J 80:229–240

Kloda A, Martinac B (2002) Common evolutionary origins of mechanosensitive ion channels in Archaea, bacteria and cell-walled Eukarya. Archaea 1:35–44

Knaysi G (1951) Elements of bacterial cytology. Comstock, Ithaca

Koch AL, Woeste S (1992) Elasticity of the sacculus of *Escherichia coli*. J Bacteriol 174:4811–4819

Koprowski P, Kubalski A (1998) Voltage-independent adaptation of mechanosensitive channels in *Escherichia coli* protoplasts. J Membr Biol 164:253–262

Koprowski P, Kubalski A (2003) C termini of the *Escherichia coli* mechanosensitive ion channel (MscS) move apart upon the channel opening. J Biol Chem 278:11237–11245

Levin G, Blount P (2004) Cysteine scanning of MscL transmembrane domains reveals residues critical for mechanosensitive channel gating. Biophys J 86:2862–2870

Levina N, Totemeyer S, Stokes NR, Louis P, Jones MA, Booth IR (1999) Protection of *Escherichia coli* cells against extreme turgor by activation of MscS and MscL mechanosensitive channels: identification of genes required for MscS activity. EMBO J 18:1730–1737

Li Y, Moe PC, Chandrasekaran S, Booth IR, Blount P (2002) Ionic regulation of MscK, a mechanosensitive channel from *Escherichia coli*. EMBO J 21:5323–5330

Li Y, Wray R, Blount P (2004) Intragenic suppression of gain-of-function mutations in the *Escherichia coli* mechanosensitive channel, MscL. Mol Microbiol 53:485–495

Macdonald AG, Martinac B (2005) Effect of high hydrostatic pressure on the bacterial mechanosensitive channel MscS. Eur Biophys J 34:434–441

Maroto R, Raso A, Wood TG, Kurosky A, Martinac B, Hamill OP (2005) TRPC1 forms the stretch-activated cation channel in vertebrate cells. Nat Cell Biol 7:179–185

Martinac B (2004) Mechanosensitive ion channels: molecules of mechanotransduction. J Cell Sci 117:2449–2460

Martinac B, Buechner M, Delcour AH, Adler J, Kung C (1987) Pressure-sensitive ion channel in *Escherichia coli*. Proc Natl Acad Sci USA 84:2297–2301

Martinac B, Adler J, Kung C (1990) Mechanosensitive ion channels of *E. coli* activated by amphipaths. Nature 348:261–263

Maurer JA, Dougherty DA (2003) Generation and evaluation of a large mutational library from the *Escherichia coli* mechanosensitive channel of large conductance, MscL – implications for channel gating and evolutionary design. J Biol Chem 278:21076–21082

McLaggan D, Jones MA, Gouesbet G, Levina N, Lindey S, Epstein W, Booth IR (2002) Analysis of the kefA2 mutation suggests that KefA is a cation-specific channel involved in osmotic adaptation in *Escherichia coli*. Mol Microbiol 43:521–536

Meyer GR, Gullingsrud J, Schulten K, Martinac B (2006) Molecular dynamics study of MscL interactions with a curved lipid bilayer. Biophys J 91:1630–1637

Miller S, Edwards MD, Ozdemir C, Booth IR (2003) The closed structure of the MscS mechanosensitive channel – cross-linking of single cysteine mutants. J Biol Chem 278:32246–32250

Mitchell P, Moyle J (1956) Osmotic function and structure in bacteria. Symp Soc Gen Microbiol 6:150–180

Moe P, Blount P (2005) Assessment of potential stimuli for mechano-dependent gating of MscL: effects of pressure, tension, and lipid headgroups. Biochemistry 44:12239–12244

Moe PC, Blount P, Kung C (1998) Functional and structural conservation in the mechanosensitive channel MscL implicates elements crucial for mechanosensation. Mol Microbiol 28:583–592

Moe PC, Levin G, Blount P (2000) Correlating a protein structure with function of a bacterial mechanosensitive channel. J Biol Chem 275:31121–31127

Munro AW, Ritchie GY, Lamb AJ, Douglas RM, Booth IR (1991) The cloning and DNA sequence of the gene for the glutathione-regulated potassium-efflux system KefC of *Escherichia coli*. Mol Microbiol 5:607–616

Nakamaru Y, Takahashi Y, Unemoto T, Nakamura T (1999) Mechanosensitive channel functions to alleviate the cell lysis of marine bacterium, *Vibrio alginolyticus*, by osmotic downshock. FEBS Lett 444:170–172

Nomura T, Sokabe M, Yoshimura K (2006) Lipid-protein interaction of the MscS mechanosensitive channel examined by scanning mutagenesis. Biophys J 91:2874–2881

Okada K, Moe PC, Blount P (2002) Functional design of bacterial mechanosensitive channels. Comparisons and contrasts illuminated by random mutagenesis. J Biol Chem 277:27682–27688

Ou X, Blount P, Hoffman RJ, Kung C (1998) One face of a transmembrane helix is crucial in mechanosensitive channel gating. Proc Natl Acad Sci USA 95:11471–11475

Park KH, Berrier C, Martinac B, Ghazi A (2004) Purification and functional reconstitution of N- and C-halves of the MscL channel. Biophys J 86:2129–2136

Patel AJ, Lazdunski M, Honore E (2001) Lipid and mechano-gated 2P domain K(+) channels. Curr Opin Cell Biol 13:422–428

Perozo E, Kloda A, Cortes DM, Martinac B (2001) Site-directed spin-labeling analysis of reconstituted Mscl in the closed state. J Gen Physiol 118:193–206

Perozo E, Cortes DM, Sompornpisut P, Kloda A, Martinac B (2002a) Open channel structure of MscL and the gating mechanism of mechanosensitive channels. Nature 418:942–948

Perozo E, Kloda A, Cortes DM, Martinac B (2002b) Physical principles underlying the transduction of bilayer deformation forces during mechanosensitive channel gating. Nat Struct Biol 9:696–703

Poolman B, Blount P, Folgering JH, Friesen RH, Moe PC, van der Heide T (2002) How do membrane proteins sense water stress? Mol Microbiol 44:889–902

Popham DL, Young KD (2003) Role of penicillin-binding proteins in bacterial cell morphogenesis. Curr Opin Microbiol 6:594–599

Powl AM, East JM, Lee AG (2005) Heterogeneity in the binding of lipid molecules to the surface of a membrane protein: hot spots for anionic lipids on the mechanosensitive channel of large conductance MscL and effects on conformation. Biochemistry 44:5873–5883

Saint N, Lacapere JJ, Gu LQ, Ghazi A, Martinac B, Rigaud JL (1998) A hexameric transmembrane pore revealed by two-dimensional crystallization of the large mechanosensitive ion channel (MscL) of *Escherichia coli*. J Biol Chem 273:14667–14670

Schleyer M, Schmid R, Bakker EP (1993) Transient, specific and extremely rapid release of osmolytes from growing cells of *Escherichia coli* K-12 exposed to hypoosmotic shock. Arch Microbiol 160:424–431

Schumann U, Edwards MD, Li C, Booth IR (2004) The conserved carboxyl-terminus of the MscS mechanosensitive channel is not essential by increases stability and activity. FEBS Lett 572:233–237

Sotomayor M, Vasquez V, Perozo E, Schulten K (2006) Ion conduction through MscS as determined by electrophysiology and simulation. Biophys J 92:886–902

Stokes NR, Murray HD, Subramaniam C, Gourse RL, Louis P, Bartlett W, Miller S, Booth IR (2003) A role for mechanosensitive channels in survival of stationary phase: regulation of channel expression by RpoS. Proc Natl Acad Sci USA 100:15959–15964

Sukharev S (2002) Purification of the small mechanosensitive channel of *Escherichia coli* (MscS): the subunit structure, conduction, and gating characteristics in liposomes. Biophys J 83:290–298

Sukharev SI, Martinac B, Arshavsky VY, Kung C (1993) Two types of mechanosensitive channels in the *Escherichia coli* cell envelope: solubilization and functional reconstitution. Biophys J 65:177–183

Sukharev SI, Blount P, Martinac B, Blattner FR, Kung C (1994) A large-conductance mechanosensitive channel in *E. coli* encoded by *mscL* alone. Nature 368:265–268

Sukharev SI, Schroeder MJ, McCaslin DR (1999a) Stoichiometry of the large conductance bacterial mechanosensitive channel of *E. coli*. A biochemical study. J Membr Biol 171:183–193

Sukharev SI, Sigurdson WJ, Kung C, Sachs F (1999b) Energetic and spatial parameters for gating of the bacterial large conductance mechanosensitive channel, MscL. J Gen Physiol 113:525–540

Sukharev S, Betanzos M, Chiang C, Guy H (2001a) The gating mechanism of the large mechanosensitive channel MscL. Nature 409:720–724

Sukharev S, Durell S, Guy H (2001b) Structural models of the MscL gating mechanism. Biophys J 81:917–936

Tsai IJ, Liu ZW, Rayment J, Norman C, McKinley A, Martinac B (2005) The role of the periplasmic loop residue glutamine 65 for MscL mechanosensitivity. Eur Biophys J 34:403–412

Tsapis A, Kepes A (1977) Transient breakdown of the permeability barrier of the membrane of *Escherichia coli* upon hypoosmotic shock. Biochim Biophys Acta 469:1–12

Van den Bogaart G, Krasnikov V, Poolman B (2007) Dual-color fluorescence-burst analysis to probe protein efflux through the mechanosensitive channel MscL. Biophys J 92:1233–1240

Vazquez-Laslop N, Lee H, Hu R, Neyfakh AA (2001) Molecular sieve mechanism of selective release of cytoplasmic proteins by osmotically shocked *Escherichia coli*. J Bacteriol 183:2399–2404

Wood JM (1999) Osmosensing by bacteria: signals and membrane-based sensors. Microbiol Mol Biol Rev 63:230–262

Wood JM (2006) Osmosensing by bacteria. Sci STKE 357:pe43

Yao X, Jericho M, Pink D, Beveridge T (1999) Thickness and elasticity of Gram-negative murein sacculi measured by atomic force microscopy. J Bacteriol 181:6865–6875

Yoshimura K, Nomura T, Sokabe M (2004) Loss-of-function mutations at the rim of the funnel of mechanosensitive channel MscL. Biophys J 86:2113–2120

Zoratti M, Petronilli V (1988) Ion-conducting channels in a Gram-positive bacterium. FEBS Lett 240:105–109

Zoratti M, Petronilli V, Szabo I (1990) Stretch-activated composite ion channels in *Bacillus subtilis*. Biochem Biophys Res Commun 168:443–450

Chapter 3
Roles of Ion Channels in the Environmental Responses of Plants

Takuya Furuichi, Tomonori Kawano, Hitoshi Tatsumi(☒), and Masahiro Sokabe

Abstract When plant cells are exposed to environmental stresses or perceive internal signal molecules involved in growth and development, ion channels are transiently activated to convert these stimuli into intracellular signals. Among the ions taken up by plant cells, Ca^{2+} plays an essential role as an intracellular second messenger in plants; the cytoplasmic free Ca^{2+} concentration ($[Ca^{2+}]_c$) is therefore strictly regulated. Signal transduction pathways mediated by changes in $[Ca^{2+}]_c$ – termed Ca^{2+} signaling – are initiated by the activation of Ca^{2+}-permeable channels in many cases. To date, a large body of electrophysiological and recent molecular biological studies have revealed that plants possess Ca^{2+} channels belonging to distinct types with different gating mechanisms, and a variety of genes for Ca^{2+}-permeable channels have been isolated and functionally characterized. Topics in this chapter focus on long-distance signal translocation in plants and the characteristics of a variety of plant Ca^{2+}-permeable channels including voltage-dependent Ca^{2+}-permeable channels, cyclic nucleotide-gated cation channels, ionotropic glutamate receptors and mechanosensitive channels. We discuss their roles in environmental responses and in the regulation of growth and development.

Department of Physiology, Nagoya University School of Medicine, 65 Tsurumai, Nagoya 466-8550, Japan, tatsumi@med.nagoya-u.ac.jp

3.1   Introduction

Land plants show a variety of behaviors and plasticities in response to environmental stresses such as drought, salinity, and attacks by pathogens or insects. When plant cells are exposed to environmental stresses or perceive internal signal molecules involved in growth and development, ion channels are transiently activated to convert these stimuli into intracellular signals. Across the plasma- and endo-membranes, the concentrations of individual ions are critically altered as a consequence of ATP-dependent pumping and transporting of ions by specific trans-membrane proteins that generate and maintain the membrane potential. When ion channels are activated by a variety of stimuli, transient changes in the cytosolic concentration of the specific ion(s) and the membrane potential are produced. Since the activities of some of the channels are regulated by the membrane potential, the initial activation of ion channels causes subsequent activations of other channels to relay and enhance the initial signal. Thus, the regulation of ion channels is indispensable for signal transduction and the appropriate final behavior of plants. Among the ions taken up by plant cells, Ca^{2+} plays an essential role as an intracellular second messenger in plants and, therefore, the cytoplasmic free Ca^{2+} concentration ($[Ca^{2+}]_c$) is strictly regulated (Muto 1993). In the steady state, $[Ca^{2+}]_c$ is kept strictly at a low level through the continuous export of Ca^{2+} by Ca^{2+}-ATPase and H^+/Ca^{2+} antiporters. When Ca^{2+}-permeable channel(s) are activated in response to a variety of external and internal stimuli, a small amount of Ca^{2+} influx through activated channels will increase $[Ca^{2+}]_c$, followed by the activation of Ca^{2+}-regulated proteins such as calmodulins (CaMs) and calcium dependent protein kinases (CDPKs). These signal transduction pathways, termed Ca^{2+} signaling, are triggered mainly by the activation of Ca^{2+}-permeable channels in plants. As illustrated in Fig. 3.1, a large body of electrophysiological studies have elucidated that plants have several Ca^{2+} channels belonging to distinct types that differ in their gating mechanisms, namely ligand-gated, voltage-dependent, and stretch-activated (mechanosensitive) channels, as shown in animal cells (Piñeros and Tester 1997). Recently, a variety of plant genes encoding Ca^{2+}-permeable channels have been identified by the efforts of genome sequencing projects for certain plants and some of them have been functionally characterized. Although they are different from those in animal cells, many of them permeate not only Ca^{2+}, but also K^+, Na^+, and other cations like those in animal cells. It has been revealed that the expression level of these genes markedly affects Ca^{2+} homeostasis in plants. In particular, Ca^{2+}-permeable channels are now known to play important roles in Ca^{2+}-signaling in some plant species. Because Ca^{2+}-signaling plays a central role in stimulus perception and signal transduction, topics in this chapter focus on the characteristics of a variety of plant Ca^{2+}-permeable channels, and their roles in environmental responses and in the regulation of growth and development.

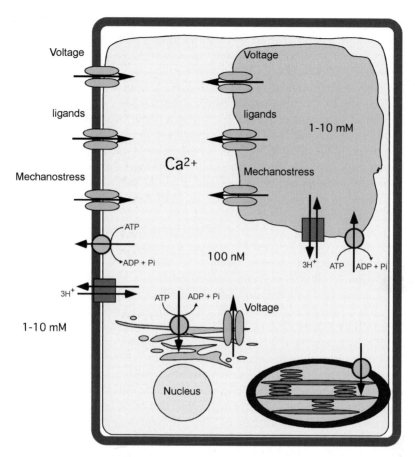

Fig. 3.1 Ca^{2+}-permeable channels in plant cells. Ca^{2+} channels responsive to voltage changes, ligands (cNMPs or Glu), and mechanical stimuli in the plasma- and endo-membranes are shown schematically. Ca^{2+}/H^+ antiporters and Ca^{2+}-ATPases are represented by *squares* and *circles*, respectively. Ca^{2+} concentrations in the cytosol, endoplasmic reticulum (*ER*), vacuole, and apoplastic space are also shown

3.2 Long-Distance Signal Translocation in Plants

For the regulation of growth and development within the entire plant body, long-distance signal transmission machineries are essential to relay information from local events such as wounding, viral infection, changes in nutritional condition, or water potential sensed by root hairs or stomata to the entire plant body to allow adaptation to these environmental stresses. Local stimuli are converted to intracellular signals, then transmitted to plant parts distant from the site of stimulus perception. As a path for systemic transportation of solutes, higher plants have developed a highly systematized vascular bundle system that plays a central role in the absorption and translocation of water, minerals, and other nutrients to support and maintain the growth of tissues and cells. Among the nutritional components, sugars such as sucrose, glucose, and fructose are essential for plant metabolism. These sugars are synthesized within chloroplasts in source tissues (mature leaves) by photosynthesis, and exported to photosynthetically less active or inactive sink tissues (e.g., roots, fruits, tubers, and immature leaves). In sink tissues, sugars are utilized for respiratory metabolism and as the substrates for synthesizing a variety of complex carbohydrates such as starch and cellulose. In addition to their involvement in metabolic processes, a role for sugars as signaling molecules has recently been highlighted (Gibson 2000; Koch 1996; Sheen et al. 1999). Sucrose, the predominant form of transported sugar in plants, specifically induces expression of genes under the control of the patatin promoter and the phloem-specific *rolC* promoter (Yokohama et al. 1994), and represses translation of *ATB2*, an *Arabidopsis thaliana* leucine zipper gene (Rook et al. 1998).

The involvement of Ca^{2+} and calmodulin in sugar-induced expression of β-amylase and sporamin was first implied by indirect evidence based on the suppressive effects of calmodulin inhibitors, including a Ca^{2+} chelator (EGTA) and Ca^{2+} channel blockers (Ohto et al. 1995). Transgenic luminiferous *Arabidopsis* and tobacco plants or suspension-cultured cells can be used to monitor changes in $[Ca^{2+}]_c$ directly; $[Ca^{2+}]_c$ increases coupled with a wide range of biotic and abiotic stimuli have been reported (Kawano et al. 1998; Knight et al. 1991). In leaves of *A. thaliana* expressing aequorin, $[Ca^{2+}]_c$ increased in response to sugar but not to non-metabolizable analogues of sugars (Furuichi et al. 2001b). This response was observed in leaves excised from autotrophically grown plants, but not in those prepared from heterotrophically grown plants, and the mRNA level of sucrose-H^+ symporters in the former was clearly higher than that in the latter. Sucrose-induced luminescence, reflecting an increase in $[Ca^{2+}]_c$ in aequorin-expressing *Arabidopsis* leaves, was suppressed by the antisense expression of the sucrose-H^+ symporters AtSUC1 and 2 (Furuichi et al. 2001a; Furuichi and Muto 2005), implying that sucrose-H^+ symporters are the key mediators of sugar-induced $[Ca^{2+}]_c$ increase. Using a two-dimensional photon-imaging system, Furuichi et al. (2001b) successfully monitored sucrose-induced transient increases in $[Ca^{2+}]_c$ in several leaves of autotrophically growing mature plants of *A. thaliana*. When 0.1 M sucrose was fed to the roots of these plants, a rapid and strong luminescence was observed in roots. The increased luminescence was followed by a weak luminescence in leaves; the luminescence moved from lower to upper leaves. The rate (i.e., velocity) of translocation of aequorin luminescence-emitting spots was roughly comparable to that of the spread of radioisotope-labeled sucrose, suggesting that the sugar signal was directly converted to a transient increase in $[Ca^{2+}]_c$ (Furuichi et al. 2001b, 2003). The application of nitrate (NO_3^-), which is also co-transported with H^+, and ammonium ion (NH_4^+), to excised leaves of *A. thaliana* also promoted a transient increase in $[Ca^{2+}]_c$, and the extent of this response was correlated with the expression levels of the corresponding transporters (Furuichi and Kawano 2006b). These observations suggest that translocation of key nutritional molecules and the subsequent $[Ca^{2+}]_c$ changes in perceptive cells play a major role in the long-distance propagation of signals required for the control of plant growth and development. Pearce et al. (1991)

were the first to find systemin, a ligand-peptide hormone from tomato leaves, which acts as a translocating signal molecule in plants. Systemin activates a defense response in the entire plant when one of the leaves is damaged. Systemin also promotes electrical responses in plants, but their roles and how the systemin-induced defense response is promoted in the entire plant are still obscure. In recent studies, some of the key components of long-distance signal translocation, such as homologous genes of glutamate receptors, were isolated and their roles have been partially characterized in plants, as described below.

A variety of signaling molecules are transferred to the entire body of animals via blood vessels with great similarity to the vascular bundle system in plants, as described above. Furthermore, animals perceive and transmit the information of environmental stresses via the nervous system. Input signals to sensory cells directly or indirectly activate a number of ion channels, and the resultant ion fluxes across the plasma membrane often cause a rapid change in the membrane potential. These electrical responses, called receptor potentials or generator potentials, will produce action potentials, which propagate towards the brain. Similarly, a large number of studies have revealed that electrical responses are also utilized in long-distance signal translocation in plants. In 1791, Luigi Galvani, one of the pioneering biologists in this field, provided the first evidence for electric signaling in plants (Galvani 1791). Another prominent scientist, Alexander von Humboldt, concluded that both animals and plants have a common bioelectrical feature, and suggested that the excitability of plant cells could be involved in long-distance signal translocation (von Humbolt 1797; Botting 1973). As illustrated in Fig. 3.2, environmental stimuli such as bacterial infection and mechanical stresses promote electrical responses, which propagate through the entire plant body. By injecting a fluorescent dye, Rhodes et al. (1996) have revealed that sieve-tube elements and companion cells of phloem are the major players in transmitting these electrical responses. These observations suggest that the plant vascular bundle system is endowed with multiple functions, corresponding to both neurons and blood vessels in animals. There are two types of electrical responses in plants, termed action potentials (APs) and variation potentials (VPs). Note that APs in plants are not identical to the well examined APs in neurons; in plants, fast and transient responses are termed as APs, and the following slow responses are termed VPs. In maize leaves, electrical stimulation that promotes APs alone lowers the concentrations of K^+ and Cl^-, implying that APs are generated mainly by K^+- and Cl^--channels (Fromm and Bauer 1994). On the other hand, transient shut-downs of H^+-pumps are thought to be involved in the generation of VPs (Julien and Frachisse 1992), which are modulated by extracellular Ca^{2+} concentration. A typical example of systemic electrical signaling in plants is a wounding response (e.g., triggered by herbivore damage), in which the electrical activity spreads from the wounded cotyledons to the entire plant, finally leading to systemic expression of proteinase inhibitor(s) that inhibit digestion by the insects (Wildon et al. 1989, 1992). The expression of proteinase inhibitor(s) in parts distant from the wounded tissue was not inhibited when chemical translocation was inhibited by chilling the petiole of the wounded leaf. These results imply that the electrical responses propagate the information without accompanying chemical

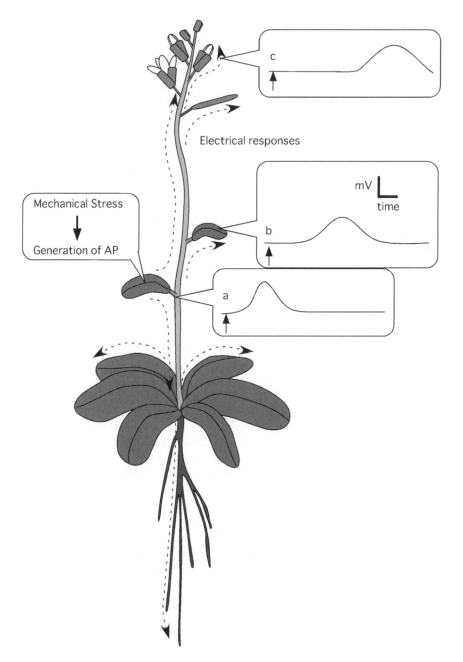

Fig. 3.2 Schematic diagram of the propagation of electrical long-distance signals in plants. Mechanical stress such as insect bites causes action potentials (*AP*) and subsequent variation potentials (*VP*), which propagate through the entire plant. The changes in the membrane potential at several points (*a*, *b*, and *c*) are shown schematically

translocation, and that electrical responses might be one of the main components for wound-induced long-distance signaling in plants.

3.3 Calcium-Permeable Channels in Plants

A large body of electrophysiological studies have elucidated that plants possess several Ca^{2+}-permeable channels with different kinetics (Fig. 3.1), but molecular identification of these channels is still in progress. To isolate the candidate Ca^{2+}-permeable channels and to examine their Ca^{2+} permeability, heterologous expression systems in yeast strains defective in the uptake of Ca^{2+} have been employed. Such yeast strains, defective in *CCH1*, a candidate for the voltage-dependent Ca^{2+} channel (VDCC) (Fischer et al. 1997) and/or *MID1*, a mechanosensitive nonselective cation channel (MSCC) (Kanzaki et al. 1999), are not capable of growth and survival in synthetic medium with lowered Ca^{2+} concentration. If the heterologously expressed protein successfully rescues the growth defect in yeast, the introduced gene may encode a possible candidate of a Ca^{2+}-permeable channel that is delivered to the plasma membrane and participates in external Ca^{2+}-entry in yeast cells. Clemens et al. (1998) have reported that a yeast disruption mutant, *mid1,* which cannot grow in low-Ca^{2+} medium, was complemented and growth restored by expressing *LCT1*, a low affinity cation channel from *A. thaliana*. This result implied that *LCT1* is a Ca^{2+}-permeable channel functioning in the plasma membrane. Recently, a variety of genes encoding presumed Ca^{2+}-permeable channels have been identified by the efforts of various genome sequencing projects in plants, and some of these channels have been functionally characterized. These channels differ from the Ca^{2+}-selective channels of animal cells mostly in their estimated secondary structures, and most of these plant channels permeate not only Ca^{2+}, but also K^+, Na^+, and other cations. Despite lower selectivity, it has been revealed that some of these channels participate in both Ca^{2+}-signaling and Ca^{2+} homeostasis in plants. Thus, in the case of some plant species, Ca^{2+}-permeable channels clearly play important roles in growth, development, and responses to environment (Fig. 3.3). The following subsections summarize the physiological characteristics and their roles in stress sensing of both the molecularly cloned Ca^{2+}-permeable channels and the molecular biologically yet-to-be identified channels with high importance to plant physiology.

3.3.1 Cyclic Nucleotide-Gated Cation Channel

Cyclic nucleotide (CN)-gated cation channels (CNGCs) play important roles in both visual (Yau and Baylor 1989) and olfactory (Schild and Restrepo 1998) signal transduction in animals. Activities of CNGCs are promoted by CN-binding and attenuated by $Ca^{2+}/$ CaM binding. To support these regulatory functions, animal CNGCs, which are composed of Shaker-units – each domain consisting of six

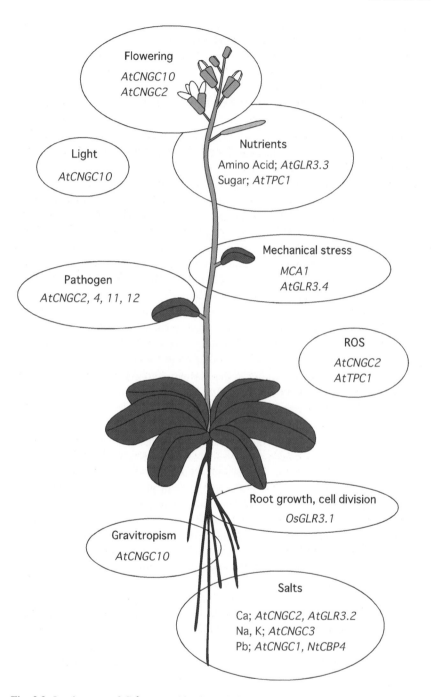

Fig. 3.3 Involvement of Ca²⁺-permeable channels in plant growth and development. The Ca²⁺-permeable channels involved in plant developmental stages and in responses to environmental stresses are presented

trans-membrane segments (S1–S6) and a pore loop (P) between S5 and S6 – are assembled with a CN-binding domain at the C-terminus and a CaM-binding domain at the N-terminus. Plant CNGCs have been cloned from barley (Schuurink et al. 1998), tobacco (Arazi et al. 1999, 2000) and *A. thaliana* (Köhler et al. 1999; Köhler and Neuhaus 2000; Leng et al. 1999). Plants CNGCs also are comprised of a Shaker unit, a CN-binding domain and a CaM-binding domain. The most distinct feature of plant CNGCs compared to animal CNGCs is that both CN-binding and CaM-binding domains overlap in their C-terminal hydrophilic tails. Due to their overlapping binding domains, CN and Ca^{2+}/CaM might competitively interact in plant CNGCs and regulate their activities.

In *A. thaliana*, CNGCs comprise a gene family with 20 genes phylogenetically divided into five groups. As in the case of growth complementation of Ca^{2+}-channel defective mutants, growth alteration of potassium channel mutants of *Escherichia coli* (LB650) and yeast (CY162) expressing plant CNGCs in lowered K^+ condition indicates that most plant CNGCs function as K^+-permeable channels. Immuno-gold detection and GFP-based detection of plant CNGCs in plant cells revealed that most plant CNGCs are localized to the plasma membrane (Borsics et al. 2007). Among the 20 Arabidopsis CNGC members, *AtCNGC2* was successfully expressed in human embryonic kidney cells (HEK293), and a CN-dependent increase in Ca^{2+} permeability was demonstrated by patch clamp recordings (Leng et al. 1999). Interestingly, the application of cAMP evoked currents in membrane patches of oocytes injected with *AtCNGC2* cRNA, demonstrating that *AtCNGC2* conducts K^+ and other monovalent cations except Na^+, while *AtCNGC1* permeates both K^+ and Na^+ (Leng et al. 2002). Animal CNGCs are also nonselective cation channels that permeate both Na^+ and K^+, and the same triplet of amino acids in the ion selectivity filter inside the channel pore is conserved. Most of the known plant CNGCs, except *AtCNGC2*, also do not discriminate between Na^+ and K^+. To resolve the unique ion selectivity of *AtCNGC2*, a site-directed mutagenesis assay revealed that specific amino acids (Asn-416 and Asp-417) within the ion selectivity filter of *AtCNGC2* facilitate K^+ conductance over that of Na^+ in a fashion similar to the well-characterized GYG triplet of K^+-selective channels (Hua et al. 2003). Phenotypic analyses have revealed that plant CNGCs are involved in a variety of environmental responses. The *AtCNGC2* mutant *dnd1* (<u>d</u>efense <u>n</u>o <u>d</u>eath) was originally isolated as a line that failed to produce a programmed cell death known as the hypersensitive response (HR) in response to avirulent *Pseudomonas syringae* pathogens. Increased expression in etiolated leaves and mature sheaths also indicated the possible involvement of *AtCNGC2* in the Ca^{2+}-influx that accompanies programmed cell death (Köhler et al. 2001). *AtCNGC4, 11*, and *12* are also involved in HR signaling (Balague et al. 2003; Yoshioka et al. 2006), indicating that plant CNGCs regulate programmed cell death in responses to pathogens. Most pathogen-related responses are associated with Ca^{2+}-flux and a transient increase in $[Ca^{2+}]_c$ (Kadota et al. 2004b), suggesting that plant CNGCs are possibly involved in these phenomena. Phenotypic features of transgenic plants also indicate that plant CNGCs participate in sensitivity to environmental cations. Tolerance to Pb^+ was improved in *AtCNGC1* knockout plants and altered in an *NtCBP4* (a CNGC in *Nicotiana tabacum*) overexpressor. Knockout plants of *AtCNGC2* unexpectedly showed hypersensitivity specifically to Ca^{2+}, but not to Na^+ and K^+ (Chan et al. 2003). Hence the seedlings of *AtCNGC3* mutants showed altered tolerance to both K^+ and Na^+ by the restricted influx of both cations, monovalent ion-dependent inhibition of germination was specifically altered by Na^+ but not K^+ (Gobert et al. 2006). Because *AtCNGC3* permeates Na^+, it might participate in Na^+ transfer from sensitive to tolerant tissues to avoid ionic toxicity. Antisense suppression of *AtCNGC10* promotes a dwarf phenotype and early flowering, which was also observed in *AtCNGC2* knockout mutants and *phyB* mutants lacking functional phytochrome B (Clough et al. 2000; Borsics et al. 2007). These lines of evidence observed in CNGC transgenic plants, as well as the physiological characteristics of these channels, imply that plant CNGCs play roles as sensors for toxic heavy metals and mediate the cation flux that triggers signaling pathways.

3.3.2   Ionotropic Glutamate Receptor

In animal cells, neurotransmitters play a key role in cell–cell signaling in nerve networks. In plants, several plant peptides, such as systemin, regulate environmental responses, growth and development, as described above. Using transgenic plants expressing the Ca^{2+}-reporting luminal protein aequorin, it was revealed that application of glutamate or glycine to Arabidopsis seedlings promotes a transient rise in $[Ca^{2+}]_c$ (Dennison and Spalding 2000), indicating that glutamate, which is an abundant nutritional component in plants, especially in phloem sap (Hayashi and Chino 1990), acts as a major neurotransmitter-like chemical in plants. As shown in the central nervous system (CNS) of animals, the potential targets of glutamate are ionotropic glutamate receptors (iGluRs), which function as glutamate-activated ion channels in fast synaptic transmission in the CNS; iGluR channels mediate fast chemical transmission across synapses by increasing the permeability to K^+, Na^+, and Ca^{2+} (Hollmann and Heinemann 1994; Dingledine et al. 1999). In contrast, the $[Ca^{2+}]_c$ increase in plants was not promoted by AMPA or NMDA, well known iGluR agonists of iGluR in animals.

Recently, cDNAs with high sequence similarities to mammalian iGluRs were isolated from *A. thaliana* (Lam et al. 1998; Lacombe et al. 2001) and *O. sativa* (Li et al. 2006). In Arabidopsis, there are 20 homologues of iGluRs, which are divided phylogenetically into three groups (Lacombe et al. 2001). Antisense suppression of AtGLR3.2 altered sensitivity to Ca^{2+} (Zhu et al. 2001), and overexpression of AtGLR3.2 showed poor growth and hypersensitivity to K^+ and Na^+ (Kim et al. 2001). In the AtGLR3.2 overexpressor, the efficiency of Ca^{2+} utilization was lowered by additive accumulation of Ca^{2+}, which led to the formation of crystals or hydroxyapatite in the cell wall to avoid Ca^{2+} toxicity. These results indicate that plant iGluRs participate in Ca^{2+}-homeostasis. All of the six amino acids commonly present in soil (glutamate, glycine, alanine, serine, asparagine and cysteine) as well as glutathione (gamma-glutamyl-cysteinyl-glycine), an abundant tripeptide in plants, have been identified as agonists of *AtGLR3.3*. In root cells, these chemicals cause a rapid depolarization of the plasma membrane, due partly to the Ca^{2+} influx (Qi et al. 2006). The agonist-induced depolarization of the plasma membrane and the $[Ca^{2+}]_c$ increase were attenuated in knockout plants of *AtGLR3.3*, indicating that *AtGLR3.3* acts as a sensor for amino acids in the rhizosphere. Although a large family of iGluR genes was found in Arabidopsis, rice *GLR3.1* is the only isolated iGluR from *O. sativa*; *GLR3.1* has been revealed as serving an important function in maintaining normal cell division and cell viability in roots. The knockout plants showed a shortened-root phenotype, due to the distorted meristematic activity and enhanced programmed cell death (Li et al. 2006), indicating that plant iGluRs coordinate plant growth and development via glutamate, which is abundant in phloem sap and capable of being translocated throughout the plant. Abiotic stressful stimuli such as touch, osmotic stress, or cold enhance the expression of *AtGLR3.4* in a Ca^{2+}-dependent manner, implying that iGluRs are required for adaptation to environmental stresses (Meyerhoff et al. 2005). In the early development of the CNS, the balance between cell proliferation and cell death is strictly regulated (Ross 1996), and glutamate and glutamate receptors are involved in these processes (Copani et al. 2001). These results suggest that plant iGluRs act as sensors for amino acids in soil and phloem sap, and participate in the regulation of plant growth and development.

3.3.3 Voltage-Dependent Ca^{2+}-Permeable Channels

Ca^{2+}-influx is an inevitable factor in cell expansion – a central process in plant development and morphogenesis. One of the most thoroughly investigated stages of plant growth that is regulated by Ca^{2+} is the elongation of roots and root hairs (Cramer and Jones 1996). The *rhd2* (root hair defective 2) mutant of *A. thaliana* is defective in Ca^{2+} uptake and root growth (Wymer et al. 1997), and it was revealed that the RHD2 protein was identical to *AtrbohC* (*A. thaliana respiratory burst oxidase homolog C*), an NADPH oxidase known to be involved in the generation of reactive oxygen species (ROS) (Foreman et al. 2003). In root cell protoplasts, hydroxyl radical (HO•), a product of NADPH oxidase, activates VDCCs (Foreman et al. 2003). Since the *rdh2* mutant lacks the NADPH oxidase activity, *rdh2* is defective in the production and accumulation of ROS in root hairs. Diphenyleneiodonium (DPI), a specific inhibitor of NADPH oxidase suppressed not only ROS production, but also root hair elongation and Ca^{2+} accumulation, indicating that ROS may stimulate the opening of Ca^{2+}-permeable channels. Pollination is also a process requiring Ca^{2+}-mediated signaling. In the case of lily pollen-tube growth, Ca^{2+} influx via VDCC regulates apical growth and a certain level of resting $[Ca^{2+}]_c$, which is maintained during germination (Shang et al. 2005). Therefore, entry of Ca^{2+} into the pollen tube via Ca^{2+} channels is necessary for fertilization. In contrast, $[Ca^{2+}]_c$ also regulates cellular ROS levels via Ca^{2+} signaling pathways since NADPH oxidases contain one Ca^{2+}-binding EF-hand motif, and the activity of NAD kinase,

which provides the substrate NADPH, is also dependent on the binding of Ca^{2+}/
CaM (Scholz-Starke et al. 2005). The relationship between Ca^{2+} signaling and
ROS production is dependent on a complicated cross-talk network system. Ca^{2+}
signaling and ROS production collaborate to control cell division and expansion
during plant growth.

Recently, some studies have demonstrated a relationship between the signal
transduction pathways for abscisic acid (ABA) and the production of ROS (Jiang
and Zhang 2001). Several other studies revealed that ABA action on stomatal
closure is mediated by the generation of ROS and $[Ca^{2+}]_c$ (Murata et al. 2001; Pei
et al. 2000). In the proposed mechanisms, ABA elicits the production of H_2O_2 and,
in turn, the resultant H_2O_2 stimulates the opening of Ca^{2+} channels, resulting in a
rapid increase in $[Ca^{2+}]_c$. The activated Ca^{2+} signaling further triggers the closure of
stomata. The inward Ca^{2+} conductance activated by ABA is a well-investigated and
important mechanism for ABA-induced stomatal closure (Pei et al. 2000; Hamilton
et al. 2000). Under drought conditions, ABA accumulates in guard cells and causes
intracellular H_2O_2 production followed by the activation of VDCCs.

3.3.4   Plant Two Pore Channel 1

The importance of VDCCs was notably implied as described, and it has been
revealed that plant VDCCs are expressed both in the plasma membrane and endo-
membranes, but the molecular identification of VDCCs has remained elusive until
recently (White et al. 2002). As a candidate for a VDCC in plants, *AtTPC1* was first
isolated by thorough homology searches of the genomic sequence of *A. thaliana*
using 30–60 bp degenerate sequences from partial amino acid sequences of several
VDCCs from animal cells (Furuichi et al. 2001a). *AtTPC1* has two conserved
homologous domains termed Shaker-units, having the highest homology with the
two pore channel 1 (*TPC1*) sequence cloned from rat (Ishibashi et al. 2000). The
overall structure is similar to half of that of the α-subunit of voltage-activated Ca^{2+}
channels. Notably, the expression of *AtTPC1* successfully complemented the growth
of a yeast *cch1* mutant in low Ca^{2+} medium, and the Ca^{2+} uptake activity was confirmed with radioactive $^{45}Ca^{2+}$. Some orthologs of *AtTPC1* have been isolated from
tobacco (Kadota et al. 2004a), rice (Hashimoto et al. 2004; Kurusu et al. 2004), and
wheat (Wang et al. 2005). In *A. thaliana* and rice, TPC1 exists as a single copy gene
in its genome and is ubiquitously expressed in the entire plant (Furuichi et al. 2001a;
Kurusu et al. 2004). In the tobacco BY-2 cultured cell line, there are two copies of
genes with high similarity (97.1% identity) and with slightly different molecular
masses (Kadota et al. 2004a), apparently detectable with a specific antibody against
AtTPC1 (T. Furuichi, unpublished result). Overexpression of *AtTPC1* altered, and
antisense suppression of *AtTPC1* attenuated, the sugar-induced influx of extracellular Ca^{2+} in Arabidopsis leaves (Furuichi et al. 2001a). The cell cycle-dependent
regulation of oxidative stress-induced $[Ca^{2+}]c$ rise and expression of NtTPC1s contribute to the cell cycle dependence of H_2O_2-induced expression of peroxidases

(Kadoto et al. 2005). In tobacco BY-2 cells, two *TPC1* orthologs, *NtTPC1-A* and *-B*, likely act as elicitor-responsive Ca^{2+} channels (Kadota et al. 2004a). Such *TPC1*-stimulating elicitors include Cryptogein (Kadota et al. 2004a, 2004b) but not the soluble cell wall preparation from *Magnaporhte grisea*, a rice blast fungus, that induces a transient increase in $[Ca^{2+}]_c$ in non-host plant material, i.e., tobacco BY-2 cells (Lin et al. 2005). Therefore, alternative Ca^{2+}-permeable channel(s) must be responsible for the *M. grisea* soluble cell-wall-component-induced Ca^{2+} influx.

In tobacco and yeast cells, physiological functions and green fluorescent protein (GFP)-based detection revealed that heterologously expressed AtTPC1 is localized (partly) in the plasma membrane. Expression of OsTPC1 in the plasma membrane was also indicated by the Ca^{2+} sensitivity of transgenic rice cells and transient expression in onion epidermal peals (Kurusu et al. 2004). In contrast, Peiter et al. (2005) reported that an AtTPC1-GFP fusion protein was expressed mainly in the vacuolar membrane when overexpressed in Arabidopsis. The transporting pathway of integral membrane proteins from the endoplasmic reticulum (ER) to the vacuole is still obscure and no characteristic signal peptides have yet been found. The difference in the localizations found in heterologous expression may be due to the absence of interacting partners. By contrast, the recycling of the plasma membrane K^+-channel, KAT1, is regulated by intrinsic sequence motifs (Mikosch et al. 2006). Further investigation of the interacting protein(s) of AtTPC1 are necessary to clarify the mechanism of localization and physiological function of TPC1 channels in intact plants.

In all plant TPC1 channels, a long hydrophilic domain (120 amino acids in AtTPC1) is present between two Shaker-units containing two EF-hand motifs, but its role(s) in channel activity are still obscure. In animal cells, downregulation of L-type Ca^{2+} channel opening by intracellular Ca^{2+} is well characterized; the EF-hand motif located in the C-terminal region of the α_{1C} subunit is required for this Ca^{2+}-dependent channel inactivation (Leon et al. 1995). Upon Ca^{2+}-binding to the EF-hand motif, the conformation of the C-terminus of the α_{1C} subunit is altered and the pore-region is closed as a consequence. When the Ca^{2+}-binding activity of one EF-hand in *AtTPC1* was disrupted by point mutagenesis, the Ca^{2+}-binding activity of *AtTPC1* was completely destroyed, and a much higher channel activity was obtained (Furuichi et al. 2004). This may indicate that Ca^{2+} binding negatively controls the gating of *AtTPC1* and consequently affects the level of glucose-induced $[Ca^{2+}]_c$ increase (Furuichi et al. 2001b).

If plant *TPC1* channels have roles and functions comparable to those of animal channels, assembling the regulatory subunits, as in the animal L-type VDCCs, may regulate TPC1 channel activity. The ion selectivity of *TPC1* channels is also unknown as yet and the pore region (p-loop) does not contain any known characteristic amino acid sequences for ion selectivity such as the EEEE-locus in the pore of L-type Ca^{2+}-selective channels (Barreiro et al. 2002) or the GYGD quartet in the pore of K^+-selective channels (Tytgat 1994). In yeast cells, the expression of *AtTPC1* promoted Rb^+ uptake, indicating that *AtTPC1* possibly permeates both Ca^{2+} and K^+ at least, but not Na^+ (Furuichi et al. 2004). A gene, *BacNaCh,* encoding a putative Ca^{2+} channel with six trans-membrane segments (S1–S6) and a hydrophobic pore region between S5 and S6 (single pore channel) has been isolated from *Bacillus halodurans* and was heterologously expressed in CHO cells for electrophysiological analysis. Contrary to expectation, *BacNaCh* was identified as a voltage-gated Na$^+$ channel with very low permeability to Ca^{2+} (Durell and Guy 2001; Ren et al. 2001). To manifest an adequate level of [Ca^{2+}]$_c$ increase, and to avoid the toxicity of Na$^+$, the ion selectivity of Ca^{2+}-permeable channels (especially Na$^+$ impermeability) is an important feature. In some cases, monomeric channel proteins possibly form a heterotetramer channel complex that shows altered kinetics, ion permeability, and voltage dependency (Paganetto et al. 2001). It is tempting to speculate that plant *TPC1* channels might be the evolutionary intermediates between the bacterial single pore Na$^+$/Ca^{2+} channel and animal L-type Ca^{2+}-selective channels (Furuichi and Kawano 2006a).

3.3.5 *Mechanosensitive Nonselective Cation Channel*

Environmental conditions affect the direction of plant growth and morphology, and these changes presumably improve the efficiency of photosynthesis, and the uptake of minerals and water by plants. Changes in the direction of the source of light and gravity, termed tropism, causes optimal bending of shoots and roots. Light controls plant movements, termed phototropism; those induced by blue light (BL) in particular have been well investigated. BL-receptors named phototropin – containing a serine/threonine kinase domain and two light sensing domains (*LOV1* and *2*) – have been isolated in recent studies (Huala et al. 1997; Kagawa et al. 2001). Involvement of Ca^{2+} signaling in BL-induced morphological responses is based on a relocation of the cytoskeleton controlled by Ca^{2+} and Ca^{2+}-binding proteins. BL-induced changes in [Ca^{2+}]$_c$ were measured by Baum et al. (1999) for the first time by using aequorin-expressing plants. Recently, Harada et al. (2003) revealed that phototropins (*phot1* and *phot2*) are the initiators for BL-induced changes in [Ca^{2+}]$_c$. Stölzle et al. (2003) reported that blue light, but not red light, activates VDCC in the plasma membrane of mesophyll cells. In the *phot1* mutant, BL-induced calcium currents were drastically reduced, and were eliminated in a *phot1 phot2* double mutant. K252a, a protein kinase inhibitor, inhibits BL-induced activation of VDCC, suggesting the involvement of photo-activation of a kinase domain of phototropins in BL-induced responses.

In the case of gravitropism – the gravity-induced bending of shoots or roots – our understanding of the mechanism is limited since exactly how plants sense the direction of gravity has not yet been fully elucidated. Gravitropism is supposed to be composed of three steps; perception of a gravity vector, transduction of the signal, and bending at responding regions (Haswell 2003). Herein, gravity perception is thought to rely upon the downward movement of amyloplasts, which are specialized plastids filled with high density starch granules found in the shoot endodermis and root tips (Sack 1997). Some *A. thaliana* mutants defective in gravitropism lack SNARE proteins – possible mediators of membrane trafficking and vesicle transport

(Tasaka et al. 1999; Kato et al. 2002; Sanderfoot and Raikhel 1999). In these mutants, most of the amyloplasts did not drop to the bottom of the cells after rotation of the plants, which showed defectiveness in gravitropism after the changes in the gravitational vector (gravistimulation). These results indicate that amyloplasts do not drop just depending on a gravity force alone, but rather that these movements are controlled by some upstream signaling events triggered by an initial gravity reception. In parallel with this mechanism, the possible involvement of Ca^{2+} signaling in gravitropism has also been implied (Sinclair and Trewavas 1997; Weisenseel and Meyer 1997) and gravistimulation-induced changes in $[Ca^{2+}]_c$ have been monitored using aequorin-expressing A. thaliana seedlings (Plieth and Trewavas 2002). When the seedlings were turned vertically around 90° to 180° as gravistimulation, a biphasic transient increase in $[Ca^{2+}]_c$ was observed and the intensity of the second peak was dependent on the degree of turning. From another point of view, changes in gravitational strength generated by the centrifugation also promote transient increases in $[Ca^{2+}]_c$ (Toyota et al. 2007). These observations suggest that plants are capable of sensing changes in both vector and strength of gravity, and that Ca^{2+}-influx is involved in these sensing machineries. A recent study has demonstrated that the gravity-induced increase in $[Ca^{2+}]_c$ is sensitive to auxin transport blockers, and the application of auxin induces an increase in $[Ca^{2+}]_c$ in a pattern similar to that induced by gravistimulation, suggesting that the gravity-induced increase in $[Ca^{2+}]_c$ is a downstream event of auxin accumulation in lower plant parts, directly promoting and inhibiting the cell elongation required for bending of stems and roots (Moore 2002). Following auxin accumulation in lower parts in the root of soybean, production of nitric oxide (NO•), which leads to enhanced cGMP synthesis, was promoted in order to inhibit cell division and fertilization (Hu et al. 2005). Other ROS, such as HO•, promote cell division and elongation in parallel, but do not induce synthesis of cGMP, indicating a specific role for NO• in plant growth and development. As discussed in Sect. 3.3.3 on Ca^{2+}-permeable channels, both ROS and cGMP activate VDCCs and CNGCs, respectively. In fact, antisense suppression of AtCNGC10 attenuates the gravitropic response in roots, although it is still unclear how AtCNGC10 participates in this response because Ca^{2+}-influx plays a central role in both gravity sensing and root elongation (Borsics et al. 2007). Further analyses are needed to reveal the channels involved in gravity-induced $[Ca^{2+}]_c$ increases.

The possible involvement of a mechanosensitive nonselective cation channel (MSCC) in the cells of the root cap in the gravity response is a model commonly assumed and classically shared among many researchers (Massa and Gilroy 2003). However, the molecular identification of plant MSCCs has seen only limited progress. The mechanosensitive channels (MSCs) MscL and MscS of E. coli were discovered (Sukharev et al. 1993), and recently MscS-related proteins have been found in, and isolated from, A. thaliana, (Haswell and Meyerowitz 2006). In A. thaliana, MscS-like proteins (MSLs) comprise a gene family of ten genes phylogenetically divided into two groups. Growth alteration of MSC mutants of E. coli (MJF465) expressing MSL3 after a hypo-osmotic shock implies that MSL3 has an MSC activity. The subcellular localization of GFP-labeled MSL3 and the phenotypic features of MSL2 and 3 knockout mutants suggest that MSLs are localized in the plastid envelop, and regulate the size and shape of plastids. Another type of MSCC, MCA1, was also isolated from *A. thaliana* by functional complementation of lethality of a yeast *mid1* mutant (Nakagawa et al. 2007). The subcellular localization of GFP-labeled MCA1 in root cells shows that MCA1 is a plasma membrane protein, and an aequorin-based analysis of $[Ca^{2+}]_c$ in the seedlings of an *MCA1* overexpressing strain revealed that MCA1 participates in the Ca^{2+}-influx in response to mechanical stresses. A knockout mutant of *MCA1* shows a unique phenotype – inability to penetrate from softer to harder agar medium – suggesting that *MCA1* is required for sensing the hardness of soil.

If MSCCs are activated by distortion of the vacuolar membrane or plasma membrane following amyloplast sedimentation, they might cause a notable increase in $[Ca^{2+}]_c$ although the linkage with auxin accumulation is still unclear. In contrast, if MSCCs are activated by an increase or decrease in the tension of the plasma membrane at interacting sites with the cell wall, then they may act as receptors of gravity, which might promote subsequent intracellular events including Ca^{2+}-induced Ca^{2+}-release. As reported by Plieth and Trewavas (2002), the gravity-induced increase in $[Ca^{2+}]_c$ seems to be a downstream event of auxin translocation, but the possible involvement of MSCCs in gravity sensing remains unclear. In addition, technically, local and restricted increases in $[Ca^{2+}]_c$ are barely detectable since the intensity of aequorin luminescence is relatively low and Ca^{2+}-sensitive fluorescent dyes are not easily applied to plant cells (Furuichi et al. 2001b). Further investigations of MSLs, MCA1, and their orthologues, as well as the improvement of Ca^{2+} imaging techniques, are required to clarify cellular mechanisms underlying gravity perception.

3.4 Conclusions

Here we provide a summary of a variety of plant Ca^{2+}-permeable channels and their roles in plant environmental responses, and regulation of growth and development (Fig. 3.3). In many cases, Ca^{2+} influx via Ca^{2+}-permeable channels triggers diverse responses crucial for plant survival and improvement in their quality of life. Our accumulating knowledge of Ca^{2+} signaling and Ca^{2+}-homeostasis may open the gates to understanding plant responses to environmental stresses, and eventually to further improvement of the growth and yield of crops in arid or salt-damaged regions that are expanding on the Earth.

References

Arazi T, Sunkar R, Kaplan B, Fromm H (1999) A tobacco plasma membrane calmodulin-binding transporter confers Ni^{2+} tolerance and Pb^{2+} hypersensitivity in transgenic plants. Plant J 20:171–182

Arazi T, Kaplan B, Fromm H (2000) A high-affinity calmodulin-binding site in a tobacco plasma-membrane channel protein coincides with a characteristic element of cyclic nucleotide-binding domains. Plant Mol Biol 42:591–601

Balague C, Lin B, Alcon C, Flottes G, Malmstrom S, Köhler C, Neuhaus G, Pelletier G, Gaymard F, Roby D (2003) HLM1, an essential signaling component in the hypersensitive response, is a member of the cyclic nucleotide-gated channel ion channel family. Plant Cell 15:365–379

Barreiro G, Guimarães CRW, Alencastro RB (2002) A molecular dynamics study of an L-type calcium channel model. Protein Eng 15:109–122

Baum G, Long JC, Jenkins GI, Trewavas AJ (1999) Stimulation of the blue light phototropic receptor NPH1 causes a transient increase in cytosolic Ca^{2+}. Proc Natl Acad Sci USA 96:13554–13559

Borsics T, Webb D, Andeme-Ondzighi C, Staehelin LA, Christopher DA (2007) The cyclic nucleotide-gated calmodulin-binding channel AtCNGC10 localizes to the plasma membrane and influences numerous growth responses and starch accumulation in *Arabidopsis thaliana*. Planta 225:563–573

Botting D (1973) Humboldt and the Cosmos. Harper and Row, New York

Chan CWM, Schorrak LM, Smith RK, Bent AF, Sussman MR (2003) A cyclic nucleotide-gated ion channel, CNGC2, is crucial for plant development and adaptation to calcium stress. Plant Physiol 132:728–731

Clemens S, Antosiewitcz DM, Ward JM, Schachtman DP, Schroeder JI (1998) The plant cDNA LCT1 mediates the uptake of calcium and cadmium in yeast. Proc Natl Acad Sci USA 95:12043–12048

Clough SJ, Fengler KA, Lippok B, Smith RK, Yu IC, Bent AF (2000) The Arabidopsis dnd1 "defense, no death" gene encodes a mutated cyclic nucleotide-gated ion channel. Proc Natl Acad Sci USA 97:9323–9328

Copani A, Uberti D, Sortino MA, Bruno V, Nicoletti F, Memo M (2001). Activation of cell-cycle-associated proteins in neuronal death: a mandatory or dispensable path? Trends Neurosci 24:25–31

Cramer GR, Jones RL (1996) Osmotic stress and abscisic acid reduce cytosolic calcium activities in roots of *Arabidopsis thaliana*. Plant Cell Environ 19:1291–1298

Dennison KL, Spalding EP (2000) Glutamate-gated calcium fluxes in Arabidopsis. Plant Physiol 124:1511–1514

Dingledine R, Borges K, Bowie D, Traynelis SF (1999) The glutamate receptor ion channels. Pharmacol Rev 51:7–61

Durell SR, Guy HR (2001) A putative prokaryote voltage-gated Ca^{2+} channel with only one 6TM motif per subunit. Biochem Biophys Res Commun 281:741–746

Fischer M, Schnell N, Chattaway J, Davies P, Dixon G, Sanders D (1997) The *Saccharomyces cerevisiae* CCH1 gene is involved in calcium influx mating. FEBS Lett 419:259–262

Foreman J, Demidchik V, Bothwell JHF, Mylona H, Torres MA, Linstead P, Costa S, Brownlee C, Jones JDG, Davies JM, Dolan L (2003) Reactive oxygen species produced by NADPH oxidase regulate plant cell growth. Nature 422:442–446

Fromm J, Bauer T (1994) Action potentials in maize sieve tubes change phloem translocation. J Exp Bot 273:463–469

Furuichi T, Muto S (2005) H^+-Coupled sugar transporter, an initiator of sugar-induced Ca^{2+}-signaling in plant cells. Z Naturforsch 60c:764–768

Furuichi T, Kawano T (2006a) Biochemistry and cell biology of calcium channels and signaling involved in plant growth and environmental responses. In: Teixeira da Silva JA (ed) Floriculture, ornamental and plant biotechnology: advances and topical issues (1st edn). Global Science Books, London, pp 26–36

Furuichi T, Kawano T (2006b) Nitrate and ammonium-induced Ca^{2+} influx in arabidopsis leaf cells. Bull Nippon Sport Sci Univ 35:227–232

Furuichi T, Cunningham KW, Muto S (2001a) A putative two pore channel AtTPC1 mediates Ca^{2+} flux in Arabidopsis leaf cells. Plant Cell Physiol 42:900–905

Furuichi T, Mori IC, Takahashi K, Muto S (2001b) Sugar-induced increase in cytosolic Ca^{2+} in *Arabidopsis thaliana* whole plants. Plant Cell Physiol 42:1149–1155

Furuichi T, Matsuhashi S, Ishioka NS, Fujimaki S, Ohtsuki S, Sekine T, Kume T, Muto S (2003) Real-time monitoring for translocation and perception of signal molecules in *Arabidopsis thaliana* mature plant. Tiara Annu Rep 2002:129–131

Furuichi T, Wüennenberg P, Osafune T, Sokabe M, Dietrich P, Hedrich R, Muto S (2004) AtTPC1, a putative voltage-dependent calcium-permeable channel, is located in the plasma membrane and contributes to cytosolic Ca^{2+}-signals induced by sucrose and reactive oxygen species. In: Abstracts of the 13th International Workshop on Plant Membrane Biology. July 2004, Montpellier, France, 114

Galvani L (1791) De viribus Electricitatis in *Motu Musculari* Commentarius. Bon Sci Art Inst Acad Commun 7:363–418

Gibson SI (2000) Plant sugar-response pathways. Part of a complex regulatory web. Plant Physiol 124:1532–15139

Gobert A, Park G, Amtmann A, Sanders D, Maathuis FJM (2006) *Arabidopsis thaliana* cyclic nucleotide gated channel 3 forms a non-selective ion transporter involved in germination and cation transport. J Exp Bot 57:791–800

Hamilton DW, Hills A, Köhler B. Blatt MR (2000) Ca^{2+} channels at the plasma membrane of stomatal guard cells are activated by hyperpolarization and abscisic acid. Proc Natl Acad Sci USA 97:4967–4972

Harada A, Sakai T, Okada K (2003) Phot1 and phot2 mediate blue light-induced transient increases in cytosolic Ca^{2+} differently in Arabidopsis leaves. Proc Natl Acad Sci USA 100:8583–8588

Hashimoto K, Saito M, Matsuoka H, Iida K, Iida H (2004) Functional analysis of a rice putative voltage-dependent Ca^{2+} channel, OsTPC1, expressed in yeast cells lacking its homologous gene CCH1. Plant Cell Physiol 45:496–500

Haswell ES (2003) Gravity perception: how plants stand up for themselves. Curr Biol 13: R761–R763

Haswell ES, Meyerowitz EM (2006) MscS-like proteins control plastid size and shape in *Arabidopsis thaliana*. Curr Biol 16:1–11

Hayashi H, Chino M (1990) Chemical composition of phloem sap from the uppermost internode of the rice plant. Plant Cell Physiol 31:247–251

Hollmann M, Heinemann S (1994) Cloned glutamate receptors. Annu Rev Neurosci 17:31–108

Hu X, Neill SJ, Tang Z, Cai W (2005) Nitric oxide mediates gravitropic bending in soybean roots. Plant Physiol 137:663–670

Hua BG, Mercier RW, Leng Q, Berkowitz GA (2003) Plants do it differently. A new basis for potassium/sodium selectivity in the pore of an ion channel. Plant Physiol 132:1353–1361

Huala E, Oeller PW, Liscum E, Han IS, Larsen E, Briggs WR (1997) Arabidopsis NPH1: a protein kinase with a putative redox-sensing domain. Science 278:2120–2123

Humboldt A von (1797) Versuche uber die gereizte Muskel- und Nervenfaser nebst Vermuthungen uber den chemischen Process des Lebens in der Tier und Pflanzenwelt. Posen Decker & Berlin, Rottmann

Ishibashi K, Suzuki M, Imai M (2000) Molecular cloning of a novel form (two-repeat) protein related to voltage-gated sodium and calcium channels. Biochem Biophys Res Commun 270:370–376

Jiang MY, Zhang JH (2001) Effect of abscisic acid on active oxygen species, antioxidative defence system and oxidative damage in leaves of maize seedlings. Plant Cell Physiol 42:1265–1273

Julien JL, Frachisse JM (1992) Involvement of the proton pump and proton conductance change in the wave of depolarization induced by wounding in *Bidens pilosa*. Can J Bot 70:1451–1458

Kadota Y, Furuichi T, Ogasawara Y, Goh T, Higashi K, Muto S, Kuchitsu K (2004a) Identification of putative voltage dependent Ca^{2+} permeable channels involved in cryptogein-induced Ca^{2+} transients and defense responses in tobacco BY-2 Cells. Biochem Biophys Res Commun 317:823–830

Kadota Y, Goh T, Tomatsu H, Tamauchi R, Higashi K, Muto S, Kuchitsu K (2004b) Cryptogein-induced initial events in tobacco BY-2 cells: pharmacological characterization of molecular relationship among cytosolic Ca^{2+} transients, anion efflux and production of reactive oxygen species. Plant Cell Physiol 45:160–170

Kagawa T, Sakai T, Suetsugu N, Oikawa K, Ishiguro S, Kato T, Tabata S, Okada K, Wada M (2001) Arabidopsis NPL1: a phototropin homolog controlling the chloroplast high-light avoidance response. Science 291:2138–2141

Kanzaki M, Nagasawa M, Kojima I, Sato C, Naruse K, Sokabe M, Iida H (1999) Molecular identification of a eukaryotic, stretch-activated nonselective cation channel. Science 285:882–886

Kato T, Morita MT, Fukaki H, Yamaguchi Y, Uehara M, Niihama M, Tasaka M (2002) SGR2, a phospholipase-like protein, and ZIG/SGR4, a SNARE, are involved in the shoot gravitropism of Arabidopsis. Plant Cell 14:33–46

Kawano T, Sahashi N, Takahashi K, Uozumi N, Muto S (1998) Salicylic acid induces extracellular generation of superoxide followed by an increase in cytosolic calcium ion in tobacco suspension culture: the earliest events in salicylic acid signal transduction. Plant Cell Physiol 39:721–730

Kadota Y, Furuichi T, Sano T, Kaya H, Gunji W, Murakami Y, Muto S, Hasezawa S, Kuchitsu K. (2005) Cell-cycle-dependent regulation of oxidative stress responses and Ca^{2+} permeable channel NtTPC1A/B in tobaccoBY-2 cells. Biochem Biophys Res Commun. 336(4): 1259–1267

Kim SA, Kwak JM, Jae SK, Wang MH, Nam HG (2001) Overexpression of the AtGluR2 gene encoding an Arabidopsis homolog of mammalian glutamate receptors impairs calcium utilization and sensitivity to ionic stress in transgenic plants. Plant Cell Physiol 42:74–84

Knight MR, Campbell AK, Smith SM, Trewavas AJ (1991) Transgenic plant aequorin reports the effects of touch and cold-shock and elicitors on cytoplasmic calcium. Nature 352:524–526

Koch KE (1996) Carbohydrate-modulated gene expression in plants. Annu Rev Plant Physiol Plant Mol Biol 47:509–540

Köhler C, Neuhaus G (2000) Characterisation of calmodulin binding to cyclic nucleotide-gated ion channels from *Arabidopsis thaliana*. FEBS Lett 4710:133–136

Köhler C, Merkle T, Neuhaus G (1999) Characterisation of a novel gene family of putative cyclic nucleotide- and calmodulin-regulated ion channels in *Arabidopsis thaliana*. Plant J 18:97–104

Köhler C, Merkle T, Roby D, Neuhaus G (2001) Developmentally regulated expression of a cyclic nucleotide-gated ion channel from Arabidopsis indicates its involvement in programmed cell death. Planta 213:327–332

Kurusu T, Sakurai Y, Miyao A, Hirochika H, Kuchitsu K (2004) Identification of a putative voltage-gated Ca^{2+} -permeable channel (OsTPC1) involved in Ca^{2+} influx and regulation of growth and development in rice. Plant Cell Physiol 45:693–702

Lacombe B, Becker D, Hedrich R, DeSalle R, Hollmann M, Kwak JM, Schroeder JI, Le Novere N, Nam HG, Spalding EP, Tester M, Turano FJ, Chiu J, Coruzzi G (2001) The identity of plant glutamate receptors. Science 292:1486–1487

Lam H-M, Chiu J, Hsieh M-H, Meisel L, Oliveira IC, Shin M, Coruzzi G (1998) Glutamate-receptor genes in plants. Nature 396:125–126

Leon M, Wang Y, Jones L, Perez-Reyes E, Wei X, Soong TW, Snutch TP, Yue DT (1995) Essential Ca^{2+}-binding motif for Ca^{2+}-sensitive inactivation of L-type Ca^{2+} channels. Science 270:1502–1506

Leng Q, Mercier RW, Yao W. Berkowitz GA (1999) Cloning and first functional characterization of a plant cyclic nucleotide-gated cation channel. Plant Physiol 121:753–761

Leng Q, Mercier RW, Hua BG, Fromm H, Berkowitz GA (2002) Electrophysiological analysis of cloned cyclic nucleotide-gated ion channels. Plant Physiol 128:400–410

Li J, Zhu S, Song X, Shen Y, Chen H, Yu J, Yi K, Liu Y, Karplus VJ, Wu P, Dend WX (2006) A rice glutamate receptor-like gene is critical for the division and survival of individual cells in the root apical meristem. Plant Cell 18:340–349

Lin C, Yu Y, Kadono T, Iwata M, Umemura K, Furuichi T, Kuse M, Isobe M, Yamamoto Y, Mastumoto H, Yoshizuka K, Kawano T (2005) Action of aluminum, novel TPC1-type channel inhibitor, against salicylate-induced and cold shock-induced calcium influx in tobacco BY-2 cells. Biochem Biophys Res Commun 332:823–830

Massa GD, Gilroy S (2003) Touch and gravitropic set-point angle interact to modulate gravitropic growth in roots. Adv Space Res 31:2195–2202

Meyerhoff O, Müller K, Roelfsema MRG, Latz A, Lacombe B, Hedrich R, Dietrich P, Becker D (2005) AtGLR3.4, a glutamate receptor channel-like gene is sensitive to touch and cold. Planta 222:418–427

Mikosch M, Hurst AC, Hertel B, Homann U (2006) Diacidic motif is required for efficient transport of the K+ channel KAT1 to the plasma membrane. Plant Physiol 142:923–930

Moore I (2002) Gravitropism: lateral thinking in auxin transport. Curr Biol 12:R482–R454

Murata Y, Pei ZM, Mori IC, Schroeder J (2001) Abscisic acid activation of plasma membrane Ca^{2+} channels in guard cells requires cytosolic NAD(P)H and is differentially disrupted upstream and downstream of reactive oxygen species production in abi1-1 and abi2-1 protein phosphatase 2C mutants. Plant Cell 13:2513–2523

Muto S (1993) Intracellular Ca^{2+} messenger system in plants. Int Rev Cytol 142:305–345

Nakagawa Y, Katagiri T, Shinozaki K, Qi Z, Tatsumi H, Furuichi T, Kishigami A, Sokabe M, Kojima I, Sato S, Kato T, Tabata S, Iida K, Terashima A, Nakano M, Ikeda M, Yamanaka T, Iida H (2007) Arabidopsis plasma membrane protein crucial for Ca^{2+} influx and touch sensing in roots. Proc Natl Acad Sci USA 104:3639–3644

Ohto M, Hayashi K, Isobe M, Nakamura K (1995) Involvement of Ca^{2+} signaling in the sugar-inducible expression of genes coding for sporamin and β-amylase of sweet potato. Plant J 7:297–307

Paganetto A, Bregante M, Downey P, Lo Schiavo F, Hoth S, Hedrich R, Gambale F (2001) A novel K^+ channel expressed in carrot roots with a low susceptibility toward metal ions. J Bioenerg Biomembr 33:63–71

Pearce G, Strydom D, Johnson S, Ryan CA (1991) A polypeptide from tomato leaves induces wound-inducible inhibitor proteins. Science 253:895–898

Pei ZM, Murata Y, Benning G, Thomine S, Klusener B, Allen GJ, Grill E, Schroeder JI (2000) Calcium channels activated by hydrogen peroxide mediate abscisic acid signalling in guard cells. Nature 406:731–734

Peiter E, Maathuis F, Mills L, Knight H, Pelloux J, Hetherington A, Sanders D (2005) The vacuolar Ca^{2+}-activated channel TPC1 regulates germination and stomatal movement. Nature 434:404–408

Piñeros M, Tester M (1997) Calcium channels in higher plant cells: selectivity, regulation and pharmacology. J Exp Bot 48:551–577

Plieth C, Trewavas AJ (2002) Reorientation of seedlings in the Earth's gravitational field induces cytosolic calcium transients. Plant Physiol 129:786–796

Qi Z, Stephens NR, Spalding EP (2006) Calcium entry mediated by GLR3.3, an Arabidopsis glutamate receptor with a broad agonist profile. Plant Physiol 142:963–971

Ren D, Navarro B, Xu H, Yue L, Shi Q, Clapham DE (2001) A prokaryotic voltage-gated sodium channel. Science 294:2372–2375

Rhodes JD, Thain JF, Wildon DC (1996) The pathway for systemic electrical signal conduction in the wounded tomato plant. Planta 200:50–57

Rook F, Geritts N, Kortstee A, van Kampe M, Borrias M, Weisbeek P, Smeekens S (1998) Sucrose-specific signalling represses translation of the Arabidopsis ATB2 bZIP transcription factor gene. Plant J 15:253–263

Ross ME (1996) Cell division and the nervous system: regulating the cycle from neural differentiation to death. Trends Neurosci 19:62–68

Sack FD (1997) Plastids and gravitropic sensing. Planta 203:S63–S68

Sanderfoot AA, Raikhel NV (1999) The specificity of vesicle trafficking: coat proteins and SNAREs. Plant Cell 11:629–642

Schild D, Restrepo D (1998) Transduction mechanisms in vertebrate olfactory receptor cells. Physiol Rev 78:429–466

Scholz-Starke J, Gambale F, Carpaneto A (2005) Modulation of plant ion channels by oxidizing and reducing agents. Arch Biochem Biophys 434:43–50

Schuurink RC, Shartzer SF, Fath A. Jones RL (1998) Characterization of a calmodulin-binding transporter from the plasma membrane of barley aleurone. Proc Natl Acad Sci USA 95:1944–1949

Shang ZL, Ma LG, Zhang HL, He RR, Wang XC, Cui SJ, Sun DY (2005) Ca^{2+} influx into lily pollen grains through a hyperpolarization-activated Ca^{2+}-permeable channel which can be regulated by extracellular CaM. Plant Cell Physiol 46:598–608

Sheen J, Zhou L, Jang J-C (1999) Sugars as signaling molecules. Curr Opin Plant Biol 2:410–418

Sinclair W, Trewavas AJ (1997) Calcium in gravitropism. A re-examination. Planta 203: S85–S90

Stölzle S, Kagawa T, Wada M, Hedrich R, Dietrich P (2003) Blue light activates calcium-permeable channels in Arabidopsis mesophyll cells via the phototropin signaling pathway. Proc Natl Acad Sci USA 100:1456–1461

Sukharev SI, Martinac B, Arshavsky VY, Kung C (1993) Two types of mechanosensitive channels in the *Escherichia coli* cell envelope: solubilization and functional reconstitution. Biophys J 65:177–183

Tasaka M, Kato T, Fukaki H (1999) The endodermis and shoot gravitropism. Trends Plant Sci 4:103–107

Toyota M, Furuichi T, Tatsumi H, Sokabe M (2007) Hypergravity stimulation induces changes in intracellular calcium concentration in Arabidopsis seedlings. Adv. Space. Res. 39(7): 1190–1197

Tytgat J (1994) Mutations in the P-region of a mammalian potassium channel (RCK1): a comparison with the Shaker potassium channel. Biochem Biophys Res Commun 203:513–518

Wang YJ, Yu JN, Chen T, Zhang ZG, Hao YJ, Zhang JS, Chen SY (2005) Functional analysis of a putative Ca^{2+} channel gene TaTPC1 from wheat. J Exp Bot 56:3051–3060

Weisenseel AJ, Meyer MH (1997) Bioelectricity, gravity and plants. Planta 203:S98–S106

White PJ, Bowen HC, Demidchik V, Nichols C, Davies JM (2002) Genes for calcium-permeable channels in the plasma membrane of plant root cells. Biochim Biophys Acta 1564:299–309

Wildon DC, Doherty HM, Eagles G, Bowles DJ, Thain JF (1989) Systemic responses arising from localized heat stimuli in tomato plants. Ann Bot 64:691–695

Wildon DC, Thain JF, Minchin PEH, Gubb IR, Reilly AJ, Skipper YD, Doherty HM, O'Donnell PJ, Bowles DJ (1992) Electrical signalling and systemic proteinase inhibitor induction in the wounded plant. Nature 360:62–65

Wymer CL, Bibikova TN, Gilroy S (1997) Cytoplasmic free calcium distributions during the development of root hairs of *Arabidopsis thaliana*. Plant J 12:427–439

Yau K-W, Baylor DA (1989) Cyclic GMP-activated conductance of retinal photoreceptor cells. Annu Rev Neurosci 12:289–327

Yokohama R, Hirose T, Fujii N, Aspuria ET, Kato A, Uchimiya H (1994) The rolC promoter of Agrobacterium rhizogenes Ri plasmid is activated by sucrose in transgenic tobacco plants. Mol Gen Genet 244:15–22

Yoshioka K, Moeder W, Kang HG, Kachroo P, Masmoudi K, Berkowitz G, Klessig DF (2006) The chimeric Arabidopsis CYCLIC NUCLEOTIDE-GATEDION CHANNEL11/12 activates multiple pathogen resistance responses. Plant Cell 18:747–763

Zhu Z, Budworth P, Han B, Brown D, Chang H-S, Zou G, Wang X (2001) Toward elucidating the global gene expression patterns of developing Arabidopsis: parallel analysis of 8,300 genes by a high-density oligonucleotide probe array. Plant Physiol Biochem 39:221–242

Chapter 4
Ion Channels, Cell Volume, Cell Proliferation and Apoptotic Cell Death

Florian Lang(✉), Erich Gulbins, Ildiko Szabo, Alexey Vereninov, and Stephan M. Huber

Abstract At some stage cell proliferation requires an increase in cell volume and a typical hallmark of apoptotic cell death is cell shrinkage. The respective alterations of cell volume are accomplished by altered regulation of ion transport including ion channels. Thus, cell proliferation and apoptosis are both paralleled by altered activity of ion channels, which play an active part in these fundamental cellular mechanisms. Activation of anion channels allows exit of Cl^-, osmolyte and HCO_3^- leading to cell shrinkage and acidification of the cytosol. K^+ exit through K^+ channels leads to cell shrinkage and a decrease in intracellular K^+ concentration. K^+ channel activity is further important for maintenance of the cell membrane potential – a critical determinant of Ca^{2+} entry through Ca^{2+} channels. Cytosolic Ca^{2+} may both activate mechanisms required for cell proliferation and stimulate enzymes executing apoptosis. The effect of enhanced cytosolic Ca^{2+} activity depends on the magnitude and temporal organisation of Ca^{2+} entry. Moreover, a given ion channel may support both cell proliferation and apoptosis, and specific ion channel blockers may abrogate both fundamental cellular mechanisms, depending on cell type, regulatory environment and condition of the cell. Clearly, further experimental effort is needed to clarify the role of ion channels in the regulation of cell proliferation and apoptosis.

Department for Physiology, University of Tübingen, Gmelinstr. 5, D 72076 Tübingen, florian.lang@uni-tuebingen.de

4.1   Introduction

Maintenance of adequate cell mass requires a delicate balance between formation of new cells by cell proliferation and their elimination by suicidal death of nucleated cells (apoptosis; Green and Reed 1998) and suicidal cell death of denucleated cells (Barvitenko et al. 2005; Rice and Alfrey 2005) such as erythrocytes (eryptosis; K.S. Lang et al. 2005).

Cell proliferation is stimulated by growth factors (Adams et al. 2004; Bikfalvi et al. 1998; Tallquist and Kazlauskas 2004), apoptosis by stimulation of receptors such as CD95 (Fillon et al. 2002; Gulbins et al. 2000; Lang et al. 1998b, 1999), somatostatin receptor (Teijeiro et al. 2002), or TNFα receptor (Lang et al. 2002a). Further triggers of apoptosis are cell density (Long et al. 2003), lack of growth factors (Sturm et al. 2004) thyroid hormones (Alisi et al. 2005), adhesion (Davies 2003; Walsh et al. 2003), choline deficiency (Albright et al. 2005), oxidants (Rosette and Karin 1996), radiation (Rosette and Karin 1996), inhibition of glutaminase (Rotoli et al. 2005), chemotherapeutics (Cariers et al. 2002; Wieder et al. 2001), energy depletion (Pozzi et al. 2002) or osmotic shock (Bortner and Cidlowski 1998; Bortner and Cidlowski 1999; Lang et al. 1998a, 2000b; Maeno et al. 2000; Michea et al. 2000; Rosette and Karin 1996).

In general, cell proliferation generates cells of a similar size as the parent cells, which requires at some point an increase in cell volume (Lang et al. 1998a). Hallmarks of apoptosis include cell shrinkage (Green and Reed 1998). Thus, both cell proliferation and suicidal cell death are paralleled by changes in cell volume that are not possible without corresponding alterations in ion channel activity.

As a matter of fact both cell proliferation and apoptosis involve activation of Cl^- channels, K^+ channels and Ca^{2+} channels. Ion channel inhibitors have been reported to interfere, at least in some cells, with cell proliferation and apoptosis. Accordingly, these channels obviously play an active role in the machinery leading to the duplication or death of a given cell.

4.2   Cell Volume Regulatory Ion Transport

Understanding the role of ion channels in the regulation of cell proliferation and apoptosis requires prior consideration of their role in cell volume regulation, i.e. regulatory cell volume increase (RVI) following cell shrinkage and regulatory cell volume decrease (RVD) following cell swelling.

RVI is accomplished by ion uptake (Lang et al. 1998b). Cell shrinkage activates the $Na^+/K^+/2Cl^-$ cotransporter and/or the Na^+/H^+ exchanger in parallel with the Cl^-/HCO_3^- exchanger (Lang et al. 1998b). The H^+ extruded by the Na^+/H^+ exchanger, and the HCO_3^- exiting though the Cl^-/HCO_3^- exchanger are replenished from CO_2 and are thus osmotically not relevant. The two carriers thus accomplish NaCl entry. Na^+ accumulated by either $Na^+/K^+/2Cl^-$ cotransport or Na^+/H^+ exchange is extruded by the Na^+/K^+ ATPase in exchange for K^+. Thus, the transporters eventually result in KCl uptake.

Shrinkage of some cells leads to activation of Na^+ channels and depolarization, which in turn dissipates the electrical gradient for Cl^- and thus leads to Cl^- entry (Wehner 2006). Some cells inhibit K^+ channels and/or Cl^- channels upon cell shrinkage to avoid cellular KCl loss (Lang et al. 1998b). Cell shrinkage is further counteracted by the cellular uptake or generation of organic osmolytes such as sorbitol, myoinositol, betaine, glycerophosphorylcholine, amino acids and the amino acid derivative taurine (Garcia-Perez and Burg 1991; Lambert 2004). The osmolytes are generated by metabolic production (Garcia-Perez and Burg 1991) or accumulated by Na^+ coupled transporters (Kwon and Handler 1995).

RVD requires release of cellular ions. In most cells ions are released by activation of K^+ channels and/or anion channels (Okada 2006; Sabirov and Okada 2004; Uchida and Sasaki 2005; Wehner 2006). Both K^+ and anion channels must be operative to accomplish KCl exit. Cell volume regulatory K^+ channels include Kv1.3, Kv1.5 and KCNE1/KCNQ1, and cell volume regulatory anion channels include ClC-2 and ClC-3 (Lang et al. 1998b). Moreover, I_{Cln} and P-glycoprotein (MDR) may participate in cell volume regulation (Jakab and Ritter 2006; Ritter et al. 2003). In some cells swelling leads to activation of unspecific cation channels mediating the entry of Ca^{2+}, which in turn activates Ca^{2+}-sensitive K^+ channels and/ or Cl^- channels (Lang et al. 1998b). Cell volume regulatory decrease could be further accomplished by activation of carriers, such as KCl-cotransport, which allows coupled exit of both ions (Adragna et al. 2004). Some cells dispose of cellular KCl via parallel activation of K^+/H^+ exchange and Cl^-/HCO_3^- exchange. The carriers accomplish KCl uptake as the H^+ and HCO_3^- taken up by those transporters eventually form CO_2, which may exit and is thus not osmotically relevant (Lang et al. 1998b). Cell swelling triggers the exit of glycerophosphorylcholine (GPC), sorbitol, inositol, betaine and taurine through as yet ill-defined anion channels or carriers (Kinne et al. 1993; Wehner et al. 2003).

4.3   Stimulation of I_{CRAC} During Cell Proliferation

Cytosolic Ca^{2+} activity plays a decisive role in the regulation of cell proliferation (Berridge et al. 1998, 2000, 2003; Parekh and Penner 1997; Santella et al. 1998; Santella 1998; Whitfield et al. 1995). Growth factors stimulate I_{CRAC} (Qian and Weiss 1997), which in turn mediates Ca^{2+} entry and subsequent Ca^{2+} oscillations in proliferating cells. The Ca^{2+} oscillations govern a wide variety of cellular functions (Berridge et al. 1998, 2000, 2003; Parekh and Penner 1997), including depolymerisation of actin filaments (Dartsch et al. 1995; Lang et al. 1992, 2000c; Ritter et al. 1997), which in turn leads to disinhibition of Na^+/H^+ exchanger and/or Na^+, K^+, $2Cl^-$ cotransporter resulting in an increase in cell volume (Lang et al. 1998a). Activation of I_{CRAC}, Ca^{2+} oscillations and depolymerisation of the actin filament network are all prerequisites of cell proliferation (Dartsch et al. 1995; Lang et al. 1992, 2000c; Ritter et al. 1997). In addition to I_{CRAC}, voltage-gated calcium channels may play a role in calcium entry during cell proliferation. Accordingly, the T-cell receptor-mediated calcium response and cytokine production may be impaired in T-lymphocytes deficient in voltage-gated calcium channel (Cav) beta-subunits (Badou et al. 2006).

4.4 Inhibition of I_{CRAC} During CD95-Induced Lymphocyte Death

Stimulation of lymphocyte apoptosis by CD95 receptor triggering is paralleled by inhibition of I_{CRAC} (Dangel et al. 2005; Lepple-Wienhues et al. 1999). The inhibition of I_{CRAC} prevents the activation of lymphocyte proliferation but may not be required for the triggering of apoptotic cell death.

4.5 Activation of Ca^{2+} Entry in Apoptosis and Eryptosis

Sustained increase of cytosolic Ca^{2+} activity triggers apoptosis in a variety of nucleated cells (Berridge et al. 2000; Green and Reed 1998; Liu et al. 2005; Parekh and Penner 1997; Parekh and Putney 2005; Spassova et al. 2004). Moreover, Ca^{2+} entry through Ca^{2+}-permeable cation channels stimulates eryptosis, i.e. the suicidal death of erythrocytes (K.S. Lang et al. 2002b, 2003; Spassova et al. 2004). The Ca^{2+}-permeable cation channels are activated by excessive cell shrinkage following hyperosmotic shock (Huber et al. 2001), by removal of intracellular and extracellular Cl^- (Duranton et al. 2002; Huber et al. 2001), by oxidative stress (Duranton et al. 2002), energy depletion (K.S. Lang et al. 2003), or infection with the malaria pathogen *Plasmodium falciparum* (Duranton et al. 2003; F. Lang et al. 2004; K.S. Lang et al. 2003). Eryptosis is similarly triggered by treatment of erythrocytes with the Ca^{2+} ionophore ionomycin (Berg et al. 2001; Bratosin et al. 2001; Daugas et al. 2001; K.S. Lang et al. 2002b, 2003). Conversely, the eryptosis following osmotic shock is blunted in the nominal absence of Ca^{2+} (K.S. Lang et al. 2003), or blockage of the cation channels with amiloride (K.S. Lang et al. 2003) or ethylisopropylamiloride (EIPA) (K.S. Lang et al. 2003). The suicidal erythrocyte cation channels are activated by prostaglandin E_2 (PGE_2), which is released upon osmotic shock (P.A. Lang et al. 2005a).

Cell volume sensitive cation channels are similarly expressed in nucleated cells, such as airway epithelial cells (Chan et al. 1992), mast cells (Cabado et al. 1994), macrophages (Gamper et al. 2000), vascular smooth muscle, colon carcinoma and neuroblastoma cells (Koch and Korbmacher 1999), cortical collecting duct cells (Volk et al. 1995), and hepatocytes (Wehner et al. 1995, 2000). Cation channels are activated by Cl^- removal in salivary and lung epithelial cells (Dinudom et al. 1995; Marunaka et al. 1994; Tohda et al. 1994). Cl^- influences the channels via a pertussis-toxin-sensitive G-protein (Dinudom et al. 1995). Whether or not those channels participate in Ca^{2+} entry and apoptosis of nucleated cells remains elusive.

4.6 Activation of K⁺ Channels in Cell Proliferation

K^+ channels have been shown to participate in the regulation of cell proliferation (Dinudom et al. 1995; Patel and Lazdunski 2004; Wang 2004). They are stimulated by growth factors (Enomoto et al. 1986; Faehling et al. 2001; Lang et al. 1991; Liu et al. 2001; O'Lague et al. 1985; Sanders et al. 1996; Wiecha et al. 1998), and are activated in a wide variety of tumour cells (DeCoursey et al. 1984; Mauro et al. 1997; Nilius and Wohlrab 1992; Pappas and Ritchie 1998; Pappone and Ortiz-Miranda 1993; Patel and Lazdunski 2004; Skryma et al. 1997; Strobl et al. 1995; Wang 2004; Zhou et al. 2003). In *ras* oncogene-expressing cells, Ca^{2+}-sensitive K^+ channels are activated by pulsatile increases of cytosolic Ca^{2+} activity leading to oscillations of cell membrane potential (F. Lang et al. 1991). Conversely, K^+ channel inhibitors may disrupt cell proliferation (for review, see Wang 2004). K^+ channel activation is thought to be particularly important for the early G1 phase of the cell cycle (Wang et al. 1998; Wonderlin and Strobl 1996). Activated K^+ channels maintain the cell membrane potential and thus establish the electrical driving force for Ca^{2+} entry through I_{CRAC} (Parekh and Penner 1997).

4.7 Inhibition of K⁺ Channels in Apoptosis

Stimulation of the CD95 receptor in Jurkat lymphocytes leads to early inhibition of Kv1.3 K^+ channels (Szabo et al. 1997, 1996, 2004) – the cell volume regulatory K^+ channel of Jurkat lymphocytes (Deutsch and Chen 1993). CD95 triggering leads to tyrosine phosphorylation of the channel protein (Gulbins et al. 1997; Szabo et al. 1996), and genetic or pharmacological knockout of the *src*-like tyrosine kinase Lck[56] abrogates the inhibition of the channel (Gulbins et al. 1997; Szabo et al. 1996). Ceramide similarly inhibits Kv1.3 and induces apoptosis in Jurkat lymphocytes (Gulbins et al. 1997). Kv1.3 is stimulated by the serum- and gluco-corticoid-inducible kinase (Lang et al. 2006), which in turn inhibits apoptosis (Aoyama et al. 2005). The early inhibition of Kv1.3 in CD95-induced apoptosis presumably contributes to the lack of early cell shrinkage despite the activation of Cl^- channels (Lang et al. 1998a). Jurkat cells only shrink approximately 1 h after CD95 triggering. Early osmotic cell shrinkage may interfere with the signalling of apoptosis (Gulbins et al. 1995). The early inhibition of Kv1.3 is, however, followed by late activation of Kv1.3 upon CD95 ligation (Storey et al. 2003).

Inhibition of K^+ channels appears to favour (Bankers-Fulbright et al. 1998; Chin et al. 1997; Han et al. 2004; Miki et al. 1997; Pal et al. 2004; Patel and Lazdunski 2004) and activation of K^+ channels to inhibit (Jakob and Krieglstein 1997; Lauritzen et al. 1997) apoptosis in several cells. Moreover, mutation of a G-protein-coupled inward rectifier K^+ channel leads to extensive neuronal cell death of *Weaver* mice (Harrison and Roffler-Tarlov 1998; Migheli et al. 1995, 1997; Murtomaki et al. 1995; Oo et al. 1996).

4.8   Stimulation of K⁺ Channels in Apoptosis

In Jurkat lymphocytes, the late execution phase is paralleled by activation of K^+ channels (Storey et al. 2003). K^+ channel activation hyperpolarizes the cell membrane and thus increases the electrical driving force for Cl^- exit. Thus, the combined activity of K^+ channels and Cl^- channels leads to cellular loss of KCl with osmotically obliged water and thus to cell shrinkage – a hallmark of apoptosis (Lang et al. 1998a). In a wide variety of other cells, apoptosis is similarly paralleled by activation of K^+ channels (Wei et al. 2004; Yu et al. 1997). Prevention of cellular K^+ release by increase of extracellular K^+ concentration or inhibition of K^+ channels (Gantner et al. 1995; P.A. Lang et al. 2003b) may disrupt the apoptotic machinery (Colom et al. 1998; P.A. Lang et al. 2003b; Prehn et al. 1997). Cellular K^+ loss may thus favour the triggering of apoptosis in a wide variety of cells (Beauvais et al. 1995; Benson et al. 1996; Bortner et al. 1997; Bortner and Cidlowski 1999, 2004; Gomez-Angelats et al. 2000; Hughes et al. 1997; Hughes and Cidlowski 1999; Maeno et al. 2000; Montague et al. 1999; Perez et al. 2000; Yurinskaya et al. 2005a, 2005b). In suicidal erythrocytes, Ca^{2+}-sensitive K^+ channels (Gardos channels) are activated by increased cytosolic Ca^{2+} activity (P.A. Lang et al. 2003a). Increase of extracellular K^+ or pharmacological inhibition of the Gardos channels blunts the cell shrinkage and eryptosis that follows exposure to the Ca^{2+} ionophore ionomycin (P.A. Lang et al. 2003a). It is not clear, however, whether the Ca^{2+}-sensitive K^+ channels support eryptosis by decreasing cytosolic K^+ activity or by inducing cell shrinkage. A decrease in cell volume leads to the release of platelet activating factor (PAF) and subsequent activation of sphingomyelinase with formation of ceramide (P.A. Lang et al. 2005b). Ceramide sensitizes the erythrocytes to the pro-apoptotic effect of increased cytosolic Ca^{2+} activity (K.S. Lang et al. 2004; P.A. Lang 2005b).

4.9   Activation of Anion Channels in Cell Proliferation

Cell proliferation has been observed to be paralleled by the activation of anion channels (Nilius and Droogmans 2001; Shen et al. 2000; Varela et al. 2004). Conversely, pharmacological inhibition of anion channels has been shown to interfere with cell proliferation (Jiang et al. 2004; Pappas and Ritchie 1998; Phipps et al. 1996; Rouzaire-Dubois et al. 2000; Shen et al. 2000; Wondergem et al. 2001). Moreover, cell proliferation is impeded in mice lacking a functional ClC-3 Cl^- channel (Wang et al. 2002). Obviously, cell proliferation needs, at some stage, transient cell shrinkage, which is accomplished by activation of Cl^- channels. As intracellular Cl^- is above electrochemical equilibrium, activation of Cl^- channels leads to Cl^- exit and depolarisation. The depolarisation drives cellular K^+ exit through K^+ channels. As a result, activation of Cl^- channels leads to exit of KCl together with osmotically obliged water and thus to cell shrinkage. (Lang et al. 1998a). Transient cell shrinkage has been shown to be required for the triggering of cytosolic Ca^{2+} oscillations in *ras* oncogene-expressing cells (Ritter et al. 1993). The Ca^{2+} oscillations depolymerise the actin filament network – an obvious prerequisite for cell proliferation (see above).

4.10 Activation of Anion and Osmolyte Channels in Apoptosis

Activation of Cl^- channels precedes the apoptosis that follows stimulation of CD95 in Jurkat cells (Szabo et al. 1998) or treatment of various cell types with TNFα or staurosporine (Maeno et al. 2000; Okada et al. 2004). At least in Jurkat cells, the same channels are activated by osmotic cell swelling, and their activation is required for regulatory cell volume decrease (Lepple-Wienhues et al. 1998). The stimulation by either CD95 triggering (Szabo et al. 1998) or cell swelling (Lepple-Wienhues et al. 1998) requires the Src-like kinase Lck[56]. This kinase is activated by both CD95 triggering (Szabo et al. 1998) and cell swelling (Lepple-Wienhues et al. 1998). Pharmacological inhibition and genetic knockout of the kinase disrupts CD95-induced apoptosis (Szabo et al. 1998) and regulatory cell volume decrease (Lepple-Wienhues et al. 1998). Stimulators of the kinase include ceramide (Gulbins et al. 1997), which is formed following activation of a sphingomyelinase upon CD95 receptor stimulation. The volume regulatory Cl^- channels are activated even in lymphocytes from patients with cystic fibrosis, which are resistant to Cl^- channel activation by protein kinase A (Lepple-Wienhues et al. 2001).

Cl^- channel blockers blunt or even disrupt CD95-induced Jurkat cell apoptosis (Szabo et al. 1998), TNFα- or staurosporine-induced apoptosis of various cell types (Maeno et al. 2000; Okada et al. 2004), apoptotic death of cortical neurons (Wei et al. 2004), antimycin-A-induced death of proximal renal tubules (Miller and Schnellmann 1993), GABA-induced enhancement of excitotoxic cell death of rat cerebral neurons (Erdo et al. 1991), cardiomyocyte apoptosis (Takahashi et al. 2005) and eryptosis (Takahashi et al. 2005).

Besides their contribution to cellular KCl loss (see above), anion channels may allow the permeation of organic osmolytes such as taurine (P.A. Lang et al. 2003b). Taurine release indeed parallels, or shortly precedes, the execution phase of apoptosis (Lang et al. 1998b; Moran et al. 2000). The loss of these organic osmolytes contributes to cell shrinkage (Lang et al. 1998a). In addition, organic osmolytes stabilise cellular proteins (Garcia-Perez and Burg 1991; Lambert 2004) and osmolyte release during apoptosis may compromise cell function by protein destabilisation. Along those lines, inhibition of inositol uptake has been shown to induce renal failure, presumably due to apoptotic death of renal tubular cells (Kitamura et al. 1998).

Cl^- channels may be permeable not only to Cl^- and organic osmolytes but also to HCO_3^-. Activation of Cl^- channels thus frequently leads to cytosolic acidification (e.g. Szabo et al. 1998) – a typical feature of cells entering apoptosis (Lang et al. 2002a; Wenzel and Daniel 2004). This acidification may enhance DNA fragmentation, since the pH optimum of DNase type II is in the acidic range (Shrode et al. 1997). Accordingly, CD95-induced apoptosis is accelerated by inhibition of Na^+/H^+

exchange, which disrupts the counter-regulation of cytosolic acidification by this carrier (Lang et al. 2000a).

4.11 Conclusions

Ion channels are activated during cell proliferation and apoptosis and play an active role in the regulation of those fundamental cellular mechanisms. Stimulation of cell proliferation may lead to early cell shrinkage through activation of Cl^- and K^+ channel activity and Ca^{2+} oscillations through activation of the Ca^{2+}-release-activated I_{CRAC} channels. The Ca^{2+} oscillations lead to depolymerisation of the actin filament network with subsequent disinhibition of Na^+/H^+ exchanger and/or $Na^+/K^+/2Cl^-$ resulting in cell swelling and cytosolic alkalinisation.

Apoptosis eventually leads to cell shrinkage, which may be accomplished by activation of K^+ and/or Cl^- channels and organic osmolyte release. Apoptosis is further paralleled by cytosolic acidification due to activation of Cl^- channels and inhibition of Na^+/H^+ exchangers. I_{CRAC} is inhibited during CD95-induced apoptosis. Apoptosis can, however, be triggered by sustained Ca^{2+} entry through Ca^{2+}-permeable cation channels.

Cl^- channels, K^+ channels and Ca^{2+}-permeable channels each participate in the machinery of both cell proliferation and suicidal cell death. The impact of an individual channel depends on further properties of the cell. Activation of K^+ channels without parallel activity of electrogenic anion transporters or Cl^- channels hyperpolarises the cell but does not shrink it. K^+ channel activity and cell membrane potential influence the cytosolic free Ca^{2+} concentration only if Ca^{2+} channels are active. Moreover, oscillating K^+ channel activity typical for proliferating cells (Lang et al. 1991; Pandiella et al. 1989) is different from the sustained K^+ channel activation typical of apoptotic cells (P.A. Lang et al. 2003a). Oscillatory Ca^{2+} channel activity with subsequent fluctuations of cytosolic Ca^{2+} concentration depolymerises the cytoskeleton (Dartsch et al. 1995; Lang et al. 1992, 2000c; Ritter et al. 1997) but is presumably too short-lived for activation of caspases (Whitfield et al. 1995) or scramblases (Dekkers et al. 2002; Woon et al. 1999). Beyond that, the amplitude of TASK-3 K^+ channel activity observed during apoptosis is one order of magnitude higher than the activity of the same channels in tumour cells (Patel and Lazdunski 2004; Wang 2004) and the Ca^{2+} entry that stimulates mitogenic transcription factors may be lower than the Ca^{2+} entry triggering apoptosis (Whitfield et al. 1995).

Thus, similar, or even identical, ion channels may be involved in the machinery of cell proliferation and apoptosis. Their effects depend on the temporal pattern and amplitude of channel activity as well as the interplay with other channels, transporters and signalling pathways. Moreover, beyond ion channels at the cell membrane, intracellular channels may participate in apoptotic signalling (e.g. O'Rourke 2004). However, the exact role of these channels also needs further elucidation. Clearly, despite the enormous body of evidence accumulated hitherto, additional experimental effort is needed to fully understand the complex interplay between channel activity and signalling of proliferating or dying cells.

Acknowledgements The authors acknowledge the meticulous preparation of the manuscript by Jasmin Bühringer. The work of the authors was supported by the Deutsche Forschungsgemeinschaft, Nr. La 315/4-3, La 315/6-1, Le 792/3-3, DFG Schwerpunkt Intrazelluläre Lebensformen La 315/11-1/-2 and Hu 781/4-3, and Bundesministerium für Bildung, Wissenschaft, Forschung und Technologie (Center for Interdisciplinary Clinical Research) 01 KS 9602, and in part (A. Vereninov) by the Russian Foundation for Basic Research, projects: RFFI no. 06–04–48060, RFFI–DFG 06–04–04000 (DFG 436 RUS 113/488/0–2R).

References

Adams TE, McKern NM, Ward CW (2004) Signalling by the type 1 insulin-like growth factor receptor: interplay with the epidermal growth factor receptor. Growth Factors 22:89–95

Adragna NC, Fulvio MD, Lauf PK (2004) Regulation of K–Cl cotransport: from function to genes. J Membr Biol 201:109–137

Albright CD, da Costa KA, Craciunescu CN, Klem E, Mar MH, Zeisel SH (2005) Regulation of chlorine deficiency apoptosis by epidermal growth factor in CWSV-1 rat hepatocytes. Cell Physiol Biochem 15:59–68

Alisi A, Demori I, Spagnuolo S, Pierantozzi E, Fugassa E, Leoni S (2005) Thyroid status affects rat liver regeneration after partial hepatectomy by regulating cell cycle and apoptosis. Cell Physiol Biochem 15:69–76

Aoyama T, Matsui T, Novikov M, Park J, Hemmings B, Rosenzweig A (2005) Serum and gluco-corticoid-responsive kinase-1 regulates cardiomyocyte survival and hypertrophic response. Circulation 111:1652–1659

Badou A, Jha MK, Matza D, Mehal WZ, Freichel M, Flockerzi V, Flavell RA (2006) Critical role for the beta regulatory subunits of Cav channels in T lymphocyte function. Proc Natl Acad Sci USA 103:15529–15534

Bankers-Fulbright JL, Kephart GM, Loegering DA, Bradford AL, Okada S, Kita H, Gleich GJ (1998) Sulfonylureas inhibit cytokine-induced eosinophil survival and activation. J Immunol 160:5546–5553

Barvitenko NN, Adragna NC, Weber RE (2005) Erythrocyte signal transduction pathways, their oxygenation dependence and functional significance. Cell Physiol Biochem 15:1–18

Beauvais F, Michel L, Dubertret L (1995) Human eosinophils in culture undergo a striking and rapid shrinkage during apoptosis. Role of K+ channels. J Leukoc Biol 57:851–855

Benson RS, Heer S, Dive C, Watson AJ (1996) Characterization of cell volume loss in CEM-C7A cells during dexamethasone-induced apoptosis. Am J Physiol 270:C1190–C1203

Berg CP, Engels IH, Rothbart A, Lauber K, Renz A, Schlosser SF, Schulze-Osthoff K, Wesselborg S (2001) Human mature red blood cells express caspase-3 and caspase-8, but are devoid of mitochondrial regulators of apoptosis. Cell Death Differ 8:1197–1206

Berridge MJ, Bootman MD, Lipp P (1998) Calcium – a life and death signal. Nature 395:645–648

Berridge MJ, Lipp P, Bootman MD (2000) The versatility and universality of calcium signalling. Nat Rev Mol Cell Biol 1:11–21

Berridge MJ, Bootman MD, Roderick HL (2003) Calcium signalling: dynamics, homeostasis and remodelling. Nat Rev Mol Cell Biol 4:517–529

Bikfalvi A, Savona C, Perollet C, Javerzat S (1998) New insights in the biology of fibroblast growth factor-2. Angiogenesis 1:155–173

Bortner CD, Cidlowski JA (1998) A necessary role for cell shrinkage in apoptosis. Biochem Pharmacol 56:1549–1559

Bortner CD, Cidlowski JA (1999) Caspase independent/dependent regulation of K(+), cell shrink-age, and mitochondrial membrane potential during lymphocyte apoptosis. J Biol Chem 274:21953–21962

Bortner CD, Cidlowski JA (2004) The role of apoptotic volume decrease and ionic homeostasis in the activation and repression of apoptosis. Pfluegers Arch 448:313–318

Bortner CD, Hughes FM Jr, Cidlowski JA (1997) A primary role for K^+ and Na^+ efflux in the activation of apoptosis. J Biol Chem 272:32436–32442

Bratosin D, Leszczynski S, Sartiaux C, Fontaine O, Descamps J, Huart JJ, Poplineau J, Goudaliez F, Aminoff D, Montreuil J (2001) Improved storage of erythrocytes by prior leukodepletion: flow cytometric evaluation of stored erythrocytes. Cytometry 46:351–356

Cabado AG, Vieytes MR, Botana LM (1994) Effect of ion composition on the changes in membrane potential induced with several stimuli in rat mast cells. J Cell Physiol 158:309–16

Cariers A, Reinehr R, Fischer R, Warskulat U, Haussinger D (2002) c-Jun-N-terminal kinase dependent membrane targeting of CD95 in rat hepatic stellate cells. Cell Physiol Biochem 12:179–186

Chan HC, Goldstein J, Nelson DJ (1992) Alternate pathways for chloride conductance activation in normal and cystic fibrosis airway epithelial cells. Am J Physiol 262:C1273–C1283

Chin LS, Park CC, Zitnay KM, Sinha M, DiPatri AJ Jr, Perillan P, Simard JM (1997) 4-Aminopyridine causes apoptosis and blocks an outward rectifier K^+ channel in malignant astrocytoma cell lines. J Neurosci Res 48:122–127

Colom LV, Diaz ME, Beers DR, Neely A, Xie WJ, Appel SH (1998) Role of potassium channels in amyloid-induced cell death. J Neurochem 70:1925–1934

Dangel GR, Lang F, Lepple-Wienhues A (2005) Effect of Sphingosin on Ca^{2+} entry and mitochondrial potential of Jurkat T cells – interaction with Bcl2. Cell Physiol Biochem 16:9–14

Dartsch PC, Ritter M, Gschwentner M, Lang HJ, Lang F (1995) Effects of calcium channel blockers on NIH 3T3 fibroblasts expressing the Ha-ras oncogene. Eur J Cell Biol 67:372–378

Daugas E, Cande C, Kroemer G (2001) Erythrocytes: death of a mummy. Cell Death Differ 8:1131–1133

Davies AM (2003) Regulation of neuronal survival and death by extracellular signals during development. EMBO J 22:2537–2545

DeCoursey TE, Chandy KG, Gupta S, Cahalan MD (1984) Voltage-gated K^+ channels in human T lymphocytes: a role in mitogenesis? Nature 307:465–468

Dekkers DW, Comfurius P, Bevers EM, Zwaal RF (2002) Comparison between Ca^{2+}-induced scrambling of various fluorescently labelled lipid analogues in red blood cells. Biochem J 362:741–747

Deutsch C, Chen LQ (1993) Heterologous expression of specific K^+ channels in T lymphocytes: functional consequences for volume regulation. Proc Natl Acad Sci USA 90:10036–10040

Dinudom A, Komwatana P, Young JA, Cook DI (1995) Control of the amiloride-sensitive Na^+ current in mouse salivary ducts by intracellular anions is mediated by a G protein. J Physiol 487:549–555

Duranton C, Huber SM, Lang F (2002) Oxidation induces a Cl^--dependent cation conductance in human red blood cells. J Physiol 539:847–855

Duranton C, Huber S, Tanneur V, Lang K, Brand V, Sandu C, Lang F (2003) Electrophysiological properties of the *Plasmodium falciparum*-induced cation conductance of human erythrocytes. Cell Physiol Biochem 13:189–198

Enomoto K, Cossu MF, Edwards C, Oka T (1986) Induction of distinct types of spontaneous electrical activities in mammary epithelial cells by epidermal growth factor and insulin. Proc Natl Acad Sci USA 83:4754–4758

Erdo S, Michler A, Wolff JR (1991) GABA accelerates excitotoxic cell death in cortical cultures: protection by blockers of GABA-gated chloride channels. Brain Res 542:254–258

Faehling M, Koch ED, Raithel J, Trischler G, Waltenberger J (2001) Vascular endothelial growth factor-A activates Ca^{2+} -activated K^+ channels in human endothelial cells in culture. Int J Biochem Cell Biol 33:337–346

Fillon S, Klingel K, Warntges S, Sauter M, Gabrysch S, Pestel S, Tanneur V, Waldegger S, Zipfel A, Viebahn R, Haussinger D, Broer S, Kandolf R, Lang F (2002) Expression of the serine/threonine kinase hSGK1 in chronic viral hepatitis. Cell Physiol Biochem 12:47–54

Gamper N, Huber SM, Badawi K, Lang F (2000) Cell volume-sensitive sodium channels upregulated by glucocorticoids in U937 macrophages. Pfluegers Arch 441:281–286

Gantner F, Uhlig S, Wendel A (1995) Quinine inhibits release of tumor necrosis factor, apoptosis, necrosis and mortality in a murine model of septic liver failure. Eur J Pharmacol 294:353–355

Garcia-Perez A, Burg MB (1991) Renal medullary organic osmolytes. Physiol Rev 71:1081–1115

Gomez-Angelats M, Bortner CD, Cidlowski JA (2000) Protein kinase C (PKC) inhibits fas receptor-induced apoptosis through modulation of the loss of K+ and cell shrinkage. A role for PKC upstream of caspases. J Biol Chem 275:19609–19619

Green DR, Reed JC (1998) Mitochondria and apoptosis. Science 281:1309–1312

Gulbins E, Schlottmann K, Brenner B, Lang F, Coggeshall KM (1995) Molecular analysis of Ras activation by tyrosine phosphorylated Vav. Biochem Biophys Res Commun 217:876–885

Gulbins E, Szabo I, Baltzer K, Lang F (1997) Ceramide-induced inhibition of T lymphocyte voltage-gated potassium channel is mediated by tyrosine kinases. Proc Natl Acad Sci USA 94:7661–7666

Gulbins E, Jekle A, Ferlinz K, Grassme H, Lang F (2000) Physiology of apoptosis. Am J Physiol Renal Physiol 279:F605–F615

Han H, Wang J, Zhang Y, Long H, Wang H, Xu D, Wang Z (2004) HERG K channel conductance promotes H_2O_2-induced apoptosis in HEK293 cells: cellular mechanisms. Cell Physiol Biochem 14:121–134

Harrison SM, Roffler-Tarlov SK (1998) Cell death during development of testis and cerebellum in the mutant mouse weaver. Dev Biol 195:174–186

Huber SM, Gamper N, Lang F (2001) Chloride conductance and volume-regulatory nonselective cation conductance in human red blood cell ghosts. Pfluegers Arch 441:551–558

Hughes FM Jr, Bortner CD, Purdy GD, Cidlowski JA (1997) Intracellular K+ suppresses the activation of apoptosis in lymphocytes. J Biol Chem 272:30567–30576

Hughes FM Jr, Cidlowski JA (1999) Potassium is a critical regulator of apoptotic enzymes in vitro and in vivo. Adv Enzyme Regul 39:157–171

Jakab M, Ritter M (2006) Cell volume regulatory ion transport in the regulation of cell migration. Contrib Nephrol 152:161–180

Jakob R, Krieglstein J (1997) Influence of flupirtine on a G-protein coupled inwardly rectifying potassium current in hippocampal neurones. Br J Pharmacol 122:1333–1338

Jiang B, Hattori N, Liu B, Nakayama Y, Kitagawa K, Inagaki C (2004) Suppression of cell proliferation with induction of p21 by Cl(−) channel blockers in human leukemic cells. Eur J Pharmacol 488:27–34

Kinne RK, Czekay RP, Grunewald JM, Mooren FC, Kinne-Saffran E (1993) Hypotonicity-evoked release of organic osmolytes from distal renal cells: systems, signals, and sidedness. Ren Physiol Biochem 16:66–78

Kitamura H, Yamauchi A, Sugiura T, Matsuoka Y, Horio M, Tohyama M, Shimada S, Imai E, Hori M (1998) Inhibition of myo-inositol transport causes acute renal failure with selective medullary injury in the rat. Kidney Int 53:146–153

Koch J, Korbmacher C (1999) Osmotic shrinkage activates nonselective cation (NSC) channels in various cell types. J Membr Biol 168:131–139

Kwon HM, Handler JS (1995) Cell volume regulated transporters of compatible osmolytes. Curr Opin Cell Biol 7:465–471

Lambert IH (2004) Regulation of the cellular content of the organic osmolyte taurine in mammalian cells. Neurochem Res 29:27–63

Lang F, Friedrich F, Kahn E, Woll E, Hammerer M, Waldegger S, Maly K, Grunicke H (1991) Bradykinin-induced oscillations of cell membrane potential in cells expressing the Ha-ras oncogene. J Biol Chem 266:4938–4942

Lang F, Waldegger S, Woell E, Ritter M, Maly K, Grunicke H (1992) Effects of inhibitors and ion substitutions on oscillations of cell membrane potential in cells expressing the RAS oncogene. Pfluegers Arch 421:416–424

Lang F, Busch GL, Ritter M, Volkl H, Waldegger S, Gulbins E, Haussinger D (1998a) Functional significance of cell volume regulatory mechanisms. Physiol Rev 78:247–306

Lang F, Madlung J, Uhlemann AC, Risler T, Gulbins E (1998b) Cellular taurine release triggered by stimulation of the Fas(CD95) receptor in Jurkat lymphocytes. Pfluegers Arch 436:377–383

Lang F, Szabo I, Lepple-Wienhues A, Siemen D, Gulbins E (1999) Physiology of receptor-mediated lymphocyte apoptosis. News Physiol Sci 14:194–200

Lang F, Madlung J, Bock J, Lukewille U, Kaltenbach S, Lang KS, Belka C, Wagner CA, Lang HJ, Gulbins E, Lepple-Wienhues A (2000a) Inhibition of Jurkat-T-lymphocyte Na$^+$/H$^+$-exchanger by CD95(Fas/Apo-1)-receptor stimulation. Pfluegers Arch 440:902–907

Lang F, Madlung J, Siemen D, Ellory C, Lepple-Wienhues A, Gulbins E (2000b) The involvement of caspases in the CD95(Fas/Apo-1)- but not swelling-induced cellular taurine release from Jurkat T-lymphocytes. Pfluegers Arch 440:93–99

Lang F, Ritter M, Gamper N, Huber S, Fillon S, Tanneur V, Lepple-Wienhues A, Szabo I, Gulbins E (2000c) Cell volume in the regulation of cell proliferation and apoptotic cell death. Cell Physiol Biochem 10:417–428

Lang F, Lang PA, Lang KS, Brand V, Tanneur V, Duranton C, Wieder T, Huber SM (2004) Channel-induced apoptosis of infected host cells – the case of malaria. Pfluegers Arch 448:319–324

Lang F, Bohmer C, Palmada M, Seebohm G, Strutz-Seebohm N, Vallon V (2006) (Patho) physiological significance of the serum- and glucocorticoid-inducible kinase isoforms. Physiol Rev 86:1151–1178

Lang KS, Fillon S, Schneider D, Rammensee HG, Lang F (2002a) Stimulation of TNF alpha expression by hyperosmotic stress. Pfluegers Arch 443:798–803

Lang KS, Roll B, Myssina S, Schittenhelm M, Scheel-Walter HG, Kanz L, Fritz J, Lang F, Huber SM, Wieder T (2002b) Enhanced erythrocyte apoptosis in sickle cell anemia, thalassemia and glucose-6-phosphate dehydrogenase deficiency. Cell Physiol Biochem 12:365–372

Lang KS, Duranton C, Poehlmann H, Myssina S, Bauer C, Lang F, Wieder T, Huber SM (2003) Cation channels trigger apoptotic death of erythrocytes. Cell Death Differ 10:249–256

Lang KS, Myssina S, Brand V, Sandu C, Lang PA, Berchtold S, Huber SM, Lang F, Wieder T (2004) Involvement of ceramide in hyperosmotic shock-induced death of erythrocytes. Cell Death Differ 11:231–243

Lang KS, Lang PA, Bauer C, Duranton C, Wieder T, Huber SM, Lang F (2005) Mechanisms of suicidal erythrocyte death. Cell Physiol Biochem 15:195–202

Lang PA, Kaiser S, Myssina S, Wieder T, Lang F, Huber SM (2003a) Role of Ca^{2+}-activated K$^+$ channels in human erythrocyte apoptosis. Am J Physiol Cell Physiol 285:C1553–C1560

Lang PA, Warskulat U, Heller-Stilb B, Huang DY, Grenz A, Myssina S, Duszenko M, Lang F, Haussinger D, Vallon V, Wieder T (2003b) Blunted apoptosis of erythrocytes from taurine transporter deficient mice. Cell Physiol Biochem 13:337–346

Lang PA, Kempe DS, Myssina S, Tanneur V, Birka C, Laufer S, Lang F, Wieder T, Huber SM (2005a) PGE(2) in the regulation of programmed erythrocyte death. Cell Death Differ 12:415–428

Lang PA, Kempe DS, Tanneur V, Eisele K, Klarl BA, Myssina S, Jendrossek V, Ishii S, Shimizu T, Waidmann M, Hessler G, Huber SM, Lang F, Wieder T (2005b) Stimulation of erythrocyte ceramide formation by platelet-activating factor. J Cell Sci 118:1233–1243

Lauritzen I, De Weille JR, Lazdunski M (1997) The potassium channel opener (−)-cromakalim prevents glutamate-induced cell death in hippocampal neurons. J Neurochem 69:1570–1579

Lepple-Wienhues A, Szabo I, Laun T, Kaba NK, Gulbins E, Lang F (1998) The tyrosine kinase p56lck mediates activation of swelling-induced chloride channels in lymphocytes. J Cell Biol 141:281–286

Lepple-Wienhues A, Belka C, Laun T, Jekle A, Walter B, Wieland U, Welz M, Heil L, Kun J, Busch G, Weller M, Bamberg M, Gulbins E, Lang F (1999) Stimulation of CD95 (Fas) blocks T lymphocyte calcium channels through sphingomyelinase and sphingolipids. Proc Natl Acad Sci USA 96:13795–13800

Lepple-Wienhues A, Wieland U, Laun T, Heil L, Stern M, Lang F (2001) A src-like kinase activates outwardly rectifying chloride channels in CFTR-defective lymphocytes. FASEB J 15:927–931

Liu XH, Kirschenbaum A, Yu K, Yao S, Levine AC (2005) Cyclooxygenase-2 suppresses hypoxia-induced apoptosis via a combination of direct and indirect inhibition of p53 activity in a human prostate cancer cell line. J Biol Chem 280:3817–3823

Liu XM, Tao M, Han XD, Fan Q, Lin JR (2001) Gating kinetics of potassium channel and effects of nerve growth factors in PC12 cells analyzed with fractal model. Acta Pharmacol Sin 22:103–110

Long H, Han H, Yang B, Wang Z (2003) Opposite cell density-dependence between spontaneous and oxidative stress-induced apoptosis in mouse fibroblast L-cells. Cell Physiol Biochem 13:401–414

Maeno E, Ishizaki Y, Kanaseki T, Hazama A, Okada Y (2000) Normotonic cell shrinkage because of disordered volume regulation is an early prerequisite to apoptosis. Proc Natl Acad Sci USA 97:9487–9492

Marunaka Y, Nakahari T, Tohda H (1994) Cytosolic (Cl$^-$) regulates Na$^+$ absorption in fetal alveolar epithelium?: roles of cAMP and Cl$^-$ channels. Jpn J Physiol 44 Suppl 2:S281–S288

Mauro T, Dixon DB, Komuves L, Hanley K, Pappone PA (1997) Keratinocyte K$^+$ channels mediate Ca^{2+}-induced differentiation. J Invest Dermatol 108:864–870

Michea L, Ferguson DR, Peters EM, Andrews PM, Kirby MR, Burg MB (2000) Cell cycle delay and apoptosis are induced by high salt and urea in renal medullary cells. Am J Physiol Renal Physiol 278:F209–F218

Migheli A, Attanasio A, Lee WH, Bayer SA, Ghetti B (1995) Detection of apoptosis in weaver cerebellum by electron microscopic in situ end-labeling of fragmented DNA. Neurosci Lett 199:53–56

Migheli A, Piva R, Wei J, Attanasio A, Casolino S, Hodes ME, Dlouhy SR, Bayer SA, Ghetti B (1997) Diverse cell death pathways result from a single missense mutation in weaver mouse. Am J Pathol 151:1629–1638

Miki T, Tashiro F, Iwanaga T, Nagashima K, Yoshitomi H, Aihara H, Nitta Y, Gonoi T, Inagaki N, Miyazaki J, Seino S (1997) Abnormalities of pancreatic islets by targeted expression of a dominant-negative KATP channel. Proc Natl Acad Sci USA 94:11969–11973

Miller GW, Schnellmann RG (1993) Cytoprotection by inhibition of chloride channels: the mechanism of action of glycine and strychnine. Life Sci 53:1211–1215

Montague JW, Bortner CD, Hughes FM, Jr., Cidlowski JA (1999) A necessary role for reduced intracellular potassium during the DNA degradation phase of apoptosis. Steroids 64:563–569

Moran J, Hernandez-Pech X, Merchant-Larios H, Pasantes-Morales H (2000) Release of taurine in apoptotic cerebellar granule neurons in culture. Pfluegers Arch 439:271–277

Murtomaki S, Trenkner E, Wright JM, Saksela O, Liesi P (1995) Increased proteolytic activity of the granule neurons may contribute to neuronal death in the weaver mouse cerebellum. Dev Biol 168:635–648

Nilius B, Droogmans G (2001) Ion channels and their functional role in vascular endothelium. Physiol Rev 81:1415–1459

Nilius B, Wohlrab W (1992) Potassium channels and regulation of proliferation of human melanoma cells. J Physiol 445:537–548

O'Lague PH, Huttner SL, Vandenberg CA, Morrison-Graham K, Horn R (1985) Morphological properties and membrane channels of the growth cones induced in PC12 cells by nerve growth factor. J Neurosci Res 13:301–321

O'Rourke B (2004) Evidence for mitochondrial K$^+$ channels and their role in cardioprotection. Circ Res 94:420–432

Okada Y (2006) Cell volume-sensitive chloride channels: phenotypic properties and molecular identity. Contrib Nephrol 152:9–24

Okada Y, Maeno E, Shimizu T, Manabe K, Mori S, Nabekura T (2004) Dual roles of plasmalemmal chloride channels in induction of cell death. Pfluegers Arch 448:287–295

Oo TF, Blazeski R, Harrison SM, Henchcliffe C, Mason CA, Roffler-Tarlov SK, Burke RE (1996) Neuron death in the substantia nigra of weaver mouse occurs late in development and is not apoptotic. J Neurosci 16:6134–6145

Pal S, He K, Aizenman E (2004) Nitrosative stress and potassium channel-mediated neuronal apoptosis: is zinc the link? Pfluegers Arch 448:296–303

Pandiella A, Magni M, Lovisolo D, Meldolesi J (1989) The effect of epidermal growth factor on membrane potential. Rapid hyperpolarization followed by persistent fluctuations. J Biol Chem 264:12914–12921

Pappas CA, Ritchie JM (1998) Effect of specific ion channel blockers on cultured Schwann cell proliferation. Glia 22:113–120

Pappone PA, Ortiz-Miranda SI (1993) Blockers of voltage-gated K channels inhibit proliferation of cultured brown fat cells. Am J Physiol 264:C1014–C1019

Parekh AB, Penner R (1997) Store depletion and calcium influx. Physiol Rev 77:901–930

Parekh AB, Putney JW Jr (2005) Store-operated calcium channels. Physiol Rev 85:757–810

Patel AJ, Lazdunski M (2004) The 2P-domain K^+ channels: role in apoptosis and tumorigenesis. Pfluegers Arch 448:261–273

Perez GI, Maravei DV, Trbovich AM, Cidlowski JA, Tilly JL, Hughes FM Jr (2000) Identification of potassium-dependent and -independent components of the apoptotic machinery in mouse ovarian germ cells and granulosa cells. Biol Reprod 63:1358–1369

Phipps DJ, Branch DR, Schlichter LC (1996) Chloride-channel block inhibits T lymphocyte activation and signalling. Cell Signal 8:141–149

Pozzi S, Malferrari G, Biunno I, Samaja M (2002) Low-flow ischemia and hypoxia stimulate apoptosis in perfused hearts independently of reperfusion. Cell Physiol Biochem 12:39–46

Prehn JH, Jordan J, Ghadge GD, Preis E, Galindo MF, Roos RP, Krieglstein J, Miller RJ (1997) Ca^{2+} and reactive oxygen species in staurosporine-induced neuronal apoptosis. J Neurochem 68:1679–1685

Qian D, Weiss A (1997) T cell antigen receptor signal transduction. Curr Opin Cell Biol 9:205–212

Rice L, Alfrey CP (2005) The negative regulation of red cell mass by neocytolysis: physiologic and pathophysiologic manifestations. Cell Physiol Biochem 15:245–250

Ritter M, Woll E, Waldegger S, Haussinger D, Lang HJ, Scholz W, Scholkens B, Lang F (1993) Cell shrinkage stimulates bradykinin-induced cell membrane potential oscillations in NIH 3T3 fibroblasts expressing the ras-oncogene. Pfluegers Arch 423:221–224

Ritter M, Woll E, Haller T, Dartsch PC, Zwierzina H, Lang F (1997) Activation of $Na^+/H^{(+)}$-exchanger by transforming Ha-ras requires stimulated cellular calcium influx and is associated with rearrangement of the actin cytoskeleton. Eur J Cell Biol 72:222–228

Ritter M, Ravasio A, Jakab M, Chwatal S, Furst J, Laich A, Gschwentner M, Signorelli S, Burtscher C, Eichmuller S, Paulmichl M (2003) Cell swelling stimulates cytosol to membrane transposition of ICln. J Biol Chem 278:50163–50174

Rosette C, Karin M (1996) Ultraviolet light and osmotic stress: activation of the JNK cascade through multiple growth factor and cytokine receptors. Science 274:1194–1197

Rotoli BM, Uggeri J, Dall'Asta V, Visigalli R, Barilli A, Gatti R, Orlandini G, Gazzola GC, Bussolati O (2005) Inhibition of glutamine synthetase triggers apoptosis in asparaginase-resistant cells. Cell Physiol Biochem 15:281–292

Rouzaire-Dubois B, Milandri JB, Bostel S, Dubois JM (2000) Control of cell proliferation by cell volume alterations in rat C6 glioma cells. Pfluegers Arch 440:881–888

Sabirov RZ, Okada Y (2004) ATP-conducting maxi-anion channel: a new player in stress-sensory transduction. Jpn J Physiol 54:7–14

Sanders DA, Fiddes I, Thompson DM, Philpott MP, Westgate GE, Kealey T (1996) In the absence of streptomycin, minoxidil potentiates the mitogenic effects of fetal calf serum, insulin-like growth factor 1, and platelet-derived growth factor on NIH 3T3 fibroblasts in a K^+ channel-dependent fashion. J Invest Dermatol 107:229–234

Santella L (1998) The role of calcium in the cell cycle: facts and hypotheses. Biochem Biophys Res Commun 244:317–324

Santella L, Kyozuka K, De Riso L, Carafoli E (1998) Calcium, protease action, and the regulation of the cell cycle. Cell Calcium 23:123–130

Shen MR, Droogmans G, Eggermont J, Voets T, Ellory JC, Nilius B (2000) Differential expression of volume-regulated anion channels during cell cycle progression of human cervical cancer cells. J Physiol 529:385–394

Shrode LD, Tapper H, Grinstein S (1997) Role of intracellular pH in proliferation, transformation, and apoptosis. J Bioenerg Biomembr 29:393–399

Skryma RN, Prevarskaya NB, Dufy-Barbe L, Odessa MF, Audin J, Dufy B (1997) Potassium conductance in the androgen-sensitive prostate cancer cell line, LNCaP: involvement in cell proliferation. Prostate 33:112–122

Spassova MA, Soboloff J, He LP, Hewavitharana T, Xu W, Venkatachalam K, van Rossum DB, Patterson RL, Gill DL (2004) Calcium entry mediated by SOCs and TRP channels: variations and enigma. Biochim Biophys Acta 1742:9–20

Storey NM, Gomez-Angelats M, Bortner CD, Armstrong DL, Cidlowski JA (2003) Stimulation of Kv1.3 potassium channels by death receptors during apoptosis in Jurkat T lymphocytes. J Biol Chem 278:33319–33326

Strobl JS, Wonderlin WF, Flynn DC (1995) Mitogenic signal transduction in human breast cancer cells. Gen Pharmacol 26:1643–1649

Sturm JW, Zhang H, Magdeburg R, Hasenberg T, Bonninghoff R, Oulmi J, Keese M, McCuskey R (2004) Altered apoptotic response and different liver structure during liver regeneration in FGF-2-deficient mice. Cell Physiol Biochem 14:249–260

Szabo I, Gulbins E, Apfel H, Zhang X, Barth P, Busch AE, Schlottmann K, Pongs O, Lang F (1996) Tyrosine phosphorylation-dependent suppression of a voltage-gated K$^+$ channel in T lymphocytes upon Fas stimulation. J Biol Chem 271:20465–20469

Szabo I, Gulbins E, Lang F (1997) Regulation of Kv1.3 during Fas-induced apoptosis. Cell Physiol Biochem 7:148–158

Szabo I, Lepple-Wienhues A, Kaba KN, Zoratti M, Gulbins E, Lang F (1998) Tyrosine kinase-dependent activation of a chloride channel in CD95-induced apoptosis in T lymphocytes. Proc Natl Acad Sci USA 95:6169–6174

Szabo I, Adams C, Gulbins E (2004) Ion channels and membrane rafts in apoptosis. Pfluegers Arch 448:304–312

Takahashi N, Wang X, Tanabe S, Uramoto H, Jishage K, Uchida S, Sasaki S, Okada Y (2005) ClC-3-independent sensitivity of apoptosis to Cl$^-$ channel blockers in mouse cardiomyocytes. Cell Physiol Biochem 15:263–270

Tallquist M, Kazlauskas A (2004) PDGF signaling in cells and mice. Cytokine Growth Factor Rev 15:205–213

Teijeiro R, Rios R, Costoya JA, Castro R, Bello JL, Devesa J, Arce VM (2002) Activation of human somatostatin receptor 2 promotes apoptosis through a mechanism that is independent from induction of p53. Cell Physiol Biochem 12:31–38

Tohda H, Foskett JK, O'Brodovich H, Marunaka Y (1994) Cl$^-$ regulation of a Ca^{2+}-activated non-selective cation channel in beta-agonist-treated fetal distal lung epithelium. Am J Physiol 266: C104–C109

Uchida S, Sasaki S (2005) Function of chloride channels in the kidney. Annu Rev Physiol 67:759–778

Varela D, Simon F, Riveros A, Jorgensen F, Stutzin A (2004) NAD(P)H oxidase-derived H$_2$O$_2$ signals chloride channel activation in cell volume regulation and cell proliferation. J Biol Chem 279:13301–13304

Volk T, Fromter E, Korbmacher C (1995) Hypertonicity activates nonselective cation channels in mouse cortical collecting duct cells. Proc Natl Acad Sci USA 92:8478–8482

Walsh MF, Thamilselvan V, Grotelueschen R, Farhana L, Basson M (2003) Absence of adhesion triggers differential FAK and SAPKp38 signals in SW620 human colon cancer cells that may inhibit adhesiveness and lead to cell death. Cell Physiol Biochem 13:135–146

Wang GL, Wang XR, Lin MJ, He H, Lan XJ, Guan YY (2002) Deficiency in ClC-3 chloride chan-
    nels prevents rat aortic smooth muscle cell proliferation. Circ Res 91:E28–E32

Wang S, Melkoumian Z, Woodfork KA, Cather C, Davidson AG, Wonderlin WF, Strobl JS
    (1998) Evidence for an early G1 ionic event necessary for cell cycle progression and survival
    in the MCF-7 human breast carcinoma cell line. J Cell Physiol 176:456–464

Wang Z (2004) Roles of K$^+$ channels in regulating tumour cell proliferation and apoptosis.
    Pfluegers Arch 448:274–286

Wehner F (2006) Cell volume-regulated cation channels. Contrib Nephrol 152:25–53

Wehner F, Sauer H, Kinne RK (1995) Hypertonic stress increases the Na$^+$ conductance of rat
    hepatocytes in primary culture. J Gen Physiol 105:507–535

Wehner F, Böhmer C, Heinzinger H, van den BF, Tinel H (2000) The hypertonicity-induced
    Na(+) conductance of rat hepatocytes: physiological significance and molecular correlate. Cell
    Physiol Biochem 10:335–340

Wehner F, Olsen H, Tinel H, Kinne-Saffran E, Kinne RK (2003) Cell volume regulation: osmo-
    lytes, osmolyte transport, and signal transduction. Rev Physiol Biochem Pharmacol
    148:1–80

Wei L, Xiao AY, Jin C, Yang A, Lu ZY, Yu SP (2004) Effects of chloride and potassium channel block-
    ers on apoptotic cell shrinkage and apoptosis in cortical neurons. Pfluegers Arch 448:325–334

Wenzel U, Daniel H (2004) Early and late apoptosis events in human transformed and non-trans-
    formed colonocytes are independent on intracellular acidification. Cell Physiol Biochem
    14:65–76

Whitfield JF, Bird RP, Chakravarthy BR, Isaacs RJ, Morley P (1995) Calcium-cell cycle regula-
    tor, differentiator, killer, chemopreventor, and maybe, tumor promoter. J Cell Biochem Suppl
    22:74–91

Wiecha J, Reineker K, Reitmayer M, Voisard R, Hannekum A, Mattfeldt T, Waltenberger J,
    Hombach V (1998) Modulation of Ca^{2+}-activated K$^+$ channels in human vascular cells by
    insulin and basic fibroblast growth factor. Growth Horm IGF Res 8:175–181

Wieder T, Essmann F, Prokop A, Schmelz K, Schulze-Osthoff K, Beyaert R, Dorken B, Daniel
    PT (2001) Activation of caspase-8 in drug-induced apoptosis of B-lymphoid cells is independ-
    ent of CD95/Fas receptor-ligand interaction and occurs downstream of caspase-3. Blood
    97:1378–1387

Wondergem R, Gong W, Monen SH, Dooley SN, Gonce JL, Conner TD, Houser M, Ecay TW,
    Ferslew KE (2001) Blocking swelling-activated chloride current inhibits mouse liver cell pro-
    liferation. J Physiol 532:661–672

Wonderlin WF, Strobl JS (1996) Potassium channels, proliferation and G1 progression. J Membr
    Biol 154:91–107

Woon LA, Holland JW, Kable EP, Roufogalis BD (1999) Ca^{2+} sensitivity of phospholipid scram-
    bling in human red cell ghosts. Cell Calcium 25:313–320

Yu SP, Yeh CH, Sensi SL, Gwag BJ, Canzoniero LM, Farhangrazi ZS, Ying HS, Tian M, Dugan
    LL, Choi DW (1997) Mediation of neuronal apoptosis by enhancement of outward potassium
    current. Science 278:114–117

Yurinskaya VE, Goryachaya TS, Guzhova TV, Moshkov AV, Rozanov YM, Sakuta GA,
    Shirokova AV, Shumilina EV, Vassilieva IO, Lang F, Vereninov AA (2005a) Potassium and
    sodium balance in U937 cells during apoptosis with and without cell shrinkage. Cell Physiol
    Biochem 16:155–162

Yurinskaya VE, Moshkov AV, Rozanov YuM, Shirokova AV, Vassilieva IO, Shumilina EV,
    Lang F, Volgareva AA, Vereninov AA (2005b) Thymocyte K$^+$, Na$^+$ and water balance during
    dexamethasone and etoposide induced apoptosis. Cell Physiol Biochem 16:15–22

Zhou Q, Kwan HY, Chan HC, Jiang JL, Tam SC, Yao X (2003) Blockage of voltage-gated K$^+$
    channels inhibits adhesion and proliferation of hepatocarcinoma cells. Int J Mol Med
    11:261–266

Chapter 5
TRPV Ion Channels and Sensory Transduction of Osmotic and Mechanical Stimuli in Mammals

Wolfgang Liedtke

Abstract In signal transduction in metazoan cells, ion channels of the transient receptor potential (TRP) family have been identified as responding to diverse external and internal stimuli, amongst them osmotic stimuli. This chapter will highlight findings on the TRP vanilloid (TRPV) subfamily – both vertebrate and invertebrate members. Of the six mammalian TRPV channels, TRPV1, 2 and 4 have been demonstrated to function in transduction of osmotic stimuli. TRPV channels have been found to function in cellular as well as systemic osmotic homeostasis in vertebrates. Invertebrate TRPV channels – five in *Caenorhabditis elegans* and two in Drosophila – have been shown to play a role in mechanosensation such as hearing and proprioception in Drosophila and nose touch in *C. elegans*, and in the response to osmotic stimuli in *C. elegans*. In a striking example of evolutionary conservation of function, mammalian TRPV4 has been found to rescue osmo- and mechano-sensory deficits of the TRPV mutant strain *osm-9* in *C. elegans*, despite the fact that the respective proteins share not more than 26% orthology.

Duke University Medical Center, Center for Translational Nuroscience, Department of Neurobiology, Department of Medicine/Division of Neurology, DUMC Box # 2900, Durham NC 27710, USA, wolfgang@neuro.duke.edu

5.1 Introduction: Response to Osmotic and Mechanical Stimuli – a Function of TRPV Ion Channels Apparent Since the "Birth" of this Subfamily

Within the transient receptor potential (TRP) superfamily of ion channels (Cosens and Manning 1969; Montell and Rubin 1989; Wong et al. 1989; Hardie and Minke 1992; Zhu et al. 1995), the TRP vanilloid (TRPV) subfamily stepped into the spotlight in 1997 (Caterina et al. 1997; Colbert et al. 1997). The spectacular find of the capsaicin-receptor TRPV1 led to subsequent research in the direction of the study of responses to ligand (capsaicin), acidity and thermal stimuli. Slightly less attention was perhaps dedicated to the other founding member, the *Caenorhabditis elegans osm-9* gene. The discovery of *osm-9* implied that TRP channels might subserve critical roles in the transduction of osmotic and mechanical stimuli. Subsequently, TRPV2, -V3, and -V4 were identified by a candidate gene approach (Caterina et al. 1999; Kanzaki et al. 1999; Liedtke et al. 2000; Strotmann et al. 2000; Wissenbach et al. 2000; Peier et al. 2002; Smith et al. 2002; Xu et al. 2002). The latter strategy also led to the identification of four additional *C. elegans ocr* genes (Tobin et al. 2002) and two *Drosophila trpv* genes: Nanchung (NAN) and Inactive (IAV; Kim et al. 2003; Gong et al. 2004). TRPV channels can be sub-divided into four branches by sequence comparison (see dendrogram in Fig. 5.1). Alluding to their function,

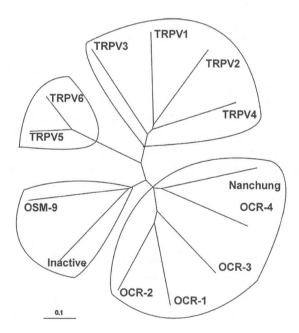

Fig. 5.1 Dendrogram of mammalian (TRPV1-6), *Caenorhabditis elegans* (OSM-9 and OCR-1 to -4) and *Drosophila melanogaster* (NAN and IAV) transient receptor potential vanilloid (TRPV) ion channels. Reproduced from Liedtke and Kim (2005) with permission

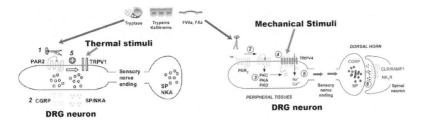

Fig. 5.2 Schematic of the "tuning" of transduction of noxious stimuli by activation of proteinase-activated-receptor-2 (PAR-2). *Left* Hyperalgesia in response to noxious *thermal* stimuli via TRPV1, *right* hyperalgesia in response to noxious *mechanical* stimuli via TRPV4. Reproduced from Amadesi et al. 2004, 2006; Grant et al. 2007 with permission

TRPV1, -V2, -V3, and -V4 have been named "thermo-TRPs"; review articles on "thermo-TRPs" are available for interested readers (Caterina and Julius 1999; Clapham 2003; Tominaga and Caterina 2004; Caterina and Montell 2005; Patapoutian 2005). TRPV5 and TRPV6 possibly function in Ca^{2+} uptake in the kidney and intestine (Hoenderop et al. 1999, 2003; Peng et al. 1999, 2003; den Dekker et al. 2003). Regarding the invertebrate TRPV channel genes, one invertebrate branch includes *C. elegans* OSM-9 and *Drosophila* IAV, and the other includes OCR-1 to -4 of *C. elegans* and *Drosophila* NAN. In cases where heterologous-expression-system data are available for TRPV channels, their non-selective conductance of cations with a (slight) preference for Ca^{2+} is apparent. This means that Ca^{2+} influx through the respective TPRPV channel is the critical signaling mechanism.

This chapter will provide some discussion on the role of mammalian and also invertebrate TRPV channels (with a focus on *C. elegans*) in signal transduction in response to osmotic as well as mechanical stimuli, because these submodalities are related via membrane tension. These "osmo- and mechano-TRPs" (Liedtke and Kim 2005) include TRPV1, -2, -4, OSM-9, OCR-2, NAN and IAV. Other TRPV channels might join this functional group within the TRP superfamily, which certainly also comprises non-TRPV channels, e.g., TRPA1 (Corey 2003; Nagata et al. 2005) or NompC (Walker et al. 2000). The available evidence will be summarized, gene by gene (Fig. 5.2), guided by the question: do TRPV ion channels function in transduction of osmotic (and mechanical) stimuli, and, if so, by which molecular mechanism?

5.2 In Vivo Findings Implicate Products of the *trpv1* Gene in Osmo-Mechano Transduction

There have been no reports on transduction of osmotic and mechanical stimuli involving TRPV1 in heterologous cellular expression systems. Genetically engineered $trpv1^{-/-}$ mice, which have previously been shown to lack thermal hyperalgesia following inflammation (Caterina et al. 2000; Davis et al. 2000), also showed an altered response of their magnocellular hypothalamic neurons to tonicity stimuli. Very recently, Reza Sharif Naeini from Charles Bourque's group reported that *trpv*1$^{-/-}$ mice failed to express an N-terminal variant of the *trpv1* gene in magnocellular neurons of the supraoptic and paraventricular nucleus of the hypothalamus (Naeini et al. 2006). As these neurons are known to secrete vasopressin, the *trpv*1$^{-/-}$ mice were found to have a profound impairment in antidiuretic hormone (ADH) secretion in response to systemic hypertonicity, and their magnocellular neurons did not show an appropriate bio-electrical response to hypertonicity. These findings led Bourque and colleagues to conclude that this *trpv1* N-terminal variant, which could not be identified at the molecular level, is likely involved as (part of) a tonicity sensor of intrinsically osmo-sensitive magnocellular neurons. In an interesting paper published soon thereafter, Ciura and Bourque reported that neurons within the organum vasculosum laminae terminalis (OVLT) – a sensory circumventricular organ in the brain, yet outside the blood-brain barrier – also express this particular *trpv1* variant, and that their osmotic sensing in the absence of *trpv1* was critically impaired (Ciura and Bourque 2006).

*trpv*1$^{-/-}$ mice also showed an abnormal response of their bladder to stretch (Birder et al. 2002). TRPV1 could be localized to sensory and autonomous ganglia neurons innervating the bladder, and also to urethelial cells. When bladder and urothel-epithelial cells were cultured, their response to mechanical stretch and hypotonicity was different from wild-type controls. Specifically, TRPV1$^+$ bladders secreted ATP upon stretch and hypotonicity, which, in turn, is known to activate nerve fibers in the urinary bladder. This response to mechanical stimulation was greatly reduced in bladders excised from *trpv*1$^{-/-}$ mice. It appears likely that this mechanism, functional in mice, also plays a role in human bladder epithelium. Intravesical instillation of TRPV1-activators is used to treat hyperactive bladder in spinal cord disease (Dinis et al. 2004; Lazzeri et al. 2004; Stein et al. 2004; Apostolidis et al. 2005). Another instance of an altered response to mechanical stimuli in *trpv*1$^{-/-}$ mice relates to the response of the jejunum to stretch (Rong et al. 2004). Afferent jejunal nerve fibers were found to respond with decreased frequency of discharge in *trpv*1$^{-/-}$ mice when compared to wild type mice. In humans, TRPV1 positive fibers were found significantly increased in the rectum in patients suffering from fecal urgency – a condition with rectal hypersensitivity in response to mechanical distension (Chan et al. 2003) – as well as in hemorrhoid tissue (di Mola et al. 2006). Expression of TRPV1$^+$ fibers in rectal biopsy samples from these patients was positively correlated with a decreased threshold to stretch; in addition, the occurrence of TRPV1$^+$ fibers was also correlated with a dysaesthesia, described as a burning sensation by the patients. Another recent study focused on possible mechanisms of signal transduction in response to mechanical stimuli in blood vessels (Scotland et al. 2004). Elevation of luminal pressure in mesenteric arteries was shown to be associated with generation of 20-hydroxyeicosatetraenoic acid, which, in turn, activated TRPV1 expressed on C-fibers leading to nerve depolarization and vasoactive neuropeptide release. With respect to nociception, using *trpv*1$^{-/-}$ mice, *trpv1* was shown to be involved in inflammatory thermal hyperalgesia, but not inflammatory mechanical hyperalgesia (Caterina and Julius 1999; Gunthorpe et al. 2002). However, a specific blocker of TRPV1 was found to reduce mechanical hyperalgesia in rats (Pomonis et al. 2003). This latter result appears contradictory in view of the obvious lack of difference between $trpv1^{-/-}$ and wild type control mice. This discrepancy could be due to a species difference between mouse and rat, or to different mechanisms affecting signaling in a $trpv1$ general knockout vs a specific temporal pharmacological blocking of TRPV1 ion channel proteins, which very likely participate in signaling multiplex protein complexes.

Taken together, loss-of-function studies using $trpv1^{-/-}$ mice clearly implicate the $trpv1$ gene as playing a significant role in transduction of osmotic and mechanical stimuli. Despite this phenotypic clarity, the details and molecular mechanisms await further investigation.

5.3   Tissue Culture Cell Data Implicate TRPV2 in Osmo-Mechanotransduction

In heterologous cellular expression systems, TRPV2 was initially described as a temperature-gated channel for stimuli >52°C (Caterina et al. 1999). Recently, TRPV2 was also demonstrated to respond to hypotonicity and mechanical stimuli (Muraki et al. 2003). Arterial smooth muscle cells from various arteries expressed TRPV2. These myocytes responded to hypotonicity with Ca^{2+} influx. This activation could be reduced by specific downregulation of TRPV2 using an antisense method. Heterologously expressed TRPV2 in CHO cells displayed a similar response to hypotonicity. These cells were also subjected to stretch by suction of the recording pipette and by stretching the cell membrane on a mechanical stimulator. Both maneuvers led to Ca^{2+} influx that was dependent on heterologous TRPV2 expression.

In aggregate, having been discovered as a "thermo-TRP", TRPV2 appears to be an "osmo-mechano-TRP" as well. However, in the absence of reports on TRPV2 null mice, this grouping is based on tissue culture data.

5.4   In Vivo Mouse- and Tissue Culture-Data Implicate the *trpv4* Gene in Osmo-Mechanotransduction, Including Hydromineral Homeostasis and Pain

CHO-immortalized tissue culture cells respond to hypotonic solution when (stably) transfected with TRPV4 (Liedtke et al. 2000). The same authors also found that HEK-293T cells expressed *trpv4* cDNA, which was cloned from these cells. However, *trpv4* cDNA was not found in other batches of HEK 293T cells, so this cell line was used for heterologous expression by other groups (Strotmann et al. 2000; Wissenbach et al. 2000). Notably, when comparing the two settings, it was obvious that the single-channel conductance of TRPV4 was different (Liedtke et al. 2000; Strotmann et al. 2000). This underscores the relevance of complementary gene expression in heterologous cellular systems for the functioning of TRPV4 in response to a basic biophysical stimulation. Also, it was found that the sensitivity of TRPV4 could be modulated by warming of the medium. Similar results were found in another investigation when expressing TRPV4 in HEK-293T cells (Gao et al. 2003; reviewed in Mutai and Heller 2003; O'Neil and Heller 2005). In addition, in this latter investigation, the cells were mechanically stretched (at isotonicity). At room temperature, there was no response to mechanical stress; however, at 37°C the response to stretch resulted in the maximum Ca^{2+} influx of all conditions tested. In two other investigations, heterologously expressed TRPV4 was found to be responsive to changes in temperature (Guler et al. 2002; Watanabe et al. 2002). Temperature change was accomplished by heating the streaming bath solution. This method of applying a temperature stimulus represents a mechanical stimulus per se. Gating of TRPV4 was found to be amplified when hypotonic solution was used as the streaming bath. In one of these investigations, temperature stimuli could not activate the TRPV4 channel in cell-detached inside-out patches (Watanabe et al. 2002).

In regards to maintenance of systemic osmotic pressure in live animals, $trpv4^{-/-}$ mice, when stressed with systemic hypertonicity, did not regulate their systemic tonicity as efficiently as did wild type controls (Liedtke and Friedman 2003). Their drinking was reduced, and systemic tonicity was significantly elevated. Continuous infusion of the ADH analogue dDAVP [desmopressin (1-desamino-8-D-arginine vasopressin)] led to systemic hypotonicity, whereas renal water readsorbtion was not changed in both genotypes. ADH synthesis in response to osmotic stimulation was reduced in $trpv4^{-/-}$ mice. Hypertonic stress led to reduced expression of c-FOS$^+$ cells in the sensory circumventricular organ OVLT, indicating an impaired osmotic activation in this brain area lacking a functional blood-brain barrier. These findings in $trpv4^{-/-}$ mice point towards a deficit in central osmotic sensing. Thus, TRPV4 is necessary for the maintenance of the tonicity equilibrium in mammals. It is conceivable that TRPV4 acts as an osmotic sensor in the central nervous system (CNS). This reported impaired osmotic regulation in $trpv4^{-/-}$ mice differs from that published in another paper. While the author's own experiments showed that $trpv4^{-/-}$ mice secrete lower amounts of ADH in response to hypertonic stimuli, the results of Mizuno et al. (2003) suggest that there is an increased ADH response to water deprivation and subsequent systemic administration of propylene glycol. The reasons for this discrepancy are not obvious. In the author's investigation, a blunted ADH response and diminished cFOS response in the OVLT of $trpv4^{-/-}$ mice upon systemic hypertonicity suggests, as one possibility, an activation of TRPV4$^+$ sensory cells in the OVLT by hypertonicity. These data imply that the $trpv4$ gene plays a significant role in the maintenance of systemic osmotic homeostasis in vivo, and imply a possible role for TRPV4 in disorders of hydromineral homeostasis.

With regards to pain-related behavior in mice, Allessandri-Haber et al. (2005) described that hypertonic and hypotonic subcutaneous solution leads to pain-related behavior in wild type mice that is not present in $trpv4^{-/-}$ mice. When sensitizing nociceptors with prostaglandin E2, the pain-related responses to hypertonic and hypotonic stimulation increased in frequency, but were greatly reduced in $trpv4^{-/-}$ mice. The in vivo behavioral data for hypertonicity could not be mirrored in acutely dissociated dorsal root ganglia (DRG) neurons upon stimulation with hypertonicity and subsequent Ca^{2+} imaging, which was, on the other hand feasible for hypotonic stimulation. Taken together, this study indicates differences in the response of mice to noxious tonicity depending on the presence/absence of TRPV4. Yet at the level of a critical transducer cell, namely the DRG sensory neuron, only hypotonicity led to a rise of intracellular Ca^{2+}, and this rise was dependent on the presence of TRPV4. These data implicate the $trpv4$ gene as playing a significant role in transduction of pain stimuli evoked or amplified by local changes in tonicity, and imply a possible role for the $trpv4$ gene in pain. This was reiterated further by two studies, also from Jon Levine's lab at UCSF. First, it was demonstrated that $trpv4$ was necessary for mechanical hyperalgesia to develop after treatment of rats with paclitaxel, a rodent pain-model of chemical deafferentiation with close similarity to human disease conditions, namely development of a painful neuropathy after treatment of paclitaxel and other taxane drugs for breast, ovarian, testicular and other cancers (Alessandri-Haber et al. 2004). Most recently, using $trpv4$ null mice generated by the author, Nicole Allessandri-Haber showed that $trpv4$ was necessary for mechanical hyperalgesia to develop after application of "inflammatory soup", which consists of several cytokines and other mediators known to be proalgesic (Alessandri-Haber et al. 2006). Finally, a group of investigators led by Nigel Bunnett, also from UCSF, determined that activation of the G-protein-coupled receptor proteinase-activated-receptor-2 (PAR-2), led to mechanical hyperalgesia in mice that was entirely dependent on TRPV4 (Grant et al. 2007). This is an intriguing finding, since activation of PAR-2 has been shown previously to sensitize thermal hyperalgesia via TRPV1 (Amadesi et al. 2004, 2006) (Fig. 5.2).

In aggregate, the $trpv4$ gene functions critically in regulation of systemic tonicity and in pain transduction of noxious osmotic stimuli in mammals. Heterologous cellular expression studies imply TRPV4 to confer responsiveness to hypotonicity (both aspects also reviewed in Voets et al. 2002; Liedtke and Kim 2005).

5.5   Recent Developments in Regards to *trpv4* Function: Regulation of TRPV4 Channels by N-glycosylation, Critical Role of TRPV4 in Cellular Volume Regulation and in Lung Injury

Another recent focus in the field of TRP ion channels is intracellular trafficking, post-translational modification and subsequent functional modulation. For TRPV4, it was reported in heterologous cells (HEK293T) that N-glycosylation between transmembrane-domain 5 and pore-loop (position 651) decreases osmotic activation via decreased plasma membrane insertion (Xu et al. 2006). Interestingly, N-glycosylation between transmembrane domains 1 and 2 had a similar effect on TRPV5, and the anti-aging hormone KLOTHO could function as beta-glucuronidase and subsequently activate TRPV5 (Chang et al. 2005). Thus, it appears feasible that KLOTHO, or related KLOTHO-like hormones, function as beta-glucuronidases regulating plasma-membrane insertion of TRPV4. How critical this mechanism is in vivo remains to be determined.

TRPV4 has also been found to play a role in maintenance of cellular osmotic homeostasis. One particular cellular defense mechanism of tonicity homeostasis is regulatory volume change, namely regulatory volume decrease (RVD) in response to hypotonicity. In a recent paper, Bereiter-Hahn's group demonstrated that immortalized CHO tissue culture cells have a poor RVD, which improved strikingly after transfection with TRPV4 (Becker et al. 2005). In yet another study, Miguel Valverde's group published findings that TRPV4 mediates the cell-swelling-induced Ca^{2+} influx into bronchial epithelial cells that triggers RVD via Ca^{2+}-dependent potassium ion channels (Arniges et al. 2004). This cell swelling response did not function in cystic fibrosis (CF) bronchial epithelia, where, on the other hand, TRPV4 could be activated by 4-α-PDD, leading to Ca^{2+} influx. This indicates that TRPV4 is downstream of the signaling step that is genetically defective in CF – the CF transmembrane resistance (CFTR) chloride conductance. These findings raise the intriguing possibility that activation of TRPV4 could be used therapeutically in CF. In yet another recent investigation, Ambudkar and colleagues found the concerted interaction of the water channel aquaporin 5 (AQP-5) with TRPV4 in hypotonic swelling-induced RVD of salivary gland epithelia (Liu et al. 2006). These findings shed light on the molecular mechanisms operative in secretory organs that secrete watery fluids. This basic physiological mechanism appears to be maintained by a concerted interaction of TRPV4 and AQP-5, which was found to be dependent on the cytoskeleton (for further discussion of the AQP-5–TRPV4 interaction, see also Sidhaye et al. 2006). In regards to volume regulation of cells in the CNS, Andrew et al. reported very recently on neuronal RVD in response to hypotonic stimulation in brain slice culture (Andrew et al. 2006). Perplexingly, the neurons were resistant to changes in tonicity, yet swelled readily when deprived of oxygen-glucose or when depolarized by potassium. This investigation once again raises the unresolved question of the molecular nature of neuronal water conductance. The behavior of the neurons appears in sharp contrast to the above AQP-5–TRPV4 interaction described for hypotonic-swelling and subsequent RVD by secretory epithelial cells. Taken together, TRPV4 also plays a role in regulatory volume decrease in response to tonicity-induced cell swelling, suggested for epithelial cells in airways and exocrine glands but not in nerve cells. This opens up the exciting possibility that TRPV4 could become a translational target in CF.

Another aspect of lung function was elucidated recently at the molecular level by Diego Alvarez from Mary Townsley's group (Alvarez et al. 2006). The lungs' alveolar septal barrier was injured in an ex-vivo lung perfusion model by stimulating TRPV4 channels (found to be expressed in the alveolar septal wall) with 4-α-PDD (4 alpha-phorbol 12,13- didecanoate) and with 14,15-epoxyeicosatrienoic acid, another known activator of TRPV4. The permeability response to 4-α-PDD was absent in $trpv4^{-/-}$ mice, whereas the lung's response to thapsigargin, a toxin known to evoke release of Ca⁺⁺ from intracellular stores, remained independent of the genotype. In aggregate, these data strongly suggest that *trpv4* plays a critical role in the injury of the alveolar septal barrier. In view of this evidence, *trpv4* is a strong candidate to function as a molecular signaling mechanism critical for acute lung injury, e.g., lung edema.

5.6 Mammalian TRPV4 Directs Osmotic Avoidance Behavior in *C. elegans*

5.6.1 Cloning of the C. elegans Gene osm-9 – the Other Founding Member of the trpv Gene Family

As referenced in the introduction, the *osm-9* mutant line was first reported in 1997 (Colbert et al. 1997). The forward genetics screen in *C. elegans* applied a confinement assay with a high-molar osmotically active substance. *osm-9* mutants did not respect this osmotic barrier, and the mutated gene was found to be a TRP channel. On closer analysis, *osm-9* mutants did not respond to aversive tonicity stimuli, they did not respond to aversive mechanical stimuli to their "nose", and they did not respond to (aversive) odorants. The OSM-9 channel protein was found to be expressed in amphid sensory neurons, the worm's cellular substrate of exteroceptive sensing of chemical, osmotic and mechanical stimuli. At the subcellular level, the OSM-9 channel was also expressed in the sensory cilia of the AWC and ASH sensory neurons. Bilateral laser ablation of the ASH neuron, referred to by some researchers as the worms' equivalent of the trigeminal ganglion or the "nociceptive" neuron (Bargmann and Kaplan 1998), has been shown to lead to a deficit in osmotic, nose touch and olfactory avoidance (Kaplan and Horvitz 1993). Next, four more TRPV channels from *C. elegans* were isolated, named OCR-1 to -4 (Tobin et al. 2002). Of these four channels, only OCR-2 was expressed in ASH. The *ocr-2* mutant phenotype was virtually identical to the *osm-9* phenotype with respect to worm "nociception", and there was genetic evidence that the two channels were necessary for proper intracellular trafficking of each other in sensory neurons, indicating an interaction between OSM-9 and OCR-2. When expressing the mammalian capsaicin receptor TRPV1 in ASH sensory neurons, neither *osm-9* nor *ocr-2* mutants could be rescued, but *osm-9 ash::trpv1* transgenic worms displayed a strong avoidance to capsaicin, which normal worms do not respond to.

5.6.2 TRPV4 Expression in ASH Rescues osm-9 Mechanical and Osmotic Deficits

Next, TRPV4 was transgenically directed to ASH amphid neurons of *osm-9* mutants. Surprisingly, TRPV4 expression in *C. elegans* ASH rescued *osm-9* mutant defects in avoidance of hypertonicity and nose touch (Liedtke et al. 2003). However, mammalian TRPV4 did not rescue the odorant avoidance defects of *osm-9*, suggesting that this function of TRPV channels differs between vertebrates and invertebrates. This basic finding of the rescue experiments in *osm-9 ash::trpv4* worms has important implications for our understanding of mechanisms of signal transduction (see Fig. 5.3).

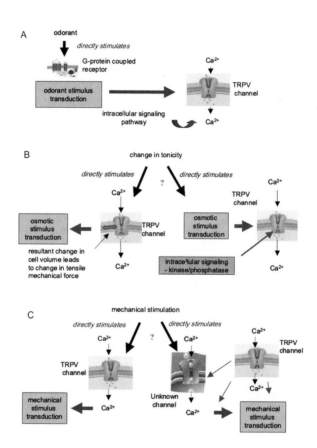

Fig. 5.3 Signal transduction in sensory (nerve) cells in response to odorant (**a**), osmotic (**b**) and mechanical (**c**) stimuli. **a** The odorant activates the TRPV ion channel via a G-protein-coupled receptor mechanism. Such a mechanism is functional in the ASH sensory neuron of *Caenorhabditis elegans* in response to, e.g., 8-octanone, a repulsive odorant cue. Intracellular signaling cascades downstream of the G-protein-coupled receptor activate the TRPV channel – OSM-9 or OCR-2. Ca^{2+} influx through the TRPV channel serves as an amplifier mechanism, which is required for this signaling pathway to elicit the stereotypical withdrawal response. **b** Two possible mechanisms for tonicity signaling. *Right* TRPV channel functions downstream of a – yet unknown – osmotic stimulus transduction mechanism, which is directly activated by a change in tonicity.

(continued)

5.6.3 Proposed TRPV4 Transduction Mechanism in osm-9 ash::trpv4 Worms

TRPV4 appeared to be integrated into the normal ASH sensory neuron signaling apparatus, since the transgene failed to rescue the respective deficits in other *C. elegans* mutants lacking in osmosensation and mechanosensation (including OCR-2, bespeaking of the specificity of the observed response). A point mutation in the pore-loop of TRPV4, M680K, eliminated the rescue, indicating that TRPV4 likely functions as a transductory ion channel. In an attempt to recapitulate the properties of the mammalian channel in the avoidance behavior of the worm, it was found that the sensitivity for osmotic stimuli and the effect of temperature on the avoidance responses of *osm-9 ash::trpv4* worms more closely resembled the known properties of mammalian TRPV4 than that of normal Caenorhabditis. TRPV4 did not rescue the odorant avoidance deficits of *osm-9* mutants. In odorant transduction, G-protein-coupled receptors function as odorant sensors, and the TRPV channel functions downstream in the signaling cascade. Moreover, TRPV4 did not function downstream of other known mutations that affect touch and osmotic avoidance in *C. elegans*.

When taken together, these findings suggest that mammalian TRPV4 was functioning as the osmotic and mechanical sensor, or at least as a component of it. It should be realized that TRPV4 was expressed functionally only in ASH, a single sensory neuron, where the mammalian protein, with a similarity to OSM-9 of approximately 25%, was trafficked correctly to the ASH sensory cilia, a distance of more than 100 μm! The rescue was specific (not for mutated *ocr-2*, not by mammalian TRPV1-capsaicin-receptor), and it respected genetically defined pathways.

The above OSM-9/TRPV4 study delivers stimulating points to be addressed in future investigations. Whereas TRPV4 restores responsiveness to hypertonicity in *C. elegans osm-9* mutants, it is gated only by hypo-osmotic stimuli in transfected

---

Fig. 5.3 (continued) This is conceptually related to the depictiond in **a**. Intracellular signaling via phosphorylation (de-phosphorylation)-dependent pathways activates the TRPV channel. For heterologous cellular expression, two groups have obtained data, contradictory in detail, that suggest phosphorylation of TRPV4 to be of relevance (Xu et al. 2003; Vriens et al. 2004b). *Left* TRPV channel is at the top of the signaling cascade, i.e., it is directly activated by a change in tonicity, which in turn can lead to altered mechanical tension of the cytoplasmic membrane. Note that the two alternatives need not be mutually exclusive. Apart from phosphorylation of the TRPV channel, which could possibly be of relevance in vivo, a direct physical linkage of the TRPV channel to the cytoskeleton, to the extracellular matrix and to the lipids of the plasma membrane in the immediate vicinity of the channel proteins has to be entertained. **c** Two possible mechanisms for mechanotransduction. *Right* An unknown mechanotransduction channel responds directly to the mechanical stimulus with Ca^{2+} influx. This activity and the subsequent signal transduction are modulated more indirectly by the TRPV channel, which acts on the unknown transduction channel, onto the biophysical properties of the membrane, and via other, yet-unknown intracellular signaling mechanisms. *Left* The TRPV channel itself functions as the mechanotransducer, i.e., it is activated directly via mechanical stimulation. Reproduced from Liedtke and Kim (2005) with permission

Table 5.1 Synopsis of the *trpv* genes covered in this chapter, in the order in which they are discussed in the text

Gene	Evidence
trpv1	**Loss-of-function studies in vivo/dissociated cells**
	trpv1$^{-/-}$ mice show abnormalities in tonicity homeostasis, response to mechanical stretch and tonicity response of bladder, bowel and vessels
	Pharmacological inhibition of TRPV1 diminishes mechanical hyperalgesia
trpv2	**Heterologous expression and loss-of-function studies in dissociated cells**
	de novo/diminished reaction to hypotonicity and mechanical stretch
trpv4	**Heterologous expression**
	de novo reaction to hypotonicity and mechanical stretch
	Loss-of-function studies in vivo
	trpv4$^{-/-}$ mice show abnormalities in tonicity homeostasis, elevated thresholds for mechanically and osmotically induced pain
	Possible regulation of channel function by N-glycosylation
	Involved in volume regulation in response to hypotonic swelling
osm-9	***Caenorhabditis elegans* mutation with defects in avoidance of osmotic, mechanical and odorant avoidance**
	Related *C. elegans* TRPV4 gene *oct-2* with identical phenotype
	Transgenic rescue by TRPV4, expression directed to one sensory neuron, of osmotic and mechanical (not odorant) defects of osm-9 mutant worms

mammalian cells. The reasons for this discrepancy are not understood. Related to this latter study, it was recently reported that TRPV2 could rescue one particular deficit of the *ocr-2* mutant, namely the dramatic downregulation of serotonin biosynthesis in the sensory ADF neuron, but mammalian TRPV2, unlike TRPV4 directing behavior in *osm-9*, did not complement the lack of the osmotic avoidance reaction of *ocr-2* (Zhang et al. 2004; Sokolchik et al. 2005). Common to these two investigations, however, is the conservation of TRPV signaling across phyla that have separated for several hundred million years of molecular evolution, despite low sequence homology!

In reference to the Drosophila TRPV channels, NAN and IAV, the interested reader is directed to original papers (Kim et al. 2003; Gong et al. 2004) and relevant reviews (Vriens et al. 2004a; Liedtke and Kim 2005).

5.7   Outlook for Future Research on TRPV Channels

In regards to TRP channels, one topic for the future is the investigation of the functional significance of protein–protein interactions of TRP(V) ion channels with to-be-discovered interaction partners (a particularly interesting example of protein–protein interactions of TRPV4 splice-variants from airway epithelia was reported recently (Arniges et al. 2006, but see also Cuajungco et al. 2006). In addition, there is the obvious potential for TRP channels as targets for translational efforts (Nilius et al. 2005), such as secretory disorders (e.g., CF), pain and hydromineral homeostasis.

Acknowledgments The author was supported by a K08 career development award of the National Institutes of Mental Health, by funding from the Whitehall Foundation (Palm Springs, FL), the Klingenstein Fund (New York, NY), Philipp Morris External Research Support. (Linthicum Heights, MD), and by Duke University (Durham, NC).

References

Alessandri-Haber N, Dina OA, Yeh JJ, Parada CA, Reichling D, Levine JD (2004) Transient receptor potential vanilloid 4 is essential in chemotherapy-induced neuropathic pain in the rat. J Neurosci 24:4444–4452

Alessandri-Haber N, Joseph E, Dina OA, Liedtke W, Levine JD (2005) TRPV4 mediates pain-related behavior induced by mild hypertonic stimuli in the presence of inflammatory mediator. Pain 118:70–79

Alessandri-Haber N, Dina OA, Joseph EK, Reichling D, Levine JD (2006) A transient receptor potential vanilloid 4-dependent mechanism of hyperalgesia is engaged by concerted action of inflammatory mediators. J Neurosci 26:3864–3874

Alvarez DF, King JA, Weber D, Addison E, Liedtke W, Townsley MI (2006) Transient receptor potential vanilloid 4-mediated disruption of the alveolar septal barrier: a novel mechanism of acute lung injury. Circ Res 99:988–995

Amadesi S, Nie J, Vergnolle N, Cottrell GS, Grady EF, Trevisani M, Manni C, Geppetti P, McRoberts JA, Ennes H, Davis JB, Mayer EA, Bunnett NW (2004) Protease-activated receptor 2 sensitizes the capsaicin receptor transient receptor potential vanilloid receptor 1 to induce hyperalgesia. J Neurosci 24:4300–4312

Amadesi S, Cottrell GS, Divino L, Chapman K, Grady EF, Bautista F, Karanjia R, Barajas-Lopez C, Vanner S, Vergnolle N, Bunnett NW (2006) Protease-activated receptor 2 sensitizes TRPV1 by protein kinase Cepsilon- and A-dependent mechanisms in rats and mice. J Physiol 575:555–571

Andrew RD, Labron MW, Boehnke SE, Carnduff L, Kirov SA (2006) Physiological evidence that pyramidal neurons lack functional water channels. Cereb Cortex 17:787–802

Apostolidis A, Brady CM, Yiangou Y, Davis J, Fowler CJ, Anand P (2005) Capsaicin receptor TRPV1 in urothelium of neurogenic human bladders and effect of intravesical resiniferatoxin. Urology 65:400–405

Arniges M, Vazquez E, Fernandez-Fernandez JM, Valverde MA (2004) Swelling-activated Ca²⁺ entry via TRPV4 channel is defective in cystic fibrosis airway epithelia. J Biol Chem 279:54062–54068

Arniges M, Fernandez-Fernandez JM, Albrecht N, Schaefer M, Valverde MA (2006) Human TRPV4 channel splice variants revealed a key role of ankyrin domains in multimerization and trafficking. J Biol Chem 281:1580–1586

Bargmann CI, Kaplan JM (1998) Signal transduction in the *Caenorhabditis elegans* nervous system. Annu Rev Neurosci 21:279–308

Becker D, Blase C, Bereiter-Hahn J, Jendrach M (2005) TRPV4 exhibits a functional role in cell-volume regulation. J Cell Sci 118:2435–2440

Birder LA, Nakamura Y, Kiss S, Nealen ML, Barrick S, Kanai AJ, Wang E, Ruiz G, De Groat WC, Apodaca G, Watkins S, Caterina MJ (2002) Altered urinary bladder function in mice lacking the vanilloid receptor TRPV1. Nat Neurosci 5:856–860

Caterina MJ, Julius D (1999) Sense and specificity: a molecular identity for nociceptors. Curr Opin Neurobiol 9:525–530

Caterina MJ, Montell C (2005) Take a TRP to beat the heat. Genes Dev 19:415–418

Caterina MJ, Schumacher MA, Tominaga M, Rosen TA, Levine JD, Julius D (1997) The capsaicin receptor: a heat-activated ion channel in the pain pathway. Nature 389:816–824

Caterina MJ, Rosen TA, Tominaga M, Brake AJ, Julius D (1999) A capsaicin-receptor homologue with a high threshold for noxious heat. Nature 398:436–441

Caterina MJ, Leffler A, Malmberg AB, Martin WJ, Trafton J, Petersen-Zeitz KR, Koltzenburg M, Basbaum AI, Julius D (2000) Impaired nociception and pain sensation in mice lacking the capsaicin receptor. Science 288:306–313

Chan CL, Facer P, Davis JB, Smith GD, Egerton J, Bountra C, Williams NS, Anand P (2003) Sensory fibres expressing capsaicin receptor TRPV1 in patients with rectal hypersensitivity and faecal urgency. Lancet 361:385–391

Chang Q, Hoefs S, van der Kemp AW, Topala CN, Bindels RJ, Hoenderop JG (2005) The beta-glucuronidase klotho hydrolyzes and activates the TRPV5 channel. Science 310:490–493

Ciura S, Bourque CW (2006) Transient receptor potential vanilloid 1 is required for intrinsic osmoreception in organum vasculosum lamina terminalis neurons and for normal thirst responses to systemic hyperosmolality. J Neurosci 26:9069–9075

Clapham DE (2003) TRP channels as cellular sensors. Nature 426:517–524

Colbert HA, Smith TL, Bargmann CI (1997) OSM-9, a novel protein with structural similarity to channels, is required for olfaction, mechanosensation, and olfactory adaptation in *Caenorhabditis elegans*. J Neurosci 17:8259–8269

Corey DP (2003) New TRP channels in hearing and mechanosensation. Neuron 39:585–588

Cosens DJ, Manning A(1969) Abnormal electroretinogram from a Drosophila mutant. Nature 224:285–287

Cuajungco M, Grimm PC, Oshima K, D'Hoedt D, Nilius B, Mensenkamp AR, Bindels RJ, Plomann M, Heller S (2006) PACSINs bind to the TRPV4 cation channel. PACSIN 3 modulates the subcellular localization of TRPV4. J Biol Chem 281:18753–18762

Davis JB, Gray J, Gunthorpe MJ, Hatcher JP, Davey PT, Overend P, Harries MH, Latcham J, Clapham C, Atkinson K, Hughes SA, Rance K, Grau E, Harper AJ, Pugh PL, Rogers DC, Bingham S, Randall A, Sheardown SA (2000) Vanilloid receptor-1 is essential for inflammatory thermal hyperalgesia. Nature 405:183–187

Den Dekker E, Hoenderop JG, Nilius B, Bindels RJ (2003) The epithelial calcium channels, TRPV5 and TRPV6: from identification towards regulation. Cell Calcium 33:497–507

Di Mola FF, Friess H, Koninger J, Selvaggi F, Esposito I, Buchler MW, di Sebastiano P (2006) Haemorrhoids and transient receptor potential vanilloid 1. Gut 55:1665–1666

Dinis P, Charrua A, Avelino A, Yaqoob M, Bevan S, Nagy I, Cruz F (2004) Anandamide-evoked activation of vanilloid receptor 1 contributes to the development of bladder hyperreflexia and nociceptive transmission to spinal dorsal horn neurons in cystitis. J Neurosci 24:11253–11263

Gao X, Wu L, O'Neil RG (2003) Temperature-modulated diversity of TRPV4 channel gating: activation by physical stresses and phorbol ester derivatives through protein kinase C-dependent and -independent pathways. J Biol Chem 278:27129–27137

Gong Z, Son W, Chung YD, Kim J, Shin DW, McClung CA, Lee Y, Lee HW, Chang DJ, Kaang BK, Cho H, Oh U, Hirsh J, Kernan MJ, Kim C (2004) Two interdependent TRPV channel subunits, inactive and Nanchung, mediate hearing in Drosophila. J Neurosci 24:9059–9066

Grant AD, Cottrell GS, Amadesi S, Trevisani M, Nicoletti P, Materazzi S, Altier C, Cenac N, Zamponi GW, Bautista-Cruz F, Barajjas Lopez C, Joseph E, Levine JD, Liedtke W, Vanner S, Vergnolle N, Geppetti P, Bunnett NW (2007) Protease-activated receptor 2 sensitizes the transient receptor potential vanilloid 4 ion channel to cause mechanical hyperalgesia in mice. J Physiol 578:715–733

Guler AD, Lee H, Iida T, Shimizu I, Tominaga M, Caterina M (2002) Heat-evoked activation of the ion channel, TRPV4. J Neurosci 22:6408–6414

Gunthorpe MJ, Benham CD, Randall A, Davis JB (2002) The diversity in the vanilloid (TRPV) receptor family of ion channels. Trends Pharmacol Sci 23:183–191

Hardie RC, Minke B (1992) The trp gene is essential for a light-activated Ca^{2+} channel in Drosophila photoreceptors. Neuron 8:643–651

Hoenderop JG, van der Kemp AW, Hartog A, van de Graaf SF, van Os CH, Willems PH, Bindels RJ (1999) Molecular identification of the apical Ca^{2+} channel in 1, 25-dihydroxyvitamin D3-responsive epithelia. J Biol Chem 274:8375–8378

Hoenderop JG, Nilius B, Bindels RJ (2003) Epithelial calcium channels: from identification to function and regulation. Pfluegers Arch 446:304–308

Kanzaki M, Zhang YQ, Mashima H, Li L, Shibata H, Kojima I (1999) Translocation of a calcium-permeable cation channel induced by insulin-like growth factor-I. Nat Cell Biol 1:165–170

Kaplan JM, Horvitz HR (1993) A dual mechanosensory and chemosensory neuron in *Caenorhabditis elegans*. Proc Natl Acad Sci USA 90:2227–2231

Kim J, Chung D, Park DY, Choi S, Shin DW, Soh H, Lee HW, Son W, Yim J, Park CS, Kernan MJ, Kim C (2003) A TRPV family ion channel required for hearing in Drosophila. Nature 424:81–84

Lazzeri M, Vannucchi MG Zardo C, Spinelli M, Beneforti P, Turini D, Faussone-Pellegrini S (2004) Immunohistochemical evidence of vanilloid receptor 1 in normal human urinary bladder. Eur Urol 46:792–798

Liedtke W, Friedman JM (2003) Abnormal osmotic regulation in trpv4$^{-/-}$ mice. Proc Natl Acad Sci USA 100:13698–13703

Liedtke W, Kim C (2005) Functionality of the TRPV subfamily of TRP ion channels: add mechano-TRP and osmo-TRP to the lexicon! Cell Mol Life Sci 62:2985–3001

Liedtke W, Choe Y, Marti-Renom MA, Bell AM, Denis CS, Sali A, Hudspeth AJ, Friedman JM, Heller S (2000) Vanilloid receptor-related osmotically activated channel (VR-OAC), a candidate vertebrate osmoreceptor. Cell 103:525–535

Liedtke W, Tobin DM, Bargmann CI, Friedman JM (2003) Mammalian TRPV4 (VR-OAC) directs behavioral responses to osmotic and mechanical stimuli in *Caenorhabditis elegans*. Proc Natl Acad Sci USA 100 [Suppl 2]:14531–14536

Liu X, Bandyopadhyay B, Nakamoto T, Singh B, Liedtke W, Melvin JE, Ambudkar I (2006) A role for AQP5 in activation of TRPV4 by hypotonicity: concerted involvement of AQP5 and TRPV4 in regulation of cell volume recovery. J Biol Chem 281:15485–15495

Mizuno A, Matsumoto N, Imai M, Suzuki M (2003) Impaired osmotic sensation in mice lacking TRPV4. Am J Physiol Cell Physiol 285:C96–C101

Montell C, Rubin GM (1989) Molecular characterization of the Drosophila *trp* locus: a putative integral membrane protein required for phototransduction. Neuron 2:1313–1323

Muraki K, Iwata Y, Katanosaka Y, Ito T, Ohya S, Shigekawa M, Imaizumi Y (2003) TRPV2 is a component of osmotically sensitive cation channels in murine aortic myocytes. Circ Res 93:829–838

Mutai H, Heller S (2003) Vertebrate and invertebrate TRPV-like mechanoreceptors. Cell Calcium 33:471–478

Naeini RS, Witty MF, Seguela P, Bourque CW (2006) An N-terminal variant of Trpv1 channel is required for osmosensory transduction. Nat Neurosci 9:93–98

Nagata K, Duggan A, Kumar G, Garcia-Anoveros J (2005) Nociceptor and hair cell transducer properties of TRPA1, a channel for pain and hearing. J Neurosci 25:4052–4061

Nilius B, Voets T, Peters J (2005) TRP channels in disease. Sci STKE 295:re8

O'Neil RG, Heller S (2005) The mechanosensitive nature of TRPV channels. Pfluegers Arch 451:193–203

Patapoutian A (2005) TRP channels and thermosensation. Chem Senses 30 [Suppl 1]:i193–i194

Peier AM, Reeve AJ, Andersson DA, Moqrich A, Earley TJ, Hergarden AC, Story GM, Colley S, Hogenesch JB, McIntyre P, Bevan S, Patapoutian A (2002) A heat-sensitive TRP channel expressed in keratinocytes. Science 296:2046–2049

Peng JB, Chen XZ, Berger UV, Vassilev PM, Tsukaguchi H, Brown EM, Hediger MA (1999) Molecular cloning and characterization of a channel-like transporter mediating intestinal calcium absorption. J Biol Chem 274:22739–22746

Peng JB, Brown EM, Hediger MA (2003) Epithelial Ca^{2+} entry channels: transcellular Ca^{2+} transport and beyond. J Physiol 551:729–740

Pomonis JD, Harrison JE, Mark L, Bristol DR, Valenzano KJ, Walker K (2003) N-(4-Tertiarybutylphenyl)-4-(3-cholorphyridin-2-yl)tetrahydropyrazine -1(2H)-carbox-amide (BCTC), a novel, orally effective vanilloid receptor 1 antagonist with analgesic properties. II. In vivo characterization in rat models of inflammatory and neuropathic pain. J Pharmacol Exp Ther 306:387–393

Rong W, Hillsley K, Davis JB, Hicks G, Winchester WJ, Grundy D (2004) Jejunal afferent nerve sensitivity in wild-type and TRPV1 knockout mice. J Physiol 560:867–881

Scotland RS, Chauhan S, Davis C, De Felipe C, Hunt S, Kabir J, Kotsonis P, Oh U, Ahluwalia A (2004) Vanilloid receptor TRPV1, sensory C-fibers, and vascular autoregulation: a novel mechanism involved in myogenic constriction. Circ Res 95:1027–1034

Sidhaye VK, Guler AD, Schweitzer KS, D'Alessio F, Caterina MJ, King LS (2006) Transient receptor potential vanilloid 4 regulates aquaporin-5 abundance under hypotonic conditions. Proc Natl Acad Sci USA 103:4747–4752

Smith GD, Gunthorpe MJ, Kelsell RE, Hayes PD, Reilly P, Facer P, Wright JE, Jerman JC, Walhin JP, Ooi L, Egerton J, Charles KJ, Smart D, Randall AD, Anand P, Davis JB (2002) TRPV3 is a temperature-sensitive vanilloid receptor-like protein. Nature 418:186–190

Sokolchik I, Tanabe T, Baldi PF, Sze JY (2005) Polymodal sensory function of the *Caenorhabditis elegans* OCR-2 channel arises from distinct intrinsic determinants within the protein and is selectively conserved in mammalian TRPV proteins. J Neurosci 25:1015–1023

Stein RJ, Santos S, Nagatomi J, Hayashi Y, Minnery BS, Xavier M, Patel AS, Nelson JB, Futrell WJ, Yoshimura N, Chancellor MB, De Miguel F (2004) Cool (TRPM8) and hot (TRPV1) receptors in the bladder and male genital tract. J Urol 172:1175–1178

Strotmann R, Harteneck C, Nunnenmacher K, Schultz G, Plant TD (2000) OTRPC4, a nonselective cation channel that confers sensitivity to extracellular osmolarity. Nat Cell Biol 2:695–702

Tobin D, Madsen DM, Kahn-Kirby A, Peckol E, Moulder G, Barstead R, Maricq AV,. Bargmann CI (2002) Combinatorial expression of TRPV channel proteins defines their sensory functions and subcellular localization in *C. elegans* neurons. Neuron 35:307–318

Tominaga M, Caterina MJ (2004) Thermosensation and pain. J Neurobiol 61:3–12

Voets T, Prenen J, Vriens J, Watanabe H, Janssens A, Wissenbach U, Boedding M, Droogmans G, Nilius B (2002) Molecular determinants of permeation through the cation channel TRPV4. J Biol Chem 277:33704–33710

Vriens J, Owsianik G, Voets T, Droogmans G, Nilius B (2004a) Invertebrate TRP proteins as functional models for mammalian channels. Pfluegers Arch 449:213–226

Vriens J, Watanabe H, Janssens A, Droogmans G, Voets T, Nilius B (2004b) Cell swelling, heat, and chemical agonists use distinct pathways for the activation of the cation channel TRPV4. Proc Natl Acad Sci USA 101:396–401

Walker RG, Willingham AT, Zuker CS (2000) A Drosophila mechanosensory transduction channel. Science 287:2229–2234

Watanabe H, Vriens J, Suh SH, Benham CD, Droogmans G, Nilius B (2002) Heat-evoked activation of TRPV4 channels in a HEK293 cell expression system and in native mouse aorta endothelial cells. J Biol Chem 277:47044–47051

Wissenbach U, Bodding M, Freichel M, Flockerzi V (2000) Trp12, a novel Trp related protein from kidney. FEBS Lett 485:127–134

Wong F, Schaefer EL, Roop BC, LaMendola JN, Johnson-Seaton D, Shao D (1989) Proper function of the Drosophila *trp* gene product during pupal development is important for normal visual transduction in the adult. Neuron 3:81–94

Xu H, Ramsey IS, Kotecha SA, Moran MM, Chong JA, Lawson D, Ge P, Lilly J, Silos-Santiago I, Xie Y, DiStefano PS, Curtis R, Clapham DE (2002) TRPV3 is a calcium-permeable temperature-sensitive cation channel. Nature 418:181–186

Xu H, Zhao H, Tian W, Yoshida K, Roullet JB, Cohen DM (2003) Regulation of a transient receptor potential (TRP) channel by tyrosine phosphorylation. SRC family kinase-dependent tyrosine phosphorylation of TRPV4 on TYR-253 mediates its response to hypotonic stress. J Biol Chem 278:11520–11527

Xu H, Fu Y, Tian W, Cohen DM (2006) Glycosylation of the osmoresponsive transient receptor potential channel TRPV4 on Asn-651 influences membrane trafficking. Am J Physiol Renal Physiol 290:F1103–F1109

Zhang S, Sokolchik I, Blanco G, Sze JY (2004) *Caenorhabditis elegans* TRPV ion channel regulates 5HT biosynthesis in chemosensory neurons. Development 131:1629–1638

Zhu X, Chu PB, Peyton M, Birnbaumer L (1995) Molecular cloning of a widely expressed human homologue for the Drosophila *trp* gene. FEBS Lett 373:193–198

Chapter 6
Mechanisms of Thermosensation in TRP Channels

Karel Talavera(✉), Thomas Voets, and Bernd Nilius

Abstract The transient receptor potential (TRP) superfamily encompasses a large number of cationic channels that are modulated by a wide variety of physical and chemical stimuli. A notorious subgroup of TRP channels, dubbed thermoTRPs, shows a dramatic dependence on temperature, which can be up to tenfold higher than that of classical ionic channels. Consequently, some thermoTRPs are thought to have a prominent role in the mechanisms of thermosensation and thermoregulation. However, the mechanisms underlying the high temperature sensitivity of thermoTRP activation are, for the most part, obscure. Only four out of the nine thermoTRPs known so far are sufficiently well characterised to allow a comprehensive model to be put forward. Temperature modulates the gating of TRPM8, TRPV1, TRPM4 and TRPM5 by shifting the voltage dependence of channel activation towards more negative potentials, which can be accounted for by a model in which voltage-dependent gating is directly affected by temperature. Heat activation of TRPV3 seems to be consistent with this mechanism, although a modification of the pore may also take place. For TRPV4,

KU Leuven; Campus Gasthuisberg, Department of Physiology, Herestraat 49, B-3000 LEUVEN, Belgium, karel.talavera@med.kuleuven.be

B. Martinac (ed.), *Sensing with Ion Channels. Springer Series in Biophysics 11*
© 2008 Springer-Verlag Berlin Heidelberg

it has been proposed that an, as yet unidentified, diffusible ligand mediates activation by heat. The mechanisms for TRPV2, TRPA1 and TRPM2 are still unknown.

6.1 Introduction

Life can be supported only through the adaptation of organisms to the ever-changing environmental conditions. Such adaptation requires the constant monitoring of the multiple variables that define the environment, such as light intensity, pressure, concentration of chemicals, and temperature. Among these, temperature is one of the most important, since it affects all physicochemical processes. Moreover, sensing not only of external but also of inner-core temperature is essential for thermoregulation, a major evolutionary step in animal adaptation. In this chapter, we review current knowledge on the mechanisms through which temperature modulates the function of transient receptor potential (TRP) channels, a family of polyvalent biological sensors that operate at the cellular level. We start by giving a short introduction to the TRP superfamily of ion channels, in which we delineate the most salient features of these channels. We then discuss the theoretical bases needed to understand the most plausible models for temperature modulation of TRP channel function. Finally, we make reference to those TRP channels for which the mechanisms of temperature modulation are far from understood. When appropriate, we make brief reference to the role of these channels in thermo-sensation and thermoregulation, but for more details the reader is referred to excellent recent reviews especially devoted to this topic (Caterina 2007; Dhaka et al. 2006; McKemy 2005; Patapoutian et al. 2003; Tominaga and Caterina 2004).

Before going into any detail, we must warn the reader that the study of thermo-sensation through the function of TRP channels is a brand new field of investigation and, unavoidably, some of the views that are discussed here might be obsolete or more refined in the near future. However, we are encouraged by important recent advances in the understanding of TRP channel function and the high impact of the subject in sensory physiology and many other branches of biology.

6.2 A Short Description of the TRP Channel Superfamily

According to their amino acid sequence, TRP channels are classified into seven subfamilies, namely TRPC, TRPV, TRPM, TRPA, TRPP, TRPML and TRPN. The TRPC (C for canonical) channels are closely related to the first TRP channel to be characterised, which was identified in the fruit fly *Drosophila melanogaster*. The TRPVs were named for the sensitivity of the founding member of this subfamily to the vanilloid compound capsaicin. Similarly, TRPMs are named for the tumor suppressor melastatin, TRPA after ANKTM1, TRPN for no-mechanoreceptor potential, TRPP for polycystic kidney disease-related protein and TRPML for mucolipin (Fig. 6.1A; Owsianik et al. 2006a).

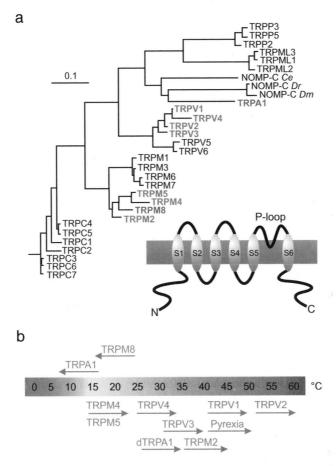

Fig. 6.1 ThermoTRPs within the transient receptor potential (TRP) ion channel superfamily. **a** Phylogenetic tree of TRP channels (Courtesy from Dr. G. Owsianik) . *Scale bar* Evolutionary distance expressed as the number of substitutions per amino acid. *Red* Heat-activated channels, *blue* cold-activated channels. *Inset* Schematic view of the topology of TRP channels in the plasma membrane. Amino and carboxy-termini are located intracellularly, whereas six segments span the membrane (*S1–S6*). The channel pore is formed by the S5 and S6 segments and the interconnecting P-loop, which contains the selectivity filter. **b** Gamma of temperature activation of the thermo-TRPs. *Arrows* Reported apparent temperature threshold for activation and the direction of increase of open probability (note that dTRPA1 and Pyrexia are *Drosophila melanogaster* thermoTRPs)

The structure of TRP channels is similar to that of voltage-gated K^+ channels (Owsianik et al. 2006a). TRP channels are made up of four identical subunits, each containing six transmembrane segments (S1–S6), with cytosolic N- and C-termini (Fig. 6.1A). These subunits are arranged around a central pore formed by the S5 and S6 segments and the interconnecting loop (Owsianik et al. 2006b). The pore allows the flux of cations according to their electrochemical gradient. This property of TRP channels makes them extremely important for the regulation of electrical and biochemical signalling in multiple cell types (Pedersen et al. 2005; Voets et al. 2005). Moreover, most TRP channels (except TRPM4 and TRPM5) allow the influx of Ca^{2+}, which is a major regulator of multiple enzymatic processes.

The opening and closing of the pore (gating) is thought to be regulated by the S1–S4 transmembrane segments as well as by the N- and C-termini, which contain multiple domains determining the interaction of the channels with cytosolic proteins and second messengers (see Owsianik et al. 2006a for further details).

Notably, compared to other ion channels, the gating of some TRP channels is highly sensitive to temperature (Fig. 6.1B). The 10-degree temperature coefficient of the gating of most ion channels, Q_{10}, is around 2–4 (Hille 2001). In contrast, nine TRP channels show Q_{10} values between 6 and 30, depending on the experimental conditions (Dhaka et al. 2006). This observation, together with the fact that these highly temperature-sensitive TRP channels (thermoTRPs for short) are expressed in thermosensitive neurons (for reviews, see Caterina 2007; Dhaka et al. 2006) and in the skin (Chung et al. 2003; Moqrich et al. 2005; Peier et al. 2002b), suggested that they may play a fundamental role in thermosensation and thermoregulation.

6.3   Mechanisms of Thermosensitivity in ThermoTRPs

In order to understand the mechanism(s) underlying the temperature dependence of thermoTRP function we may formulate two fundamental questions: Why is the temperature dependence so pronounced compared to other channels and enzymes? And where does it come from? All biological processes have some degree of temperature dependence, as do the underlying physicochemical processes, according to the fundamental laws and principles of thermodynamics and statistical physics. Thus, we should be convinced that, physically speaking, there is nothing really special about thermoTRPs, and that their surprising properties should arise from some scaled-up behaviour, reminiscent of other, less temperature-sensitive channel types.

6.3.1   Some Theoretical Basics of Ion Channel Thermodynamics

First, the transition of a channel between two of its states, e.g. from closed to open, can be viewed as a chemical reaction (Hille 2001). The opening of a channel involves a change in protein conformation, which, although complex, is not qualitatively different from, for example, the transitions between the boat and chair conformations of cyclohexane, which we may have learnt about in high school. Thus, the influence of temperature on channel transitions can be studied using the formalism that applies to chemical reactions, such as that of Arrhenius or, the more modern, Eyring's rate theory (Hille 2001).

Within this view, conversion of reactants into products or, in our case, the transition of a channel from the closed to the open state is determined by the interactions

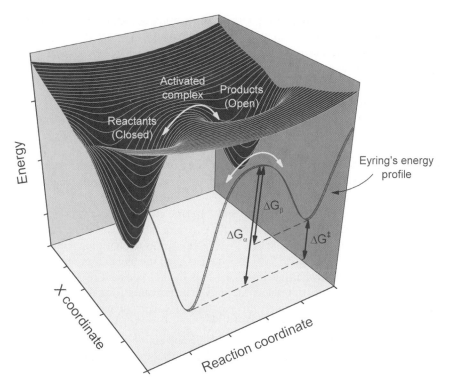

Fig. 6.2 Energy landscape associated with a hypothetical chemical reaction. Energy is plotted as function of a coordinate X and another coordinate that goes in the direction of the reaction pathway. The two energy wells are stable configurations that correspond to reactants and products of the chemical reaction. These states might represent the states of an ion channel that follows a closed–open gating scheme. In Eyring's rate theory, the multidimensional energy landscape is substituted by a unidimensional free energy profile following the reaction coordinate

between the particles of the system and occurs through the movement of the system via a landscape of energy that is characteristic for these interactions. For the hypothetical example shown in Fig. 6.2, the two energy wells may correspond to the stable conformations of the system, one representing the reactants or the closed state of the channel and the other representing the products or the open state of the channel.

Most chemical reactions, particularly conformational changes of proteins, involve multiple degrees of freedom, which results in multidimensional and complex energy landscapes. In Eyring's theory, this energy landscape is simplified to a unidimensional Gibb's free energy profile (Fig. 6.2). The system thus transits from one state to the other along the so-called reaction coordinate, which follows the minimal energy path of the original complex energy landscape (Fig. 6.2). For the transition to occur, the system must reach a configuration called "activated complex", by acquiring an energy that must be larger than the difference between the energy of the activation complex and the energy of the departing state. This energy can be provided by thermal agitation (internal kinetic energy), which in turn can be transferred from the surroundings to the system under analysis through random (thermal) interactions. In that sense, temperature is the parameter that indicates how much of this energy is provided by the surroundings[1].

The main achievement of Eyring's theory is that it allows the rate constants of any chemical reaction to be calculated as a function of temperature, given a knowledge of the free energy profile. Consider a voltage-independent channel that follows a simple closed-open gating scheme. The equation describing the gating can be written as:

$$Closed \underset{\beta(T)}{\overset{\alpha(T)}{\rightleftharpoons}} Open \tag{6.1}$$

where the rates for opening (α) and closing (β) are only temperature-dependent and can be determined from the equations:

$$\alpha(T) = \kappa \frac{kT}{h} \exp\left(-\frac{\Delta G_\alpha}{RT}\right) \quad \text{and} \quad \beta(T) = \kappa \frac{kT}{h} \exp\left(-\frac{\Delta G_\beta}{RT}\right) \tag{6.2}$$

where k is the Boltzmann constant (1.38×10^{-23} JK^{-1}), T is the absolute temperature, h is Planck's constant (6.63×10^{-34} Js), ΔG_α and ΔG_β are the free energy changes (per mole) associated with each transition, and R is the gas constant (8.31 JK^{-1}). The transmission coefficient, κ, reflects the maximal probability for the transition to occur. For the case of ion channels this parameter is unknown and usually taken as 1 for the sake of simplicity.

ΔG_α and ΔG_β are determined by the enthalpic and entropic contributions in each transition through the equations: $\Delta G_\alpha = \Delta H_\alpha - T\Delta S_\alpha$ and $\Delta G_\beta = \Delta H_\beta - T\Delta S_\beta$. Thus,

$$\alpha(T) = \frac{kT}{h} \exp\left(-\frac{\Delta H_\alpha}{RT} + \frac{\Delta S_\alpha}{R}\right) \quad \text{and} \quad \beta(T) = \frac{kT}{h} \exp\left(-\frac{\Delta H_\beta}{RT} + \frac{\Delta S_\beta}{R}\right) \tag{6.3}$$

The temperature dependence of the probability of finding the channel in the open state, $P_o(T)$, is given by:

$$P_o(T) = \frac{\alpha(T)}{\alpha(T) + \beta(T)} = \frac{1}{1 + \exp\left(\dfrac{\Delta G^\ddagger}{RT}\right)} \tag{6.4}$$

$$\text{or } P_o(T) = \frac{1}{1 + \exp\left(\dfrac{\Delta H^\ddagger}{RT} - \dfrac{\Delta S^\ddagger}{R}\right)}, \tag{6.5}$$

where $\Delta G^\ddagger$, $\Delta H^\ddagger$ and $\Delta S^\ddagger$ are the differences of Gibb's free energy, enthalpy and entropy between the open and the closed states, respectively.

---

[1] Note that we implicitly assume that the surroundings have enough heat capacity to function as a thermostat and that it is in equilibrium with the system. This condition is perfectly met in a typical patch-clamp experiment in which the cells are perfused with relatively large volumes of extracellular solution.

Thus, it is clear that the open probability of a channel that follows such a simple scheme of activation is sensitive to temperature. Assuming that $\Delta H^{\ddagger}$ and $\Delta S^{\ddagger}$ do not depend on temperature, the derivative of the open probability with respect to the temperature is:

$$\frac{dP_O(T)}{dT} = \frac{\Delta H^{\ddagger}}{RT^2} P_O^{\,2} \exp\left(\frac{\Delta H^{\ddagger}}{RT} - \frac{\Delta S^{\ddagger}}{R}\right) \tag{6.6}$$

from which it is clear that the sign of $\Delta H^{\ddagger}$ determines whether the channel is cold- ($\Delta H^{\ddagger} < 0$) or heat-activated ($\Delta H^{\ddagger} > 0$).

6.3.2   Thermodynamics of Channel Gating in the Presence of an External Field: Voltage-Gated Channels

The rate of the transitions in a system may be influenced not only by interactions between the inner components, but also by external force fields. In the case of ion channels, these force fields can be determined, for example, by the tension of the membrane and/or the electric potential across the cell membrane. These external potentials add up to the Gibb's free energy profile and, depending on their behaviour along the reaction coordinate, they may modify the energy profile and favour either the forward or the backward transition of the reaction. For example, for an ion channel that activates by membrane depolarisation, the application of a positive membrane potential will tilt the free energy profile in favour of the channel opening. As a result, the rate of opening will increase and the rate of closing will decrease.

For voltage-gated channels following a closed–open gating mechanism, the corresponding equation is:

$$Closed \underset{\beta(V,T)}{\overset{\alpha(V,T)}{\rightleftarrows}} Open \tag{6.7}$$

where the rate constants for channel opening and closing are voltage- and temperature-dependent and can be described by:

$$\alpha(V,T) = \frac{kT}{h} \exp\left(-\frac{\Delta H_{\alpha}}{RT} + \frac{\Delta S_{\alpha}}{R} + \frac{zF\delta V}{RT}\right) \tag{6.8}$$

$$\text{and } \beta(V,T) = \frac{kT}{h} \exp\left(-\frac{\Delta H_{\beta}}{RT} + \frac{\Delta S_{\beta}}{R} - \frac{zF(1-\delta)V}{RT}\right) \tag{6.9}$$

where z is the effective gating valence of the voltage sensor given in elementary charge units ($e_0 = 1.602 \cdot 10^{-19}$ C), F is Faraday's constant ($9.64 \cdot 10^4$ Cmol^{-1}) and δ accounts for the coupling between the local electric potential sensed by the gating charge and the membrane potential V. Note that the voltage term in each rate has the form of an enthalpic contribution, which is reminiscent of the relationship of enthalpy with the application of external forces applied to the system.

The open probability is therefore:

$$P_O(T,V) = \frac{1}{1+\exp\left(\dfrac{\Delta G^{\ddagger} - ZFV}{RT}\right)} = \frac{1}{1+\exp\left(\dfrac{\Delta H^{\ddagger}}{RT} - \dfrac{\Delta S^{\ddagger}}{R} - \dfrac{ZFV}{RT}\right)} \qquad (6.10)$$

and its partial derivative with respect to temperature is:

$$\frac{\partial P_O(T,V)}{\partial T} = \frac{\left(\Delta H^{\ddagger} - zFV\right)}{RT^2} P_O^{\,2} \exp\left(\frac{\Delta H^{\ddagger}}{RT} - \frac{\Delta S^{\ddagger}}{R} - \frac{zFV}{RT}\right) \qquad (6.11)$$

From this equation it is clear that the sign of ($\Delta H^{\ddagger}$–zFV) determines whether the channel is heat- or cold-activated (for more details, see Nilius et al. 2005c). Interestingly, this indicates that there is a voltage ($V = \Delta H^{\ddagger}/zF$) at which the channel changes from cold- to heat-activated. However, as we will see below, for the voltage-gated TRP channels studied so far, this voltage has very large values of no physiological relevance.

The time constant of current relaxation as a function of temperature and voltage is given by:

$$\tau(T,V) = \frac{\alpha(T,V)}{\alpha(T,V) + \beta(T,V)} \qquad (6.12)$$

which is equivalent to

$$\tau(T,V) = \frac{h/kT}{\exp\left(-\dfrac{\Delta H_{\alpha}}{RT} + \dfrac{\Delta S_{\alpha}}{R} + \dfrac{zF\delta V}{RT}\right) + \exp\left(-\dfrac{\Delta H_{\beta}}{RT} + \dfrac{\Delta S_{\beta}}{R} - \dfrac{zF(1-\delta)V}{RT}\right)} \qquad (6.13)$$

6.3.3 The Principle of Temperature-Dependent Gating in TRPV1 and TRPM8

The history of thermoTRPs started with the identification of TRPV1 as a heat-activated capsaicin receptor (Caterina et al. 1997). Ten years after this important breakthrough, the family of thermoTRPs continues to grow, with most of its members behaving as heat-activated channels. Notably, the discovery of TRPM8 as a cold-activated menthol receptor generalised the role of TRP channels as cellular thermosensors (McKemy et al. 2002; Peier et al. 2002a). Moreover, it imposed the challenging question of how two relatively closely related ion channels can be activated by opposed thermal stimuli. As often occurs in Science, the answer to this question was found while looking for something else. Almost simultaneously, several laboratories identified TRPM4 and TRPM5 – members of the TRPM subfamily – as voltage-gated channels (Hofmann et al. 2003; Launay et al. 2002; Nilius et al. 2003a; Prawitt et al. 2003). While testing whether the gating of the closely related TRPM8 was voltage-dependent, Voets et al. (2004) found that this channel is activated by membrane depolarisation. Moreover, they also realised that the effect of temperature on channel opening is voltage-dependent. A series of experiments using voltage steps[2] at different temperatures revealed that, at high temperature, activation of TRPM8 occurs at very positive potentials, whereas cooling induces a shift of the voltage dependence of activation towards more physiological (negative) potentials. Interestingly, for the heat-activated TRPV1, it was an increase rather than a decrease in temperature that caused a negative shift of the activation curve. For both channels, these shifts were very large, but graded. In order to quantify this effect, the voltage dependence of channel activation was characterised by a classical Boltzmann function of the form:

$$P_O(V) = \frac{1}{1 + \exp\left(-(V - V_{act})/s_{act}\right)}, \tag{6.14}$$

where V_{act} is the voltage for half-maximal activation and s_{act} the slope factor of the activation. For TRPM8, cooling shifted V_{act} to more negative potentials at a rate of 7.3 mV/°C, whereas for TRPV1, heating shifted V_{act} at a rate of 9.1 mV/°C (Fig. 6.3b, d).

This finding has two important implications. First, it indicates that the thermal modulation of TRPM8 and TRPV1 does not involve a critical phenomenon involving abrupt phase transitions. And second, it implies that the concept of a temperature threshold for channel activation, so commonly used in the TRP literature, has no rigorous foundation, at least not for those thermoTRPs showing the TRPM8/ TRPV1 type of mechanism[3]. Given the importance that has been given to this concept, especially when used to look for a correlation between the "thermal threshold" of TRP currents and the threshold of the responses of thermosensory cells, we believe that it is worth discussing its real value. The method most often used to estimate this "threshold" consists of determining the value of the temperature at which the curves fitting the temperature dependence of background and TRP currents intercept each other. This method has an important shortcoming: the point of interception will depend on the relative size of the background currents, which may be different for each expression system[4]. Besides the fact that any "threshold measure" inherently depends on voltage, the other problem of the "thermal threshold"

---

[2] Given the initial belief that all TRP channels were not voltage-gated, voltage ramp protocols became the most popular method used to elicit TRP currents. The use of step protocols, which allow the steady-state and kinetic properties to be determined, is now recognised as being essential for the correct analysis of the gating properties of many TRP channels (Nilius et al. 2003a, 2005c).

[3] So far, these thermoTRPs are TRPM8, TRPV1 (Voets et al. 2004, 2005), TRPM4, TRPM5 (Talavera et al. 2005) and probably also TRPV3 (Chung et al. 2003) (see below).

[4] The reader familiar with voltage-gated channels may realise that the concept of threshold for voltage-dependent activation is subjected to the same limitation. Although estimated values may be of some utility for some applications, they are totally useless for, e.g., the determination of the actual current contribution at very low, but physiologically relevant, values of open probability.

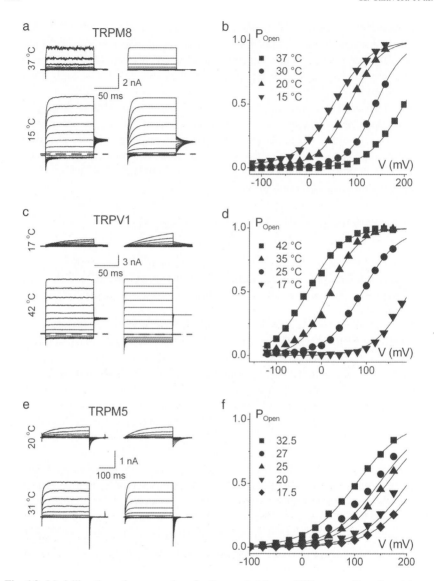

Fig. 6.3 Modelling the gating properties of voltage-gated thermoTRPs. **a, c, e** Experimental current traces (*left*) of TRPM8, TRPV1 and TRPM5 recorded at different temperatures compared to traces simulated with models for these channels (*right*) (see text). **b, d, f** Open probability of TRPM8, TRPV1 and TRPM5 as a function of the membrane potential at various temperatures. Note that for TRPM8, heating shifts the activation curve to positive potentials, whereas the opposite occurs for TRPV1 and TRPM5

is conceptual. The use of this parameter would be justified only if it were demonstrated that thermally induced activation is due to an abrupt phenomenon, e.g. a phase transition-type process. For these reasons, we believe that threshold values may be of no actual use, or at least should be considered with extreme care, when trying to understand the thermosensory role of TRPs in native cells.

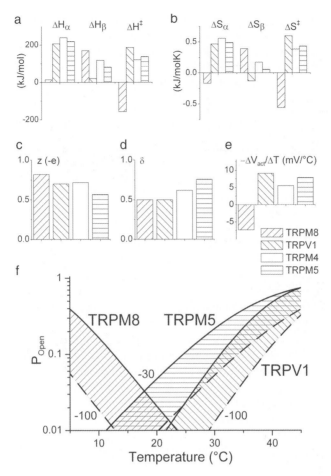

Fig. 6.4 Comparison of the thermodynamic properties of TRPM8, TRPV1, TRPM4 and TRPM5. **a, b** Enthalpies and entropies associated with the opening (ΔH_α and ΔS_α) and closing (ΔH_β and ΔS_β) transitions, and differences in enthalpy and entropy between the open and the closed states ($\Delta H^{\ddagger}$ and $\Delta S^{\ddagger}$). **c, d, e** Effective valence of the gating charge (z), coupling electric factor (δ) and shift of V_{act} per temperature degree ($-\Delta V_{act}/\Delta T$). **f** Diagram of activity of TRPM8, TRPV1 and TRPM5 (Ca^{2+} 500 nM) as a function of temperature. The *shaded bands* correspond to open probabilities within the range -100 mV (*dashed lines*) to -30 mV (*continuous lines*)

How then should we describe the effects of temperature on the gating of thermoTRPs? Obviously, the best answer to this question should come only after we understand the underlying mechanism. Somewhat unexpectedly, a simple closed–open scheme of channel gating (see above) could be used to describe the voltage- and temperature-dependence of macroscopic currents of TRPM8 and TRPV1 (Fig. 6.3). This allowed estimation of the gating and thermodynamic parameters that determine the most prominent functional features of these channels (Fig. 6.4). In addition, the model provides a highly instrumental mathematical formalism that can be used to understand the role of thermoTRPs in sensory cells.

Characterisation of the gating kinetics of TRPM8 revealed that cold-activation results from the closing rate being tenfold more sensitive to temperature than the rate of channel opening, which is determined by the large difference in enthalpy associated with each transition (Fig. 6.4). The opposite occurs in TRPV1. The marked temperature-dependent shift of the voltage dependence can be explained easily in terms of the model. The voltage for half-maximal activation can be expressed as: $V_{act} = \dfrac{1}{zF}(\Delta H^{\ddagger} - T\Delta S^{\ddagger})$. This implies that the activation curve of the channel shifts with changes in temperature according to the factor $-\Delta S^{\ddagger}/zF$. Thus, if $\Delta S^{\ddagger} < 0$ the channel is cold-activated, whereas if $\Delta S^{\ddagger} > 0$ it is heat-activated. Interestingly, small values of z result in large shifts in the activation curve.

Notably, this analysis was previously applied to other voltage-gated channels, but a comparison between model parameters indicate that TRPM8 and TRPV1 combine relatively larger absolute value of $\Delta S^{\ddagger}$ with a low gating valence z, which explains the large shifts in V_{act} upon changing temperature. "Classical" voltage-gated channels have a much larger z and, accordingly, display much less pronounced temperature sensitivity (Correa et al. 1992; Tiwari and Sikdar 1999; van Lunteren et al. 1993).

But, what do these thermodynamic parameters actually mean? For a system under constant pressure, such as in a typical patch-clamp experiment, the difference in enthalpy is the maximal thermal energy that it is possible to extract from the transition. On the other hand, the entropy quantifies how many microscopic conformations, or microstates, can be realised with a certain amount of energy of the system. Thus, $\Delta S^{\ddagger}$ gives a measure of the variation of the disordered status of the channel in the different conformations. Here we may agree with the reader that these "cold" concepts are far from being useful, even for the best channel biophysicist. Fortunately, the literature on protein thermodynamics offers a more practical interpretation of these parameters. The difference in enthalpy is commonly associated with the energy transferred during the formation or rupture of hydrogen bonds and salt bridges between different parts of the protein. The difference in entropy, however, has a main contribution from the movement of water occurring during the transition. These notions might also be applicable to the thermodynamics of TRP channel gating. Given the current lack of necessary structural information, we are still far from being able to interpret the available experimental data in such terms. However, considering recent advances in the study of the structure of ion channels (Long et al. 2005), it is probably worth keeping these ideas in mind.

For the moment, we can perform mutagenesis studies to gain some insight into the structure–function relationship of thermoTRPs. In this respect, the only comprehensive study of the structural determinants of the thermal modulation of TRPM8 and TRPV1 is that of Brauchi et al. (2006). These authors reported that exchange of the C-terminus is sufficient to confer heat-induced activation on TRPM8 and cold-induced activation on TRPV1. Based on these results, the authors argued for a model in which the temperature and voltage sensitivities of these channels are structurally dissociated and that opening of the channel is regulated through an allosteric mechanism (Brauchi et al. 2004). They suggested that voltage sensitivity is associated to the putative voltage sensor in the S4 transmembrane domain and that the temperature sensor is contained within the C-terminus (Brauchi et al. 2006). This model differs fundamentally from the closed–open kinetic scheme proposed by Voets et al. (2004). This discrepancy seems to arise from the different methods used to determine the voltage dependence of the open probability. Indeed, Brauchi et al. (2004) used the tail current protocol, which may have yielded unreliable values of open probability given the high rate of channel deactivation at high temperatures (Voets et al. 2007).

In addition, it must be pointed out that the behaviour of the chimeric TRPM8/TRPV1 channels observed by Brauchi et al. (2006) does not necessary imply an allosteric mechanism and distinct voltage and temperature sensors. Alternatively, the simple closed–open model may provide an easy explanation. This model implies that the movement of the putative voltage sensor of TRPM8 and TRPV1 is subjected to the influence of thermal energy and that it functions as the thermosensitive element of the channel. In this scenario, it can be easily envisaged that the exchange of the C-terminus did not affect the putative voltage and temperature sensor in the S4 segment, but rather the enthalpic and entropic components of the gating transitions. In this regard, elucidation of the nature of the voltage sensor in TRP channels and its relation with thermosensing seems to be the most urgent issue. Notably, recent data indicate that the gating charge of TRPM8 is carried by basic residues in the putative S4 segment and in the S4–S5 linker and that the neutralisation of these residues leads to alterations in the modulation of channel gating by temperature (Voets et al. 2007).

6.3.4   Heat-Induced Activation of TRPM4 and TRPM5: Sweet Confirmation of the Principle

TRPM4 and TRPM5 are Ca^{2+}-impermeable non-selective cation channels activated by intracellular Ca^{2+} and membrane depolarisation (Hofmann et al. 2003; Launay et al. 2004; Liu and Liman 2003; Nilius et al. 2003a, 2005a, 2005b; Prawitt et al. 2003; Ullrich et al. 2005). Given the similarities between the voltage dependence of TRPM4 and TRPM5 and that of TRPM8 and TRPV1, it was of interest to determine the effects of temperature on the gating properties of TRPM4 and TRPM5. Not surprisingly, in the presence of intracellular Ca^{2+}, both channels displayed strong temperature-dependent activation. As in the case of TRPM8 (Voets et al. 2004) and TRPV1 (Tominaga et al. 1998; Voets et al. 2004), thermal modulation of TRPM4 and TRPM5 was observed in excised patches, indicating that the underlying mechanism is supported by the environment of the close vicinity of the channels. An unexpected result was the observation that TRPM4 and TRPM5 are heat-activated channels, closely resembling the behaviour of TRPV1 (Fig. 6.3e). This contrasts with the relatively higher sequence similarity with TRPM8 (Fig. 6.1a). Thus, one may conclude that the structural determinants for the strong temperature dependence are conserved over TRPM and TRPV subfamilies, in contrast to those that determine whether these channels are heat- or cold-activated.

In order to gain more insight into the mechanism of temperature modulation of TRPM4 and TRPM5, it was necessary to determine the effects of temperature on both the Ca^{2+} and the voltage dependencies of channel activation. For TRPM5, heating induced similar shifts of the voltage dependence of activation towards negative potentials at different intracellular Ca^{2+} concentrations (~7 mV/°C, Fig. 6.3f). Moreover, for TRPM4, it was possible to demonstrate that the effective intracellular Ca^{2+} concentration for channel activation showed little temperature dependence, with a Q_{10} of 1.32. From the voltage and temperature dependence of the steady-state and kinetic properties of TRPM4 and TRPM5 currents in the presence of saturating intracellular Ca^{2+} concentrations, it was possible to obtain a thermodynamic model for these channels (Fig. 6.4). From these results, it was concluded that TRPM4 and TRPM5 follow the same principle of temperature-dependent gating as that formulated for TRPM8 and TRPV1.

Notably, TRPM4 and TRPM5 have not been found in thermosensory cells and therefore they might not play any role in thermosensation. However, it was found that the strong temperature sensitivity of TRPM5 seems to underlie enhanced sweetness perception at high temperatures (Talavera et al. 2005).

Figure 6.4f shows the open probability as a function of temperature for TRPM8, TRPV1 and TRPM5. Note that, although high open probability occurs in different ranges of temperature for the different channels, the concept of thermal threshold for activation has no rigorous foundation.

6.4 Most ThermoTRPs are Little Understood

Five out of the nine thermoTRPs known so far lack a mechanistic explanation for their modulation by temperature. In the following, we summarise what is known about them in this respect.

6.4.1 ThermoTRPVs are Still Hot

Besides TRPV1, three other TRPVs: TRPV2, TRPV3 and TRPV4 (but not TRPV5 and TRPV6 (N. Nilius, J. Prenen, unpublished)) are strongly activated by heating.

6.4.1.1 TRPV2

Soon after the discovery of TRPV1, TRPV2 was identified as a vanilloid receptor-like channel (VRL-1), which was sharply activated by temperatures above ~52°C[5]

---

[5] Note that for the thermoTRPs that are still poorly understood, the temperature threshold for activation remains the only parameter useful to (rather crudely) characterize the temperature range of channel activation.

in heterologous (Caterina et al. 1999) and native (Bender et al. 2005; Jahnel et al. 2003) expression systems. The mechanism of thermal activation of this channel is an absolute mystery. The only thing we know about it is that thermal activation, but not stimulation with sub-threshold temperatures, sensitises this channel for subsequent stimulation (Caterina et al. 1999).

6.4.1.2 TRPV3

TRPV3 was described almost simultaneously by three different groups as a heat-activated channel activated in the physiological temperature range (Peier et al. 2002b; Smith et al. 2002; Xu et al. 2002). Like TRPV1 and TRPV2, TRPV3 is sensitised by the repetitive application of thermal (Peier et al. 2002b; Xu et al. 2002) and chemical (Chung et al. 2005, 2004) stimuli.

Interestingly, TRPV3 shows several features reminiscent of the behaviour of TRPV1, TRPM8, TRPM4 and TRPM5. First, TRPV3 is activated by membrane depolarisation (Chung et al. 2004; Xu et al. 2002) and can be activated by heat in excised patches (Chung et al. 2004). Second, it shows weak voltage sensitivity, with a slope factor of 31.7 mV (Chung et al. 2005), corresponding to 0.8 elementary charges of effective gating valence in the putative voltage sensor. Finally, Chung et al. (2005) reported that exposure to high temperatures induced a biphasic increase of TRPV3 currents during repetitive thermal stimulations, with the initial phase being characterised by a progressive shift of the voltage for half-maximal activation to more negative potentials.

Taken together, these characteristics suggest that the principle of temperature-dependent gating operating in other voltage-gated thermoTRPs (see above and Talavera et al. 2005; Voets et al. 2004) are applicable also to TRPV3. However, other features of the heat-induced activation of TRPV3 are not compatible with such a simple mechanism. Indeed, Chung et al. (2005) observed that, during the transition between the first and the second phases of heat-induced stimulation, TRPV3 undergoes changes in cationic selectivity, and in the sensitivity to ruthenium red and 2,2-diphenyltetrahydrofuran. Given that mutation of the putative selectivity filter affected the transition to the second phase of activation, the authors proposed a mechanism in which the occurrence of this phase is related to a modification of the TRPV3 pore.

6.4.1.3 TRPV4

Heterologous and native TRPV4 currents are noticeably enhanced by heating above ~25–34°C (Chung et al. 2003; Güler et al. 2002; Watanabe et al. 2002b). TRPV4 is the only non voltage-gated thermoTRP, thus its thermosensitivity should not rely on the same mechanism operating in TRPM8, TRPV1, TRPM4 and TRPM5. Notably, TRPV4 cannot be activated by heat in inside-out patches (Chung et al. 2003; Watanabe et al. 2002a). This observation suggests that heat-induced activation of this channel occurs via a diffusible endogenous ligand, whose production or interaction with TRPV4 is modulated by temperature. TRPV4 is one of the paradigms of promiscuous stimulation (Nilius et al. 2003b, 2004). Besides heat, this channel can be activated by hypotonic cell swelling (Liedtke et al. 2000; Nilius et al. 2001; Strotmann et al. 2000; Wissenbach et al. 2000) via the phospholipase A_2-arachidonic acid pathway (Vriens et al. 2004, 2005; Watanabe et al. 2003) and phorbol ester derivatives (Watanabe et al. 2002a). Notably, these modes of stimulation seem to be independent of each other (Vriens et al. 2004), although they obviously show some degree of convergence due to the saturation of the increase in channel open probability.

Structure–function approaches revealed that some of the three N-terminal ankyrin binding domains and the aromatic side chain in residue 555 (murine sequence) are compulsory for modulation by temperature (Watanabe et al. 2002b).

6.4.2   TRPA1 Channels: Close Cousins with Different Thermosensation

The TRPA subfamily contains only one mammalian member: TRPA1. This channel exhibits at least 14 N-terminal ankyrin repeats (Story et al. 2003), an unusual structural feature that is relevant to the proposed mechano-sensor role of the channel (Lee et al. 2006; Nagata et al. 2005). Some reports (Bandell et al. 2004; Story et al. 2003) indicate that TRPA1 is activated by noxious cold (<17°C), although such a mode of stimulation has subsequently been disputed (Bautista et al. 2006; Jordt et al. 2004; Nagata et al. 2005). Interestingly, the homologous channels found in the fruit fly *D. melanogaster*, named dTRPA1 and pyrexia, are activated by heating (see also Rosenzweig et al. 2005). Hopefully, this striking difference will help determine the structural bases of thermosensation in these channels.

Notably, TRPA1 seems to manifest voltage-dependent gating with low voltage sensitivity, reminiscent of the properties of other voltage-gated thermoTRPs (Y. Karashima, B. Nilius, unpublished data; T. Jegla, personal communication). It remains to be investigated whether the principle of temperature-dependent gating originally proposed for TRPM8 and TRPV1 also operates in TRPA1 channels.

6.4.3   Last, But Not Least: TRPM2

TRPM2 is the latest acquisition in the thermoTRP subgroup of TRP channels. These channels are strongly activated by heat, with an optimal temperature around mammalian core body temperature (Togashi et al. 2006). With regard to the possible mechanism of activation, the lack of voltage-dependent gating at room- and warm-temperatures argues against a similarity with TRPM8, TRPV1, TRPM4 and TRPM5. Interestingly, heating induced an increase in the relative permeability to

Ca^{2+} over that of monovalent cations, reminiscent of the effects on the secondary heat-induced activation of TRPV3.

6.5   Concluding Remarks

Temperature sensing of TRP channels comprises the whole physiological range between ~5°C and ~55°C. Remarkably, this range comprises the temperature range detected by "cold" and "warm" fibres in early electrophysiological measurements on sensory fibres (Hensel and Zotterman 1951a, 1951b). With the advent of thermoTRP channels, we now know the identity of the molecular thermometers in our sensory system and understand, at least for some of these temperature sensors, the fundamental gating principle coupling thermodynamics and sensory physiology. Finally, it is hoped that the still intrinsic mystery of why "cold" and "warm" channels respond to changes in temperature with opposite changes in entropy and enthalpy will be unravelled by the long-awaited crystallisation and analysis of the three-dimensional structures of thermoTRPs.

Acknowledgements We thank all members of the Leuven TRP laboratory for stimulating discussions. This work was supported by the Human Frontiers Science Programme (HFSP Research Grant Ref. RGP 32/2004), the Belgian Federal Government, the Flemish Government, and the Onderzoeksraad KU Leuven (GOA 2004/07, F.W.O. G. 0136.00; F.W.O. G.0172.03, Interuniversity Poles of Attraction Program, Prime Ministers Office IUAP (P5/05), Excellentiefinanciering EF/95/010).

References

Bandell M, Story GM, Hwang SW, Viswanath V, Eid SR, Petrus MJ, Earley TJ, Patapoutian A (2004) Noxious cold ion channel TRPA1 is activated by pungent compounds and bradykinin. Neuron 41:849–857

Bautista DM, Jordt SE, Nikai T, Tsuruda PR, Read AJ, Poblete J, Yamoah EN, Basbaum AI, Julius D (2006) TRPA1 mediates the inflammatory actions of environmental irritants and proalgesic agents. Cell 124:1269–1282

Bender FL, Mederos YSM, Li Y, Ji A, Weihe E, Gudermann T, Schafer MK (2005) The temperature-sensitive ion channel TRPV2 is endogenously expressed and functional in the primary sensory cell line F-11. Cell Physiol Biochem 15:183–194

Brauchi S, Orio P, Latorre R (2004) Clues to understanding cold sensation: thermodynamics and electrophysiological analysis of the cold receptor TRPM8. Proc Natl Acad Sci USA 101:15494–15499

Brauchi S, Orta G, Salazar M, Rosenmann E, Latorre R (2006) A hot-sensing cold receptor: C-terminal domain determines thermosensation in transient receptor potential channels. J Neurosci 26:4835–4840

Caterina MJ (2007) Transient receptor potential ion channels as participants in thermosensation and thermoregulation. Am J Physiol Regul Integr Comp Physiol 292:R64–R76

Caterina MJ, Schumacher MA, Tominaga M, Rosen TA, Levine JD, Julius D (1997) The capsaicin receptor: a heat-activated ion channel in the pain pathway. Nature 389:816–824

Caterina MJ, Rosen TA, Tominaga M, Brake AJ, Julius D (1999) A capsaicin-receptor homologue with a high threshold for noxious heat. Nature 398:436–441

Chung MK, Guler AD, Caterina MJ (2005) Biphasic currents evoked by chemical or thermal activation of the heat-gated ion channel, TRPV3. J Biol Chem 280:15928–15941

Chung M-K, Lee H, Caterina MJ (2003) Warm temperatures activate TRPV4 in mouse 308 keratinocytes. J Biol Chem 278:32037–32046

Chung MK, Lee H, Mizuno A, Suzuki M, Caterina MJ (2004) 2-Aminoethoxydiphenyl borate activates and sensitizes the heat-gated ion channel TRPV3. J Neurosci 24:5177–5182

Correa AM, Bezanilla F, Latorre R (1992) Gating kinetics of batrachotoxin-modified Na^+ channels in the squid giant axon. Voltage and temperature effects. Biophys J 61:1332–1352

Dhaka A, Viswanath V, Patapoutian A (2006) TRP ion channels and temperature sensation. Annu Rev Neurosci 29:135–161

Güler A, Lee H, Shimizu I, Caterina MJ (2002) Heat-evoked activation of TRPV4 (VR-OAC) J Neurosci 22:6408–6414

Hensel H, Zotterman Y (1951a) Action potentials of cold fibres and intracutaneous temperature gradient. J Neurophysiol 14:377–385

Hensel H, Zotterman Y (1951b) The effect of menthol on the thermoreceptors. Acta Physiol Scand 24:27–34

Hille B (2001) Ionic channels of excitable membranes. Sinauer, Sunderland, MA

Hofmann T, Chubanov V, Gudermann T, Montell C (2003) TRPM5 is a voltage-modulated and Ca^{2+}-activated monovalent selective cation channel. Curr Biol 13:1153–1158

Jahnel R, Bender O, Munter LM, Dreger M, Gillen C, Hucho F (2003) Dual expression of mouse and rat VRL-1 in the dorsal root ganglion derived cell line F-11 and biochemical analysis of VRL-1 after heterologous expression. Eur J Biochem 270:4264–4271

Jordt SE, Bautista DM, Chuang HH, McKemy DD, Zygmunt PM, Hogestatt ED, Meng ID, Julius D (2004) Mustard oils and cannabinoids excite sensory nerve fibres through the TRP channel ANKTM1. Nature 427:260–265

Launay P, Fleig A, Perraud AL, Scharenberg AM, Penner R, Kinet JP (2002) TRPM4 is a Ca^{2+}-activated nonselective cation channel mediating cell membrane depolarization. Cell 109:397–407

Launay P, Cheng H, Srivatsan S, Penner R, Fleig A, Kinet JP (2004) TRPM4 regulates calcium oscillations after T cell activation. Science 306:1374–1377

Lee G, Abdi K, Jiang Y, Michaely P, Bennett V, Marszalek PE (2006) Nanospring behaviour of ankyrin repeats. Nature 440:246–249

Liedtke W, Choe Y, Marti-Renom MA, Bell AM, Denis CS, Sali A, Hudspeth AJ, Friedman JM, Heller S (2000) Vanilloid receptor-related osmotically activated channel (VR-OAC), a candidate vertebrate osmoreceptor. Cell 103:525–535

Liu D, Liman ER (2003) Intracellular Ca^{2+} and the phospholipid PIP2 regulate the taste transduction ion channel TRPM5. Proc Natl Acad Sci USA 100:15160–15165

Long SB, Campbell EB, Mackinnon R (2005) Voltage sensor of Kv1.2: structural basis of electromechanical coupling. Science 309:903–908

McKemy DD (2005) How cold is it? TRPM8 and TRPA1 in the molecular logic of cold sensation. Mol Pain 1:16

McKemy DD, Neuhausser WM, Julius D (2002) Identification of a cold receptor reveals a general role for TRP channels in thermosensation. Nature 416:52–58

Moqrich A, Hwang SW, Earley TJ, Petrus MJ, Murray AN, Spencer KS, Andahazy M, Story GM, Patapoutian A (2005) Impaired thermosensation in mice lacking TRPV3, a heat and camphor sensor in the skin. Science 307:1468–1472

Nagata K, Duggan A, Kumar G, Garcia-Anoveros J (2005) Nociceptor and hair cell transducer properties of TRPA1, a channel for pain and hearing. J Neurosci 25:4052–4061

Nilius B, Prenen J, Wissenbach U, Bodding M, Droogmans G (2001) Differential activation of the volume-sensitive cation channel TRP12 (OTRPC4) and volume-regulated anion currents in HEK-293 cells. Pfluegers Arch Eur J Physiol 443:227–233

Nilius B, Prenen J, Droogmans G, Voets T, Vennekens R, Freichel M, Wissenbach U, Flockerzi V (2003a) Voltage dependence of the Ca^{2+}-activated cation channel TRPM4. J Biol Chem 278:30813–30820

Nilius B, Watanabe H, Vriens J (2003b) The TRPV4 channel: structure–function relationship and promiscuous gating behaviour. Pfluegers Arch 446:298–303

Nilius B, Vriens J, Prenen J, Droogmans G, Voets T (2004) TRPV4 calcium entry channel: a paradigm for gating diversity. Am J Physiol Cell Physiol 286:C195–C205

Nilius B, Prenen J, Janssens A, Owsianik G, Wang C, Zhu MX, Voets T (2005a) The selectivity filter of the cation channel TRPM4. J Biol Chem 280:22899–228906

Nilius B, Prenen J, Tang J, Wang C, Owsianik G, Janssens A, Voets T, Zhu MX (2005b) Regulation of the Ca^{2+} sensitivity of the nonselective cation channel TRPM4. J Biol Chem 280:6423–6433

Nilius B, Talavera K, Owsianik G, Prenen J, Droogmans G, Voets T (2005c) Gating of TRP channels: a voltage connection? J Physiol 567:33–44

Owsianik G, D'Hoedt D, Voets T, Nilius B (2006a) Structure–function relationship of the TRP channel superfamily. Rev Physiol Biochem Pharmacol 156:61–90

Owsianik G, Talavera K, Voets T, Nilius B (2006b) Permeation and selectivity of trp channels. Annu Rev Physiol 68:685–717

Patapoutian A, Peier AM, Story GM, Viswanath V (2003) ThermoTRP channels and beyond: mechanisms of temperature sensation. Nat Rev Neurosci 4:529–539

Pedersen SF, Owsianik G, Nilius B (2005) TRP channels: an overview. Cell Calcium 38:233–252

Peier AM, Moqrich A, Hergarden AC, Reeve AJ, Andersson DA, Story GM, Earley TJ, Dragoni I, McIntyre P, Bevan S, Patapoutian A (2002a) A TRP channel that senses cold stimuli and menthol. Cell 108:705–715

Peier AM, Reeve AJ, Andersson DA, Moqrich A, Earley TJ, Hergarden AC, Story GM, Colley S, Hogenesch JB, McIntyre P, Bevan S, Patapoutian A (2002b) A heat-sensitive TRP channel expressed in keratinocytes. Science 296:2046–2049

Prawitt D, Monteilh-Zoller MK, Brixel L, Spangenberg C, Zabel B, Fleig A, Penner R (2003) TRPM5 is a transient Ca^{2+}-activated cation channel responding to rapid changes in $[Ca^{2+}]_i$. Proc Natl Acad Sci USA 100:15166–15171

Rosenzweig M, Brennan KM, Tayler TD, Phelps PO, Patapoutian A, Garrity PA (2005) The Drosophila ortholog of vertebrate TRPA1 regulates thermotaxis. Genes Dev 19:419–424

Smith GD, Gunthorpe J, Kelsell RE, Hayes PD, Reilly P, Facer P, Wright JE, Jerman JC, Walhin JP, Ooi L, Egerton J, Charles KJ, Smart D, Randall AD, Anand P, Davis JB (2002) TRPV3 is a temperature-sensitive vanilloid receptor-like protein. Nature 418:186–190

Story GM, Peier AM, Reeve AJ, Eid SR, Mosbacher J, Hricik TR, Earley TJ, Hergarden AC, Andersson DA, Hwang SW, McIntyre P, Jegla T, Bevan S, Patapoutian A (2003) ANKTM1, a TRP-like channel expressed in nociceptive neurons, is activated by cold temperatures. Cell 112:819–829

Strotmann R, Harteneck C, Nunnenmacher K, Schultz G, Plant TD (2000) OTRPC4, a nonselective cation channel that confers sensitivity to extracellular osmolarity. Nat Cell Biol 2:695–702

Talavera K, Yasumatsu K, Voets T, Droogmans G, Shigemura N, Ninomiya Y, Margolskee RF, Nilius B (2005) Heat activation of TRPM5 underlies thermal sensitivity of sweet taste. Nature 438:1022–1025

Tiwari JK, Sikdar SK (1999) Temperature-dependent conformational changes in a voltage-gated potassium channel. Eur Biophys J 28:338–345

Togashi K, Hara Y, Tominaga T, Higashi T, Konishi Y, Mori Y, Tominaga M (2006) TRPM2 activation by cyclic ADP-ribose at body temperature is involved in insulin secretion. EMBO J 25:1804–1815

Tominaga M, Caterina MJ (2004) Thermosensation and pain. J Neurobiol 61:3–12

Tominaga M, Caterina MJ, Malmberg AB, Rosen TA, Gilbert H, Skinner K, Raumann BE, Basbaum AI, Julius D (1998) The cloned capsaicin receptor integrates multiple pain-producing stimuli. Neuron 21:531–543

Ullrich ND, Voets T, Prenen J, Vennekens R, Talavera K, Droogmans G, Nilius B (2005) Comparison of functional properties of the Ca^{2+}-activated cation channels TRPM4 and TRPM5 from mice. Cell Calcium 37:267–278

Van Lunteren E, Elmslie KS, Jones SW (1993) Effects of temperature on calcium current of bullfrog sympathetic neurons. J Physiol 466:81–93

Voets T, Droogmans G, Wissenbach U, Janssens A, Flockerzi V, Nilius B (2004) The principle of temperature-dependent gating in cold- and heat-sensitive TRP channels. Nature 430:748–754

Voets T, Talavera K, Owsianik G, Nilius B (2005) Sensing with TRP channels. Nat Chem Biol 1:85–92

Voets T, Owsianik G, Janssens A, Talavera K, Nilius B (2007) TRPM8 voltage sensor mutants reveal a mechanism for integrating thermal and chemical stimuli. Nat Chem Biol 3:174–182

Vriens J, Watanabe H, Janssens A, Droogmans G, Voets T, Nilius B (2004) Cell swelling, heat, and chemical agonists use distinct pathways for the activation of the cation channel TRPV4. Proc Natl Acad Sci USA 101:396–401

Vriens J, Owsianik G, Fisslthaler B, Suzuki M, Janssens A, Voets T, Morisseau C, Hammock BD, Fleming I, Busse R, Nilius B (2005) Modulation of the Ca2 permeable cation channel TRPV4 by cytochrome P450 epoxygenases in vascular endothelium. Circ Res 97:908–915

Watanabe H, Davis JB, Smart D, Jerman JC, Smith GD, Hayes P, Vriens J, Cairns W, Wissenbach U, Prenen J, Flockerzi V, Droogmans G, Benham CD, Nilius B (2002a) Activation of TRPV4 channels (hVRL-2/mTRP12) by phorbol derivatives. J Biol Chem 277:13569–13577

Watanabe H, Vriens J, Suh SH, Benham CD, Droogmans G, Nilius B (2002b) Heat-evoked activation of TRPV4 channels in a HEK293 cell expression system and in native mouse aorta endothelial cells. J Biol Chem 277:47044–47051

Watanabe H, Vriens J, Prenen J, Droogmans G, Voets T, Nilius B (2003) Anandamide and arachidonic acid use epoxyeicosatrienoic acids to activate TRPV4 channels. Nature 424:434–438

Wissenbach U, Bodding M, Freichel M, Flockerzi V (2000) Trp12, a novel Trp related protein from kidney. FEBS Lett 485:127–134

Xu HX, Ramsey IS, Kotecha SA, Moran MM, Chong JHA, Lawson D, Ge P, Lilly J, Silos Santiago I, Xie Y, DiStefano PS, Curtis R, Clapham DE (2002) TRPV3 is a calcium-permeable temperature-sensitive cation channel. Nature 418:181–186

Chapter 7
TRPC Family of Ion Channels and Mechanotransduction

Owen P. Hamill(✉), and Rosario Maroto

Abstract Here we review recent evidence that indicates members of the canonical transient receptor potential (TRPC) channel family form mechanosensitive (MS) channels. The MS functions of TRPCs may be mechanistically related to their better known functions as store-operated (SOCs) and receptor-operated channels (ROCs). In particular, mechanical forces may be conveyed to TRPC channels through

Department of Physiology & Biophysics, University of Texas Medical Branch at Galveston, Galveston, TX 77555, USA, ohamill@utmb.edu "conformational coupling" and/or "Ca^{2+} influx factor" mechanisms that are proposed to transmit information regarding the status of internal Ca^{2+} stores to SOCs located in the plasma membrane. Furthermore, all TRPCs are regulated by receptors coupled to phospholipases (e.g., PLC and PLA_2) that may themselves display mechanosensitivity and modulate channel activity via their generation of lipidic second messengers (e.g., diacylglycerol, lysophospholipids and arachidonic acid). Accordingly, there may be several nonexclusive mechanisms by which mechanical forces may regulate TRPC channels, including direct sensitivity to bilayer deformations (e.g., involving changes in lipid packing, bilayer thickness and/or lateral pressure profile), physical coupling to internal membranes and/or cytoskeletal proteins, and sensitivity to lipidic second messengers generated by MS enzymes. Various strategies that can be used to separate out different MS gating mechanisms and their possible role in each of the TRPCs are discussed.

7.1   Introduction

Mechanotransduction (MT) is a fundamental physiological process by which mechanical forces are transduced into electrical, ionic and/or biochemical signals. MT can span a time scale of milliseconds as in the case of a fast sensory process (e.g., in hearing and touch) to days and even years as in the case of the growth and reorganization of tissues (e.g., skin, muscle and the endothelia) in response to mechanical loading or mechanical stress. Because the plasma membrane forms the interface with the external physical world, it is continually subject to mechanical deformations arising from tissue stretch, compression, gravity, interstitial fluid pressure, fluid shear stress, and also from cytoskeleton (CSK)-generated contractile and tractile forces (Howard et al. 1988; Sachs 1988; Hamill and Martinac 2001; Perbal and Driss-Ecole 2003; Wang and Thampatty 2006; Pickard 2007). Furthermore, the membrane bilayer may be subject to local mechanical distortions caused by the insertion of lipidic second messenger molecules [e.g., diacylglycerol (DAG), arachidonic acid (AA) and lysophospholipids (LPLs)] that act by altering local packing thickness and/or lateral pressure profile, and thereby influence membrane protein conformations with consequences similar to those of global membrane deformations (Martinac et al. 1990; Hamill and Martinac 2001; Perozo et al. 2002; Kung 2005; Martinac 2007; Markin and Sachs 2007; Powl and Lee 2007). It is therefore not surprising to find that a wide range of integral and membrane-associated proteins are specialized to sense and transduce membrane distortions into different homeostatic responses. Here we focus on the seven members of the mammalian canonical transient receptor potential (TRPC1–7) channel family that provide an illustration of how very closely related membrane proteins have evolved different mechanisms for sampling their global and local mechanical environment.

7.2   Distinguishing Direct from Indirect MS Mechanisms

Because TRPCs are gated by a variety of stimuli including direct lipid bilayer stretch as well as by lipidic second messengers that are generated by membrane-associated enzymes that may themselves be mechanosensitive (MS), it is important to establish criteria that may be used to distinguish direct from indirect MS mechanisms of TRPC channel activation. Below we list some tests that may be useful in making this discrimination.

7.2.1   Stretch Activation of Channels in the Patch

The most convenient way of demonstrating an MS channel is to use a patch clamp to apply pressure or suction after formation of the giga-seal while simultaneously measuring single channel current activity (Hamill et al. 1981; Guharay and Sachs 1984). The cell-free membrane patch configurations (inside-out and outside-out) can also be used to determine if MS channel activity is retained when the cytoplasmic membrane face is perfused with solutions deficient in soluble second messengers (e.g., Ca^{2+}, cAMP, ATP). However, MS enzymes that generate membrane delimited second messengers may retain their activity following patch excision. Similarly, critical elements of the CSK involved in gating MS channels may be preserved in cell-free membrane patches (Ruknudin et al. 1991). Alternative approaches for testing CSK involvement may involve testing for MS channel activity in membrane patches formed on CSK-deficient membrane blebs induced by ATP depletion or by high ionic strength solution (Zhang et al. 2000; Honoré et al. 2006), and determining how agents that disrupt the CSK elements (e.g., cytochalasins and colchicine) affect the activity of MS channels (Guharay and Sachs 1984; Small and Morris 1994; Honoré et al. 2006).

7.2.2   Osmotic Swelling and Cell Inflation

Osmotic stress can also be used to test if a channel is MS either by swelling the cell while recording from a cell-attached patch or from the whole cell (Hamill 1983; Christensen 1987; Sackin 1989; Cemerikic and Sackin 1993; Vanoye and Reuss 1999; Spassova et al. 2006; Numata et al. 2007). The advantage of this approach is that the action of inhibitors and activators can be consistently recognized when assessed on whole cell vs patch currents. However, osmotic cell swelling also activates a number of membrane-associated enzymes, including Src kinase and phospholipase A2 (PLA_2) (Lehtonen and Kinnunen 1995; Cohen 2005a). Furthermore, although some stretch-sensitive channels are sensitive to osmotic cell swelling and direct cell inflation, others are not (Levina et al. 1999; Vanoye and Reuss 1999; Zhang and Hamill 2000a, 2000b). One basis for this difference is that some cells may possess large excess membrane reserves in the form of folds microvilli and caveola (e.g., *Xenopus* oocytes and skeletal muscle) that can buffer rapid increases in bilayer tension (Zhang and Hamill 2000a, 2000b; Hamill 2006). Conversely, not all channels activated by cell swelling are activated by membrane stretch when applied to the patch (Ackerman et al. 1994; Strotmann et al. 2000).

7.2.3 Gating Kinetics

Another criterion that may be useful in distinguishing direct from indirect mechanisms is the delay time in activation in response to pressures steps applied to the patch with a fast pressure-clamp (McBride and Hamill 1993). Channels that are directly MS should be limited only by the conformational transitions of the channel protein, and may as a consequence show only brief delays (i.e., in the sub-ms or ms range) in their activation and deactivation (McBride and Hamill 1992, 1993). In comparison, channels dependent on enzymatic reactions and/or diffusion of second messenger may be expected to show much longer delays in opening and closing (e.g., ≥ 1 s). These kinetic measurements are best made in the cell-attached or cell-free patch using the pressure clamp to apply pressure steps (1–5 ms rise time) in order to measure the latency in activation and deactivation of the channels (McBride and Hamill 1992, 1993, 1995, 1999; Besch et al. 2002). Rapid activation and deactivation kinetics have been reported for the mechanosensitive Ca^{2+}-permeable cation channel (MscCa) that is formed by TRPCs in *Xenopus* oocytes (Hamill and McBride 1992; McBride and Hamill 1992, 1993) and the expressed TRAAK channel [i.e., a TWIK (tandem of pore domains in a weak inward rectifier K^+ channel)-Related Arachidonic Acid stimulated K^+ channel; Honoré et al. 2006]. In contrast, both the MscK expressed in snail neurons, which may also be a two-pore-domain K^+-channel-like TRAAK (Vandorpe and Morris 1992), and the cation channel formed by TRPC6 show long activation delays of 5–10 s (Small and Morris 1994; Spassova et al. 2006). However, because the delays can be abolished by mechanical and/or chemical CSK disruption it seems more likely that the delays reflect CSK constraint of the bilayer, which prevents rapid transmission of tension to the channel (Small and Morris 1994; Hamill and McBride 1997; Spassova et al. 2006). No studies measuring possible delays in pressure activation of TRPs suspected of being indirectly MS have been performed to date. In the case of the activation of TRPV4, which has been functionally linked to MS PLA_2 generation of AA and its subsequent metabolism to 5′, 6′-epoxyeicosatrienoic acid (5′6′-EET) by cytochrome P450 epoxygenase activity (Vriens et al. 2004; Watanabe et al. 2003), whether the apparent lack of stretch sensitivity was overlooked because of long delays and slow channel activation still needs to be determined (Strotmann et al. 2000).

7.2.4 The Use of MS Enzyme Inhibitors

A further strategy for implicating potential MS enzymatic steps in channel activation is to test specific enzyme inhibitors on channel activity. For example, bromoenol lactone (BEL) can be used as a selective blocker of the Ca^{2+}-independent phospholipase A2 (iPLA$_2$), PP2 (4-amino-5-(4-chlorophenyl)-7-(t-butyl) pyrazolo[3,4,d] pyrimidine) is a Src tyrosine kinase blocker, and U73122 can be used to block phospholipase C (PLC) (reviewed by Hamill and Maroto 2007). Using this approach, it has been reported that the stretch sensitivity of a 30 pS cation channel measured in cell-attached patches formed on arterial smooth muscle is abolished by perfusion of the whole muscle cell with U73122, indicating that the channel may derive its stretch sensitivity from a MS iPLA$_2$ (Park et al. 2003). However, a different study reported that U73122 was ineffective in blocking the stretch sensitivity of a similar 30 pS cation channel measured in inside-out patches isolated from CHO cells transfected with hTRPC6 (Spassova et al. 2006). In another example that involves the excess Ca^{2+}-influx that occurs in dystrophic muscle, and has been proposed to be mediated by TRPC-dependent MS and/or store-operated channels (SOC) (Vandebrouck et al. 2002; Ducret et al. 2006), it was shown that the Ca^{2+} influx could be abolished by BEL (Boittin et al. 2006) and potentiated by the bee venom melittin, a potent activator of PLA$_2$ (Lindahl et al. 1995; Boittin et al. 2006).

7.2.5 Reconstitution of MS Channel Activity in Liposomes

The most unequivocal method for distinguishing direct from indirect mechanosensitivity is to examine whether the detergent-solubilized channel protein retains stretch sensitivity when reconstituted in pure liposomes. So far this criterion has been applied to several MS channels in prokaryotes and MscCa expressed in the frog oocyte (Sukharev et al. 1993, 1994; Sukharev 2002; Kloda and Martinac 2001a, 2001b; Maroto et al. 2005). This approach also offers the potential of definitive evidence on whether lipidic second messengers (e.g., DAG, AA, LPLs and 5′6′-EET) activate the channel by binding directly to the channel protein and/or the surrounding lipid without the requirement of additional proteins and/or enzymatic steps. Furthermore, the same approach may be used to determine whether multi-protein component MS signaling complexes can be functionally reconstituted (e.g., TRPV4/PLA$_2$/P450 and TRPCs/PLC).

7.3   Extrinsic Regulation of Stretch Sensitivity

It seems highly unlikely that the stretch sensitivity of different membrane channels will be accounted for by a single structural domain analogous to the S-4 voltage sensor domain common to voltage-gated Na^+, K^+ and Ca^{2+} channels (Hille 2001). This is because even the relatively simple peptide channels gramicidin and alamethicin, which possess dramatically different structures and gating mechanisms, exhibit stretch sensitivity (Opsahl and Webb 1994; Hamill and Martinac 2001; Martinac and Hamill 2002). Furthermore, stretch sensitivity is not a fixed channel property, but rather can undergo significant changes with changing extrinsic conditions. For example, mechanical and/or chemical disruption of the CSK can either enhance or abolish the stretch sensitivity of specific channels (Guharay and Sachs 1984; Hamill and McBride 1992, 1997; Small and Morris 1994; Patel and Honoré 2001; Hamill 2006); changes in bilayer thickness (Martinac and Hamill 2002), membrane voltage (Gu et al. 2001; Morris and Juranka 2007), or dystrophin expression (Franco-Obregon and Lansman 2002) can switch specific MS channels between being stretch-activated to being stretch-inactivated; specific lipids (Patel and Honoré 2001; Chemin et al. 2005), nucleotides (Barsanti et al. 2006a, and references therein) and increased internal acidosis (Honoré et al. 2002; Barsanti et al. 2006b) can convert MS channels into constitutively open 'leak' channels. The basis for many of these changes involves changes in the way the bilayer, CSK and/or extracellular matrix conveys mechanical forces to the channel protein. The practical consequence of this plasticity is that the specific conditions associated with reconstitution and/or heterologous expression may alter the stretch sensitivity of the channel.

7.4   Stretch Sensitivity and Functional MT

Although stretch sensitivity measured in the patch can be used to demonstrate a channel protein is MS at the biophysical level, it cannot prove that the channel functions as a physiological mechanotransducer because conditions associated with the giga-seal formation can increase the stretch sensitivity of the membrane patch (Morris and Horn 1991; Zhang and Hamill 2000b; Hamill 2006). Indeed, many structurally diverse voltage- and receptor-gated channels (e.g., Shaker, L-type Ca^{2+} channels, NMDA-R, S-type K^+ channels), as well as the simple model peptide channels alamethicin and gramicidin A, display stretch sensitivity in patch recordings (Opsahl and Webb 1994; Paoletti and Ascher 1994; Martinac and Hamill 2002; Morris and Juranka 2007). In order to demonstrate functionality one also needs to show that blocking the channel (pharmacologically and/or genetically) inhibits a mechanically induced cellular/physiological process.

7.5   General Properties of TRPCs

The designation TRP originated with the discovery of a *Drosophila* mutant that showed a transient rather than a sustained receptor potential in response to light (Cosens and Manning 1969). This response was subsequently shown to involve a PLC-dependent Ca^{2+}-permeable cation channel (Minke et al. 1975; Montell and Rubin 1989; Minke and Cook 2002). Beginning in the mid 1990s, seven mammalian TRP homologs were identified that, together with the *Drosophila* TRP, make up the canonical TRP (TRPC) subfamily. Other TRP subfamilies include TRPV (vanilloid), TRPA (ankyrin), TRPP (polycystin), TRPM (melastatin), TRPML (mucolipid), TRPN (NOMPC) and TRPY (yeast); these combine with TRPC to form the TRP superfamily (Montell 2005; Nilius and Voets 2005; Ramsey et al. 2006; Saimi et al. 2007; Nilius et al. 2007). In addition to the TRPCs, specific members of the other TRP subfamilies have also been implicated in MT so that the MS mechanisms discussed here may also apply to these channels (Walker et al. 2000; Palmer et al. 2001; Zhou et al. 2003; Muraki et al. 2003; Nauli and Zhou 2004; O'Neil and Heller 2005; Voets et al. 2005, Saimi et al. 2007; Numata et al. 2007).

The proposed transmembrane topology of TRPCs is reminiscent of voltage-gated channels – sharing six transmembrane spanning helices (TM1–6), cytoplasmic N- and C-termini and a pore region between TM5 and TM6 – but lacking the positively charged residues in the TM4 domain that forms the voltage sensor. TRPC channels also share an invariant sequence in the C-terminal tail called a TRP box (E-W-K-F-A-R), as well as 3–4 N-terminal ankyrin repeats. Although the ankyrin repeats may act as gating springs for MS channels (Howard and Bechstedt 2004) and the positively charged residues in the TRP box may interact directly with the membrane phospholipids, phosphatidylinositol 4,5-bisphosphate (PIP_2) (Rohács et al. 2005) their exact roles remain to be verified (Vazquez et al. 2004a; Owsianik et al. 2006). The TRPCs share very little sequence identity in the region that is C terminal of the TRP box, except for the common feature of calmodulin (CaM)- and inositol 1,4,5-trisphosphate receptor (IP_3R)-binding domains that have been implicated in Ca^{2+} feedback inhibition and activation by store depletion, respectively (Kiselyov et al. 1998; Vaca and Sampieri 2002; Bolotina and Csutora 2005). Based on sequence homology, the TRPCs have been subdivided into the major subgroups of TRPC1/4/5 (showing 65% homology) and TRPC3/6/7 (showing 70–80% homology). TRPC2 is grouped alone because it forms a functional channel in rodents but not in humans (i.e., it is a pseudogene in humans because of the presence of multiple stop codons within its open reading frame).

7.5.1   TRPC Expression

TRPCs are widely expressed in mammalian tissues with some cell types expressing all seven members and others expressing only one or two (Riccio et al. 2002b; Goel et al. 2006; Antoniotti et al. 2006; Hill et al. 2006). Cells that express only one

TRPC may prove particularly useful models for dissecting out specific TRPC functions. However, to justify this role it is necessary to verify that their selective expression is reflected at both the transcriptional and protein levels. This is important because low turnover proteins may require little mRNA, and high mRNA levels need not translate into high membrane protein levels (Andersen and Seilhamer 1997). Another caveat is that TRPC expression patterns can vary significantly during development (see Strübing et al. 2003), and with culture conditions (e.g., presence or absence of growth factors). For example, TRPC1 expression is upregulated by (1) serum deprivation, which leads to increased proliferation of pulmonary arterial smooth muscle cells (Golovina et al. 2001), (2) tumor necrosis factor α, which enhances endothelial cell death (Paria et al. 2003), and (3) vascular injury in vivo, which contributes to human neoitimal hyperplasia (Kumar et al. 2006). Also, TRPC6 expression in pulmonary arterial smooth muscle cells is enhanced in idiopathic pulmonary hypertension and by platelet-derived growth factor (Yu et al. 2003, 2004). Compared with mammalian cells there is less information on TRPC expression in lower vertebrates. For example, although a TRPC1 homologue has been identified in *Xenopus* oocytes, a systematic study of expression of other TRPs in lower vertebrates has not yet been carried out (Bobanović et al. 1999).

7.5.2   TRPC Activation and Function: Mechanisms of SOC and ROC

Studies of TRPC activation and function are complicated by their polymodal activation and splice variants that display different activation mechanisms (see Ramsey et al. 2006). However, all TRPCs are regulated by PLC-coupled receptors (i.e., G-protein-coupled receptors or tyrosine kinase receptors). PLC hydrolyzes a component of the bilayer, PIP_2, into two distinct messengers – the soluble inositol 1,4,5-trisphosphate ($InsP_3$) that activates the IP_3R in the endoplasmic reticulum (ER) to release Ca^{2+} from internal stores – and the lipophilic DAG, which may regulate TRPs indirectly via protein kinase C (PKC) or by interacting directly with TRPCs in a membrane delimited manner (Hofmann et al. 1999; Delmas et al. 2002; Clapham 2003; Ahmmed et al. 2004; Ramsey et al. 2006). Although all TRPCs could be classified as receptor-operated channels (ROCs, but see Janssen and Kwan 2007), they are more often subdivided into either (1) SOCs, based on their sensitivity to Ca^{2+} store depletion, or (2) ROCs based on both their activation by DAG, AA or their byproducts, and their insensitivity to Ca^{2+} store depletion (Hofmann et al. 1999; Shuttleworth et al. 2004). To be classified as a SOC, the channel should be gated by a variety of procedures that share only the common feature of reducing Ca^{2+} stores, which may or may not depend on IP3R signaling (see Parekh and Putney 2005). Unfortunately, there have been conflicting reports for all seven TRPCs on whether they function as SOCs, ROCs or both. Furthermore, the nature of the mechanism(s) that activates SOCs remains controversial, with at least two main classes of mechanism in contention. One mechanism depends upon a soluble

Ca^{2+} influx factor (CIF) that is released from depleted Ca^{2+} stores and diffuses to the plasma membrane where it activates the SOC, possibly by releasing inhibitory CaM from $iPLA_2$, which generates LPLs and AA (see Bolotina and Csutora 2005) Direct support for this CIF-CaM-$iPLA_2$-LPL model has come from the demonstration that functional $iPLA_2$ is required for SOC activation, displacement of CaM from $iPLA_2$ activates SOC, and the direct application of LPLs (but not AA) to inside-out patches activates SOCs (Smani et al. 2003, 2004). On the other hand, the generality of this model has been questioned by the finding that BEL, an $iPLA_2$ inhibitor, does not block thapsigarin-induced Ca^{2+} entry in RBL 2H3 or bone-marrow-derived mast cells (Fensome-Green et al. 2007).

The second main SOC mechanism involves conformational coupling (i.e., "CC" mechanism) between the SOC and a molecule located in the ER that transmits information regarding $[Ca^{2+}]$ levels in internal stores. This mechanism has received its strongest support with the discovery of STIM (stromal interaction molecule), a resident ER protein with a putative Ca^{2+} binding domain in the ER lumen. Following Ca^{2+} store depletion, STIM has been shown to undergo rapid clustering and increased interactions with elements of the plasma membrane (Liou et al. 2005; Roos et al. 2005). Furthermore, it has been demonstrated that the STIM1 carboxylterminus activates native SOC, Ca^{2+} release-activated currents (I_{CRAC}) and TRPC1 channels (Huang et al. 2006). It may turn out that both CIF and CC mechanisms can operate in a redundant manner to activate SOCs, with the CIF mechanism conferring indirect MS on SOC via a MS $iPLA_2$ (Lehtonen and Kinnunen 1995), and the CC mechanism allowing for direct transmission of mechanical force via a direct STIM–SOC physical connection.

7.5.3   TRPC–TRPC Interactions

Assuming that a cell expresses all seven TRPC subunits, and that four TRPC subunits are required to form a channel (i.e., as a homotetramer or heterotetramer), then there could be as many as 100 different TRPC channels with different neighbor subunit interactions. However, this number would be much lower if only certain TRPC–TRPC combinations are permitted. Two different models have been proposed to underlie the permissible TRPC interactions: the homotypic model, which allows subunits interactions only within each major subgroup – with TRPC1/4/5 combinations forming SOCs and TRPC3/6/7 combinations forming ROCs – (Hofmann et al. 2002; Sinkins et al. 2004); and the heterotypic model, which permits interactions both within and between members of the two subgroups. In a specific heterotypic model developed by Villreal and colleagues it is proposed that TRPC1, 3 and 7 combine to form SOCs (i.e., without participation of TRPC4 and TRPC6) while TRPC3, 4, 6 and 7 combine to form ROCs (i.e., without TRPC1 participation) (Zagranichnaya et al. 2005; Villereal 2006). In this case, the TRPC1 role is limited to forming a SOC and TRPC4 and TRPC6 are limited to forming ROCs. However, in contradiction of an exclusive ROC role for TRPC4, it has been reported that SOC currents in adrenal cells are abolished by TRPC4 anti-sense treatment (Phillip et al. 2000) and that endothelial cells isolated from TRPC4 knockout mice lack SOC activity (Freichel et al. 2003, 2004). In contrast to the exclusive roles of TRPC1, 4 and 6, TRPC3 and TRPC7 can participate in forming both SOCs and ROCs (Zagranichnaya et al. 2005). The validity of the different models has yet to be resolved. However, whereas the homotypic model has been based mostly on gain-of-function results from TRPC overexpression studies, the heterotypic model has been based mostly on loss-of-function results from TRPC suppression studies. At least one complication with the overexpression studies is related to the finding that different levels of specific TRPC expression can influence the function displayed in the transfected cell. In particular, low TRPC3 expression results in increased SOC activity, while high TRPC3 expression results in increased ROC activity (Vazquez et al. 2003). This variation may occur because high expression levels favor TRPC3 homotetrameric channels, whereas low TRPC3 expression allows for heterotetrameric channels with incorporation of endogenous subunits as well as exogenous TRPC (Brereton et al. 2001; Vazquez et al. 2003). Differences in channel function may also arise depending upon whether the cell is permanently or transiently transfected, presumably because stable transfection provides added time for adaptive changes in endogenous protein expression (Lièvremont et al. 2004).

7.5.4   TRPC Interactions with Scaffolding Proteins

TRPCs also interact with a variety of regulatory and scaffolding proteins that may add further diversity and segregation of the channels (Ambudkar 2006). In particular, it has been shown that several TRPCs assembly into multi-protein and lipid signaling complexes that result in physical and functional interactions between the plasma membrane, and CSK and ER resident proteins. These interactions may also allow for mechanical forces to be conveyed via a tethered mechanism to gate the channel (Guharay and Sachs 1984; Howard et al. 1988; Hamill and Martinac 2001; Matthews et al. 2007; Cantiello et al. 2007). Alternatively, the interactions may also serve to constrain the development or transmission of bilayer tension to the TRPC channel and thereby "protect" it from being mechanically activated (Small and Morris 1994; Hamill and McBride 1997). For all TRPCs, the C-terminal coiled-coil domains and the N-terminal ankyrin repeats have the potential to mediate protein-CSK interactions. All TRP family members also encode a conserved proline rich sequence LP(P/X)PFN in their C termini that is similar to the consensus binding site for Homer, a scaffold protein that has been shown to facilitate TRPC1 interaction with IP_3R – disruption of which has been proposed to promote SOC activity (Yuan et al. 2003). In particular, TRPC1 mutants lacking Homer protein binding sites show diminished interaction between TRPC1 and IP_3R and the TRPC1 channels are constitutively active. Moreover, co-expression of a dominant-negative form of Homer increases basal TRPC1 channel activity (Yuan et al. 2003). Another protein, I-mfa, which inhibits helix-loop-helix transcription factors, also binds to TRPC1 and blocks SOC function (Ma et al. 2003). TRPC1 also possesses a dystrophin domain within its C-terminus (Wes et al. 1995) that may allow for interaction with dystrophin – the major CSK protein in skeletal muscle – and this could possibly explain why the absence of dystrophin in Duchenne muscular dystrophic muscle results in TRPC1 channels being abnormally gated open (see Sect. 7.6.1.4). TRPC1 also shows a putative caveolin-1-binding domain that may promote its functional recruitment into lipid rafts and increase SOC activity (Lockwich et al. 2000; Brazier et al. 2003; Ambudkar 2006). As mentioned previously, TRPC1 also interacts with STIM, the putative ER Ca^{2+} sensor molecule that regulates SOC function (Huang et al. 2006). Junctate – another IP_3R-associated protein – interacts with TRPC2, 3 and 5, but apparently not with TRPC1, to regulate their SOC/ROC function (Treves et al. 2004; Stamboulian et al. 2005). In pulmonary endothelial cells, TRPC4 is localized to cell–cell adhesions in cholesterol-rich caveolae and has been shown to interact with the spectrin CSK via the protein 4.1 (Cioffi et al. 2005; Torihashi et al. 2002). Furthermore, either deletion of the putative 4.1 protein binding site on the TRPC4 C-terminus or addition of peptides that competitively bind to that site are able to reduce SOC activity. Another site for TRPC4–CSK interaction involves the PSD-95/disc large protein/zona occludens 1 (PDZ) binding domain located at the TRPC4 distal C-terminus, which binds to the Na^+/H^+ exchange regulatory factor (NHERF) scaffolding protein (Mery et al. 2002; Tang et al. 2000). TRPC6 interacts with the stomatin-like protein podocin, which may modulate its mechano-operated channel (MOC) function in the renal slit diaphragm (Reiser et al. 2005). Interestingly, another stomatin homolog, MEC-2, was proposed to link the putative MS channel to the microtubular CSK in *Caenorhabditis elegans* neurons (Huang et al. 1995) but most recently has been implicated, along with podocin, in regulating MS channel function by forming large protein–cholesterol complexes in the plasma membrane (Huber et al. 2006).

In summary, TRPCs undergo dynamic interactions with various scaffolding proteins that may act to inhibit or promote a particular mode of channel activation. Any one of these interactions may be important in modulating MS of TRPCs by focusing mechanical force on the channel or constraining the channel and/or bilayer from responding to mechanical stretch. It may be that the right combination of TRPC proteins and accessory proteins are needed to produce channels that are not constitutively active but are responsive to factors associated with store depletion and/or mechanical stimulation.

7.5.5 TRPC Single Channel Conductance

Single channel conductance provides the best functional fingerprint of a specific channel, and is superior to identification by whole cell current properties that depend upon multiple factors including single TRPC channel conductance, gating and membrane insertion as well as functional coupling with other channel classes (i.e., voltage- and Ca^{2+}-activated channels). For example, whole cell currents generated by expression and co-expression of TRPC1/4/5 and/or TRPC3/6/7 subgroup members show I–V relations with dramatically different rectifications (Lintschinger et al. 2000; Strübing et al. 2001). However, these differences may reflect voltage-dependent changes in any one or a combination of the above parameters. Unfortunately, compared with studies of whole cell TRPC generated currents, there have been relatively few studies of the single channel activity that is enhanced by TRPC overexpression or reduced by TRPC suppression. Furthermore, no study to date has distinguished unequivocally between channel currents arising from TRPC homomers or heteromers. To make this distinction one needs to transfect with mutant subunits that produce predictable and measurable changes in channel conductance (or channel block) depending on the subunit stoichiometry within the channel complex (see Hille 2001). Another practical issue for the comparison of different TRPC channel conductance values has been the lack of standardized recording conditions (i.e., pipette solutions with the same composition, and measured over the same voltage range). Nevertheless, a survey of the TRPC single channel values indicates roughly the following order: TRPC3 (65 pS) > TRPC5 (50 pS) > ~ TRPC4 ~ TRPC6 (~30 pS) ≥ TRPC1 (3–20 pS) for estimates made from cell-attached recordings with 100–150 mM Na^+/Cs^+, 1–4 mM Ca^{2+}/Mg^{2+} between −40 and −100 mV (Hofmann et al. 1999; Hurst et al. 1998; Kiselyov et al. 1998; Yamada et al. 2000; Vaca and Sampieri 2002; Liu et al. 2003; Bugaj et al. 2005; Maroto et al. 2005; Inoue et al. 2006; Saleh et al. 2006). The only available estimates for TRPC2 (42 pS) and TRPC7 (60 pS) were made with no divalents (Zufall et al. 2005; Perraud et al. 2001). One basis for the low conductance of TRPC1 compared with TRPC3, 4, 5, 6 and 7 is that TRPC1 lacks the negatively charged aspartate or glutamate residues at analogous positions to D633 in TRPC5 and the other TRPCs, which is situated nine residues from the end of the TM6 domain (Obukhov and Nowycky 2005). Removal of external Ca^{2+} (or Mg^{2+}) has been reported to increase TRPC1 (but not TRPC6) channel conductance and, according to some reports, cause a positive shift in TRPC1 current reversal potential (e.g., Vaca and Sampieri 2002; Maroto et al. 2005; Spassova et al. 2006). The heterogeneity in TRPC1-associated conductance measurements (i.e., 3–20 pS) may also indicate that its conductance is altered when it combines with other subunits. For example, the homomeric TRPC5 channel has a conductance of ~50 pS but the TRPC1/TRPC5 heteromer is reduced to ~10 pS (Strübing et al. 2001). In this case TRPC1 may cause structural distortion of the putative D633 ring formed by the TRPC5 monomeric assembly. The intracellular Mg^{2+} block of TRPC5 at physiological potentials that is relieved at positive potentials also appears to be mediated by D633 (Obukhov and Nowycky 2005). TRPC4 and TRPC6 may have similar voltage-dependent activities because both channels possess aspartate at positions equivalent to D633, and anionic rings at this location may space the properties of TRPC4 and TRPC6. It may also turn out that different TRPCs display multiconductance states some of which are favored by specific conditions. In any case, the conductance values listed above can serve as a baseline for future measurements of the purified/reconstituted TRPCs.

7.5.6   TRPC Pharmacology

Pharmacological tools available to study TRPCs are limited, with different agents reported to block, stimulate or have no effect on different TRPCs (Xu et al. 2005; Ramsey et al. 2006). For example, SKF-96365 blocks TRPC3- and TRPC6-mediated whole cell currents (at ~5 μM), and is considered a more selective ROC- than SOC-blocker. In contrast, 2-aminoethoxydiphenyl borate (2-APB) blocks TRPC1 (80 μM), TRPC5 (20 μM) and TRPC6 (10 μM) but not TRPC3 (75 μM), and is considered a more selective SOC- than ROC-blocker. In the case of Gd^{3+} (and La^{3+}), TRPC1 and TRPC6 are blocked but TRPC4 and TRPC5 are potentiated at 1–10 μM (Jung et al. 2003), while flufenamate blocks TRPC3, TRPC5 and TRPC7 (100 μM) but potentiates TRPC6. Amiloride, which is known to block different MscCa, has yet to be tested on TRPC channels (Lane et al. 1991, 1992; Rüsch et al. 1994). The newest anti-MscCa agent, the tarantula venom peptide GsmTX-4 (Suchyna et al. 1998, 2004; Gottlieb et al. 2004; Jacques-Fricke et al. 2006) has more recently been shown to block TRPC channels in mammalian cells but does not abolish MscCa activity in *Xenopus* oocytes at 5 μM concentration (Hamill 2006; Spassova et al. 2006). At this stage it would be highly useful to carry out a systematic screen of the various agents reported to target MscCa and/or TRPC, including gentamicin, GsmTX-4, amiloride, 2-APB, amiloride, and SFK-96365 on ROCs as well as SOCs (Flemming et al. 2003).

7.6   Evidence of Specific TRPC Mechanosensitivity

There are several lines of evidence indicating specific TRPCs are MS, with the main evidence pointing towards TRPC1, TRPC4 and TRPC6. TRPC1 is generally considered to form a SOC that can be directly activated by LPLs, whereas TRPC4 and TRPC6 appear to form ROCs activated by AA and DAG, respectively. Here we consider whether the same mechanisms underlying SOC and ROC activity and sensitivity to lipidic second messengers is also the basis for their mechanosensitivity.

7.6.1   TRPC1

TRPC1 was the first identified vertebrate TRP homolog (Wes et al. 1995; Zhu et al. 1995), and initial heterologous expression of human TRPC1 in CHO and sf9 cells showed enhanced SOC currents (Zitt et al. 1996). However, a subsequent study indicated that hTRPC1 expression in sf9 cells induced a constitutively active nonselective cation channel that was not sensitive to store depletion (Sinkins et al. 1998). This early discrepancy raises the possibility that store sensitivity (and perhaps stretch sensitivity) may depend upon a variety of conditions (e.g., expression levels, presence of endogenous TRPCs and state of phosphorylation). For example, TRPC1 has multiple serine/threonine phosphorylation sites in the putative pore-forming region and the N- and C-termini, and at least one report indicates that PKC_α-dependent phosphorylation of TRPC1 can enhance Ca^{2+} entry induced by store depletion (Ahmmed et al 2004). Despite the early discrepant reports concerning TRPC1 and SOC function, many studies now point to TRPC1 forming a SOC (Liu et al. 2000, 2003; Xu and Beech 2001; Kunichika et al. 2004; for reviews see Beech 2005; Beech et al. 2003), and in cases where TRPC1 overexpression has not resulted in enhanced SOC (Sinkins et al. 1998; Lintschinger et al. 2000; Strübing et al. 2001) it has been argued that TRPC1 was not trafficked to the membrane (Hofmann et al. 2002). This does not seem to be the case for hTRPC1 when expressed in the oocyte (Brereton et al. 2000). In any case, direct involvement of TRPC1 in forming the highly Ca^{2+}-selective I_{CRAC} seems to be reduced by the recent finding that a novel protein family (i.e., CRAM1 or Orai1) forms I_{CRAC} channels (Peinelt et al. 2006; but see Mori et al. 2002; Huang et al. 2006).

7.6.1.1   A TRPC1 Homologue Expressed in *Xenopus* Oocytes

In 1999, xTRPC1 was cloned from *Xenopus* oocytes and shown to be ~90% identical in sequence to hTRPC1 (Bobanović et al. 1999). An anti-TRPC1 antibody (T1E3) targeted to an extracellular loop of the predicted protein was generated and shown to recognize an 80 kDa protein. Immunofluorescent staining indicated an irregular "punctuate" expression pattern of xTRPC1 that was uniformly evident over the animal and vegetal hemispheres. A subsequent patch clamp study also indicated that MscCa was uniformly expressed over both hemispheres (Zhang and Hamill 2000a). This uniform surface expression is in contrast to the polarized expression of the ER and phosphatidylinositol second messenger systems that are more abundant in the animal hemisphere (Callamaras et al. 1998; Jaconi et al. 1999). These results indicate that neither TRPC1 nor MscCa are tightly coupled to ER internal Ca^{2+} stores and IP_3 signaling. Originally, it was speculated that the punctuate expression of TRPC1 might reflect discrete channel clusters, but it might also indicate that these channels are localized to the microvilli that make up > 50% of the membrane surface area (Zhang et al. 2000). In another study testing the idea that xTRPC1 forms a SOC, Brereton et al. (2000) found that antisense oligonucleotides targeting different regions of the xTRP1 sequence did not inhibit IP_3-, or thapsigargin-stimulated Ca^{2+} inflow (but see Tomita et al. 1998). Furthermore, overexpression of hTRPC1 did not enhance basal or IP_3-stimulated Ca^{2+} inflow (Brereton et al. 2000). On the other hand, they did see enhancement of a LPA-stimulated Ca^{2+} influx. Interestingly, LPA also enhances a mechanically induced Ca^{2+} influx in a variety of cell types (Ohata et al. 2001). Based on this apparent lack of TRPC1-linked SOC activity, Brereton et al. (2000) proposed that TRPC1 might form the endogenous cation channel activated by the marine toxin, maitotoxin (MTX). However, in another study directly comparing the properties of the endogenous MTX-activated conductance measured in normal liver cells and the enhanced MTX-activated conductance measured in hTRPC1-transfected liver cells, Brereton et al. (2001) found that the endogenous conductance showed a higher selectivity for Na^+ over Ca^{2+}, and a higher sensitivity to Gd^{3+} block ($K_{50\% block}$ = 1 μM vs 3 μM) compared with the enhanced conductance. These differences may indicate that other endogenous TRPC subunits combine with TRPC1 to form the endogenous MTX-activated conductance, whereas the enhanced MTX-activated conductance is formed exclusively by hTRPC1 homotetramers (Brereton et al. 2001). Finally, unlike in hTRPC1-transfected oocytes, hTRPC1-transfected rat liver cells did show an increased thapsigarin-induced Ca^{2+} inflow (Brereton et al. 2000, 2001).

7.6.1.2 MTX and TRPCs

Evidence from several studies indicates that oocyte MTX-activated conductance may be mediated by MscCa (Bielfeld-Ackermann et al. 1998; Weber et al. 2000; Diakov et al. 2001). In particular, both display the same cation selectivity, both are blocked by amiloride and Gd^{3+}, both are insensitive to flufenamic and niflumic acid, and both have a single channel conductance of ~25 pS (i.e., when measured in symmetrical 140 mM K^+ and 2 mM external Ca^{2+}). Because MTX is a highly amphipathic molecule (Escobar et al. 1998), it may activate MscCa by changing bilayer mechanics, as has been proposed for other amphipathic agents that activate or modulate MS channel activity (Martinac et al. 1990; Kim 1992; Hamill and McBride 1996; Casado and Ascher 1998, Perozo et al. 2002).

7.6.1.3 TRPC1 and Volume Regulation

To directly test whether TRPC1 might be MS, Chen and Barritt (2003) selectively suppressed TRPC1 expression in rat liver cells and measured the cellular response to osmotic cell swelling. Liver cells are known to express MscCa (Bear 1990), and previous studies had shown that osmotic swelling of epithelial cells activates an MscCa-dependent Ca^{2+} influx that stimulates Ca^{2+}-activated K^+ efflux accompanied by Cl^-/H_2O efflux and regulatory volume decrease (RVD; Christensen 1987). However, contrary to expectations, hypotonic stress actually caused a greater swelling and faster RVD in the TRPC1 suppressed liver cells than in the control liver cells (Chen and Barritt 2003). This may occur because TRPC1 suppression results in a compensatory overexpression of other transport mechanisms that enhance both cell swelling and RVD. It should also be recognized that cell swelling does not always activate MscCa. For example, although hypotonic solution activates a robust Ca^{2+}-independent Cl^- conductance in Xenopus oocytes that should contribute to RVD, it fails to activate the endogenous MscCa (Ackerman et al. 1994; Zhang and Hamill 2000a).

7.6.1.4 TRPC1 in Muscular Dystrophy

Both TRPC1 and MscCa are expressed in skeletal muscle and both have been implicated in the muscular degeneration that occurs in Duchenne muscular dystrophy (DMD). In particular, muscle fibers from the *mdx* mouse (i.e., an animal model of DMD) show an increased vulnerability to stretch-induced membrane wounding (Yeung and Allen 2004; Allen et al. 2005) that has been linked to elevated $[Ca^{2+}]_i$ levels caused by increased Ca^{2+} leak channel activity (Fong et al. 1990) and/or abnormal MscCa activity (Franco and Lansman 1990). Based on the observation that the channel activity was increased by thapsigargin-induced store depletion, it was proposed that the channel may also be a SOC belonging to the TRPC family (Vandebrouck et al. 2002, see also Hopf et al. 1996). To test this idea, *mdx* and normal muscle were transfected with anti-sense oligonucleotides designed against the most conserved TRPC regions. The transfected muscles showed a significant reduction in expression of TRPC-1 and -4 but not -6 (all three TRPCs are expressed in normal and *mdx* muscle) and a decrease in Ca^{2+} leak channel activity. Previous studies indicate that MscCa behaves more like a Ca^{2+} leak channel in *mdx* mouse muscle patches (Franco-Obregon and Lansman 2002) and in some *Xenopus* oocyte patches (Reifarth et al. 1999). In a more recent study it has been confirmed that SOC and MscCa in *mdx* mouse muscle display the same single channel conductance and sensitivity to block by Gd^{3+}, SKF96365, 2APB and GsMTx-4 (Ducret et al. 2006). The presence of a dystrophin domain on the C-terminus of TRPC1 (Wes et al. 1995) could explain the shift in MscCa gating mode in *mdx* muscle that lacks dystrophin (Franco-Obregon and Lansman 2002, but see Suchyna and Sachs 2007). However, the findings that TRPC6 and TRPV2 form stretch-sensitive cation channels and are expressed in normal and *mdx* mouse skeletal muscle raises the possibility that several TRPs may contribute to MscCa activity in normal and DMD muscle (Kanzaki et al. 1999; Vandebrouck et al. 2002; Iwata et al. 2003; Muraki et al. 2003; Spassova et al. 2006).

7.6.1.5 TRPC1 Interaction with Polycystins

Further clues pointing to a MS role for TRPC1 relates to the demonstration that TRPC1 interacts with the putative MS channel TRPP2 when they are co-expressed in HEK-293 (Tsiokas et al. 1999; Delmas 2004). TRPP2 is a member of the TRPP family (polycystin) and has been shown to form a Ca^{2+}-permeable cation channel that is mutated in autosomal dominate polycystic kidney disease (ADPKD) (Nauli et al. 2003; Nauli and Zhou 2004; Giamarchi et al. 2006; Cantiello et al. 2007). TRPP2 was originally designated polycystin kidney disease 2 (PKD2) and shown to combine with PKD1, a membrane protein with a large extracellular N-terminal domain that seemed well suited for acting as an extracellular sensing antenna for mechanical forces. Both TRPP2 and PKD1 are localized in the primary cilium of renal epithelial cells that is considered essential for detecting laminar fluid flow (Praetorius and Spring 2005). However, the osmosensitive TRPV4 is also expressed in renal epithelial cells and may also associate with TRPP2 (Giamarchi et al. 2006). It remains to be determined whether TRPC1 combines with TRPP2 in renal epithelial cells and whether knock-out of TRPC1 and/or TRPV4 blocks fluid flow detection.

7.6.1.6 TRPC1 in Mechanosensory Nerve Endings

If TRPC1 is a mechanosensory channel, it might be expected to be found in specialized mechanosensory nerve endings. To address this issue, Glazebrook et al. (2005) used immunocytochemical techniques to examine the distribution of TRPC1 and TRPC3–7 in the soma, axons and sensory terminals of arterial mechanoreceptors, and found that TRPC1, 3, 4 and 5 (but not TRPC6 and TRPC7) were expressed in the peripheral axons and the mechanosensory terminals. However, only TRPC1 and TRPC3 extended into the low threshold mechanosensory complex endings, with TRPC4 and TRPC5 limited mainly to the major branches of the nerve. Although these results are consistent with TRPC1 (and possibly TRPC3) involvement in baroreception, it was concluded that, because it was not present in all fine terminals, TRPC1 was more likely involved in modulation rather than direct MT. However, it is not clear that all fine endings are capable of transduction. Furthermore, other putative MS proteins (i.e., β and γ ENaC subunits) are expressed in baroreceptor nerve terminals (Drummond et al. 1998), in which case different classes of MS channels (i.e., ENaC and TRPC) may mediate MT in different mechanosensory nerves.

7.6.1.7 TRPC1 Involvement in Wound Closure and Cell Migration

For a cell to migrate there must be coordination between the mechanical forces that propel the cell forward and the mechanisms that promote retraction of the cell rear. The first study to implicate TRPC1 in cell migration was by Moore et al. (1998). They proposed that shape changes induced in endothelial cells by activation of TRPC1 were a necessary step for angiogenesis and cell migration. In another study, it was demonstrated that TRPC1 overexpression promoted, while TRPC1 suppression inhibited, intestinal cell migration as measured by wound closure assay (Rao et al. 2006). Based on the proposal that MscCa regulates fish keratocyte cell migration (Lee et al. 1999), and identification of TRPC1 as an MscCa subunit (Maroto et al. 2005), the role of TRPC1 in migration of the highly invasive/metastatic prostate tumor cell line PC-3 has been tested. TRPC1 activity was shown to be essential for PC-3 cell migration and, in particular, Gd^{3+}, GsMTx-4, anti-TRPC1 antibody and siRNA targeting of TRPC1 were shown to block PC-3 migration by inhibiting the Ca^{2+} dynamics that coordinated cell migration (R. Maroto et al., manuscript submitted). However, again TRPC1 may not be the only TRP channel involved in this function since TRPC6 and TRPM7 have recently been reported to be stretch-activated channels (Spassova et al. 2006; Numata et al. 2007). Irrespective of the exact molecular identity of MscCa, it seems that this channel may be a more promising target for blocking tumor cell invasion and metastasis than integrins and metalloproteinases. This is because when a tumor cell switches from mesenchymal to amoeboid migration mode it appears to remain dependent upon Ca^{2+} influx via MscCa, whereas it becomes relatively independent of integrin and metalloproteinase activity (for review, see Maroto and Hamill 2007).

7.6.1.8   Reconstitution of xTRPC1 in Liposomes

Perhaps the most direct evidence for an MS role for TRPC1 comes from studies in which the proteins forming the oocyte MscCa were detergent-solubilized, fractionated by FPLC, reconstituted in liposomes and assayed for MscCa activity using patch recording (Maroto et al. (2005). A specific protein fraction that ran with a conductivity of $16\,mS\ cm^{-1}$ was shown to reconstitute the highest MscCa activity, and silver-stained gels indicated a highly abundant 80 kDa protein. Based on previous studies that identified xTRPC1 and hTRPC1 as forming an ~80 kDa protein when expressed in oocytes (Bobanović et al. 1999; Brereton et al. 2000), immunological methods were used to demonstrate that TRPC1 was present in the MscCa active fraction. Furthermore, heterologous expression of hTRPC was shown to increase the MscCa activity expressed in the transfected oocyte, whereas TRPC1-antisense reduced endogenous MscCa activity (Maroto et al. 2005). Despite the almost tenfold increase in current density in the TRPC1-injected oocyte, the channel activation and deactivation kinetics in the two patches were similar, at least in some patches. On the other hand, in some cases the kinetics of the TRPC1-dependent channels show delayed activation and deactivation kinetics (Hamill and Maroto 2007). The basis for this heterogeneity in kinetics of TRPC1 channels remains unclear but may reflect local differences in the underlying CSK and/or bilayer or even the MscCa subunit composition that occurs with TRPC1 overexpression. Maroto et al. (2005) also demonstrated that hTRPC1 expression in CHO cells results in increased MscCa activity, consistent with a ~fivefold greater increase in channel density. Furthermore, the presence of endogenous MscCa activity is consistent with previous reports that indicate CHO cells express TRPC1 along with TRPC2, 3, 4, 5 and 6 (Vaca and Sampieri 2002).

7.6.2   TRPC2

So far there have been no studies addressing the possibility that TRPC2 is an MS channel. However, evidence does indicate that TRPC2 may function either as a ROC or a SOC depending upon cell type (Vannier et al. 1999; Gailly and Colson-Van Schoor 2001; Chu et al. 2004; Zufall et al. 2005). For example, because TRPC2$^{-/-}$ mice fail to display gender discrimination, the channel has been implicated in pheromone detection in the rodent vomeronasal organ (VNO) (Liman et al. 1999; Zufall et al. 2005). Furthermore, because a DAG-activated channel in VNO neurons is down-regulated in TRPC2$^{-/-}$ mice and TRPC2 is localized in sensory microvilli that lack Ca^{2+} stores, it would seem that TRPC2 functions as a ROC rather than a SOC, at least in VNO neurons (Spehr et al 2002; Zufall et al. 2005). On the other hand, in erythroblasts, and possibly sperm, TRPC2 has been reported to be activated by store depletion. In both cell types, long splice variants of TRPC2 were detected (Yildrin et al. 2003), whereas VNO neurons express a short splice variant (Chu et al. 2002; Hofmann et al. 2000). In hemaetopoiesis, erthyropoietin is proposed to modulate Ca^{2+} influx via TRPC2 in possible combination with TRPC6 (Chu et al. 2002, 2004). In sperm, TRPC2 may participate in the acrosome reaction based on its inhibition by a TRPC2 antibody (Jungnickel et al. 2001). However, the fact that TRPC2$^{-/-}$ mice display normal fertility raises serious doubts regarding this role (Stamboulian et al. 2005).

7.6.3   TRPC3

TRPC3 co-localizes with TRPC1 in specialized mechanosensory nerve endings, indicating that these two TRPCs may combine to form an MS channel (see Sect. 7.6.1.6). Because TRPC3 is activated by the DAG analog 1-oleoyl-2-acetylglycerol (OAG) in a direct manner like TRPC6 (Hofmann et al. 1999), it would seem likely that it may also be sensitive to direct membrane stretch like TRPC6 (Spassova et al. 2006). However, TRPC3, unlike TRPC6, can also contribute to forming SOCs (Zitt et al. 1997; Hofmann et al. 1999; Kamouchi et al. 1999; Trebak et al. 2002; Vasquez et al. 2001; Liu et al. 2005; Groschner and Rosker 2005; Zagranichnaya et al. 2005; Kawasaki et al. 2006), and whether TRPC3 forms a SOC or a ROC has been shown to depend on levels of TRPC3 expression, indicating that subunit stoichiometry may determine activation mode (Vasquez et al. 2003; Putney et al. 2004). Finally, suppression of TRPC3 in cerebral arterial smooth muscle, while suppressing pyridine receptor-induced depolarization, does not appear to alter pressure increased depolarization and contraction, which therefore might be dependent on TRPC6 alone (Reading et al. 2005).

7.6.4   TRPC4

There is disagreement on whether TRPC4 functions as a SOC and/or ROC (Philipp et al. 1998; Tomita et al. 1998; McKay et al. 2000; Plant and Shaefer 2005). However, at least two studies by the Villreal group indicate that TRPC4 forms a ROC activated by AA rather than by DAG as in the case of TRPC3/6/7 and TRPC2 (Wu et al. 2002; Zagranichnaya et al. 2005). In particular, using siRNA and antisense strategies to reduce endogenous TRPC4 expression, TRPC4 was shown to be required for AA-induced Ca^{2+} oscillations but not for SOC function. This AA activation may have implications for the mechanosensitivity of TRPC4 since AA has been show to activate/modulate a variety of MS channels by directly altering the mechanical properties of the bilayer surrounding the channel (Kim 1992; Hamill and McBride 1996; Casado and Ascher 1998; Patel and Honoré 2001). Since AA is produced by PLA_2, which is itself MS (Lehtonen and Kinnunen 1995), TRPC4 may derive its mechanosensitivity from this enzyme in addition to possibly being directly sensitive to bilayer stretch. Studies of TRPC4$^{-/-}$ mice indicate that TRPC4 is an essential determinant of endothelial vascular tone and endothelial permeability as well neurotransmitter release from central neurons (reviewed by Freichel et al. 2004).

7.6.5 TRPC5

The human TRPC5 encodes a protein that is very similar to TRPC4 in its first ~700 amino acids but shows more variability in final C-terminal ~200 amino acids (Sossey-Alaoui et al. 1999; Zeng et al. 2004). Both TRPC5 and TRPC4 differ from other TRPCs in terms of possessing a C-terminal VTTRL motif that binds to PDZ domains of the scaffolding proteins EBP50 (NHERF1). However, co-expression and deletion experiments have shown that the VTTRL motif is not necessary for TRPC5 activation although it may mediate the EBP50 modulatory effects on TRPC5 activation kinetics (Obukhov and Nowycky 2004). TRPC5 (and 4) differ from the other TRPCs in that La^{3+} and Gd^{3+} cause potentiation at micromolar concentrations and block only at higher concentrations (Schaefer et al. 2000; Strübing et al. 2001; Jung et al. 2003). On this basis alone, TRPC5 and TRPC4 homotetramers would seem to be excluded from forming MscCa because Gd^{3+} has usually been reported to block MscCa at $1–10\,\mu M$ (Yang and Sachs 1989; Hamill and McBride 1996). 2-APB blocks TRPC5 as well as the activating effect of Gd^{3+} possibly by directly occluding the Gd^{3+} activation site (Xu et al. 2005). TRPC5 (and TRPC4) also differ from TRPC3/6/7 in that they are not activated directly by DAG (Hofmann et al. 1999; Schaefer et al. 2000; Venkatachalam et al. 2003). However, TRPC5 is activated by LPLs including LPC when applied to excised membrane patches, but not by the fatty acid AA (Flemming et al. 2006; Beech 2006). This latter result would seem to contradict the idea that TRPC4 forms the AA-activated ROC, ARC, unless the two closely related TRPCs differ significantly in their AA sensitivity (Zagranichnaya et al. 2005).

The most intriguing functional evidence implicating TRPC5 as a putative MscCa comes from the demonstration that TRPC5, like MscCa, functions as negative regulator of neurite outgrowth (Calabrese et al. 1999; Greka et al. 2003; Hui et al. 2006; Jacques-Fricke et al. 2006; Pellegrino and Pelligrini 2007). In particular, MscCa blockers, including gentamicin, Gd^{3+} and GsmTX-4, potentiate neurite outgrowth (Calabrese et al. 1999; Jacques-Fricke et al. 2006) as does expression of a TRPC5 dominant-negative pore mutant. In contrast, overexpression of TRPC5 suppresses neurite outgrowth (Greka et al. 2003; Hui et al. 2006). Although it is tempting to suggest that TRPC5 may form MscCa in neurites, the stretch sensitivity of TRPC5 and its sensitivity to block by GsmTX-4 needs to be directly tested. Furthermore, because neurite outgrowth is potentiated by ruthenium red (a TRPV4 blocker) and suppressed by the specific TRPV4 agonist 4α-phorbol 12, 12-didecanoate, it has been suggested that TRPV4 forms the MscCa (Jacques-Fricke et al. 2006). Furthermore, in contrast to its proposed role in suppressing cell motility, TRPC5, possibly in combination with TRPC1, has also been implicated in mediating sphingosine 1-phosphate-stimulated smooth muscle cell migration (Xu et al. 2006).

7.6.6 TRPC6

The general consensus is that TRPC6 forms a ROC that is directly activated by DAG, and is insensitive to activation by IP_3 and Ca^{2+} store depletion (Boulay et al. 1997; Hofmann 1999; Estacion et al. 2004; Zagranichnaya et al. 2005; Zhang et al. 2006). Although TRPC6 is a member of the TRPC3/6/7 subfamily it shows distinct functional and structural properties. Functionally, while TRPC6 forms only a ROC, TRPC3 and TRPC7 appear capable of participating in forming both ROCs and SOCs (Zagranichnaya et al. 2005). Structurally, whereas TRPC6 carries two extracellular glycosylation sites, TRPC3 carries only one (Dietrich et al. 2003). Furthermore, exogenously expressed TRPC6 shows low basal activity compared with TRPC3, and elimination of the extra glycosylation site that is missing in TRPC3, transforms TRPC6 into a constitutively active TRPC-3 like channel. Conversely, engineering of an additional glycosylation site in TRPC3 markedly reduces TRPC3 basal activity. It will be interesting to determine how these manipulations alter the apparent MS functions of TRPC6 described below.

7.6.6.1 TRPC6 Role in Myogenic Tone

TRPC6 is proposed to mediate the depolarization and constriction of small arteries and arterioles in response to adrenergic stimulation (Inoue et al. 2001, 2006; Jung et al. 2002) and elevation of intravascular pressure consistent with TRPC6 forming a MOC as well as a ROC (Welsh et al. 2000, 2002). The cationic current activated by pressure in vascular smooth muscle is suppressed by antisense-DNA to TRPC6 (Welsh et al. 2000). Furthermore, because cation entry was stimulated by OAG and inhibited by PLC inhibitor (Park et al. 2003), it was proposed that TRPC6 forms a MS channel that is activated indirectly by pressure according to the pathway:

$\uparrow$Intravascular pressure $\rightarrow$ $\uparrow$PLC $\rightarrow$$\uparrow$[DAG] $\rightarrow$$\uparrow$TRPC $\rightarrow$$\uparrow$[$Ca^{2+}$]$\rightarrow$$\uparrow$ myogenic tone.

In this scheme it is PLC rather than TRPC6 that is MS and, since all TRPCs are coupled to PLC-dependent receptors, this would imply that all TRPC could display some degree of mechanosensitivity. However, while there are reports that PLC can be mechanically stimulated independent of external Ca^{2+} (Rosales et al. 1997; Mitchell et al. 1997; Moore et al. 2002), there are more cases that indicate that the mechanosensitivity of PLC derives from stimulation by Ca^{2+} influx via MscCa (Matsumoto et al. 1995; Ryan et al. 2000; Ruwhof et al. 2001). In this case, it becomes important to demonstrate that TRPC6 can be mechanically activated in the absence of external Ca^{2+} (e.g., using Ba^{2+}). There is other evidence to indicate that TRPC6 may be coupled to other MS enzymes. For example, TRPC6 is similar to TRPV4 in that it is activated by 20-hydroxyeicosatetraenoic acid (20-HETE), which is the dominate AA metabolite produced by cytochrome P-450 w-hydroxylase enzymes (Basora et al. 2003). TRPC6 may also be activated by Src family protein tyrosine kinase-mediated tyrosine phosphorylation (Welsh et al. 2002). Indeed, PP2 a specific inhibitor of Src PTKs, abolishes TRPC6 (and TRPC3) activation and strongly inhibits OAG-induced Ca^{2+} entry (Soboloff et al. 2005). OAG may operate solely through TRPC6 homomers, whereas vasopressin may act on OAG-insensitive TRPC heteromers (e.g., formed by TRPC1 and TRPC6). At least consistent with this last possibility is evidence of co-immunoprecipitation between TRPC1 and TRPC6 (Soboloff et al. 2005). A further complication is that DAG-dependent activation of PKC appears to stimulate the myogenic channels based on their block by the PKC inhibitor chelerythrine (Slish et al. 2002), whereas PKC activation seems to inhibit TRPC6 channels, which would seem more consistent with direct activation by DAG/OAG (Soboloff et al. 2005).

Despite the above evidence implicating TRPC6 as the "myogenic" channel, TRPC6-deficient mice show enhanced rather that reduced myotonic tone and increased rather than reduced responsiveness to constrictor agonist in small arteries. These effects result in both a higher elevated mean arterial blood pressure and a shift in the onset of the myogenic tone towards lower intravascular pressures, again opposite to what would be expected if TRPC6 were critical for myoconstriction (Dietrich et al. 2005). Furthermore, isolated smooth muscle from TRPC6$^{-/-}$ mice shows increased basal cation entry and more depolarized resting potentials, but both effects are blocked if the muscles are also transfected with siRNA targeting TRPC3. Based on this latter observation, it was suggested that constitutively active TRPC3 channels are upregulated in TRPC6$^{-/-}$ mice. However, the TRPC3 subunits are unable to functionally replace the lost TRPC6 function that involves suppression of high basal TRPC3 activity (i.e., the TRPC3/6 heteromer is a more tightly regulated ROC and/or MOC). In summary, although evidence indicates that TRPC6 may be a pressure- or stretch-sensitive channel and contribute to MOC, the TRPC6 knockout mouse indicates a phenotype that cannot be explained if TRPC6 alone forms the vasoconstrictor channel. It may also be relevant that another study could find no evidence that Gd^{3+}-sensitive MscCa contributes to myogenic tone in isolated arterioles from rat skeletal muscle (Bakker et al. 1999).

In the most direct study concerning TRPC6 mechanosensitivity, a stretch-activated channel current with a conductance of 25 pS (measured at +60 mV) was activated in cell-attached patches formed on HEK293 cells transfected with hTRPC6 with a significant delay (~5 s) in turn on and turn off following a brief (2 s) pressure pulse (Spassova et al (2006). Although these long delays could indicate an indirect mechanism of stretch activation, possibly involving MS PLC (see Sect. 7.2.3), it was found that treatment of cells with cytochalasin D reduced the delays and increased stretch sensitivity, which is more consistent with the actin CSK acting as a mechanical constraint that acts to delay the transmission of tension to the bilayer. It was also found that either hypoosmotic cell swelling or application of OAG to TRPC6-transfected cells activated whole cell cation conductance that was not blocked by the PLC inhibitor U73122, apparently ruling out an indirect mechanism involving MS PLC as was previously implied (Park et al. 2003).

7.6.6.2   TRPC6 Role in Kidney Disease

Autosomal dominant focal segmental glomerulosclerosis (FSGS) is a kidney disease that leads to progressive renal kidney failure characterized by leakage of plasma proteins like albumin into the urine (proteinuria). Recently, mutations in TRPC6 were associated with familial FSGS and implicated in aberrant calcium signaling that leads to podocyte injury (Winn et al. 2005; Reiser et al. 2005). Furthermore, two of the mutants were demonstrated to be gain-of-function mutations that produce larger ROCs than the ROC currents measured in wild type TRPC6-expressing HEK-293 cells. Ultra-filtration of plasma by the renal glomeruli is mediated mainly by the podocyte, which is an epithelial cell that lies external to the glomerular basement membrane (GBM) and lines the outer endothelium of the capillary tuft located inside the Bowman's capsule. The podocyte covers the GBM and forms interdigitating foot processes that are connected by slit diaphragms, which are ultra-thin membrane structures that form a zipper-like structure at the center of the slit with pores smaller than albumin (Tryggvason and Wartovaara 2004; Kriz 2005). The podocyte-specific proteins nephrin and podocin are localized in the slit diaphragm, and the extracellular domains of nephrin molecules of neighboring foot processes interact to form the zipper structure. Podocin, a member of the stomatin family, is a scaffolding protein that accumulates in lipid rafts and interacts with the cytoplasmic domain of nephrin (Durvasula and Shankland 2006). Both nephrin and podocin have been shown to be mutated in different familial forms of FSGD. Furthermore, TRPC6 interacts with both nephrin and podocin, and a nephrin-deficiency in mice leads to overexpression and mislocalization of TRPC6 in podocytes as well as disruption of the slit diaphragm (Reiser et al. 2005). Mechanical forces play an important role in ultra-filtration, both in terms of the high transmural distending forces arising from the capillary perfusion pressure, as well as the intrinsic forces generated by the contractile actin network in the foot process that control, in a Ca-dependent manner, the width of the filtration slits. As a consequence, TRPC6 may act as the central signaling component mediating pressure-induced constriction of the slit.

In summary, two quite diverse physiological functions, myogenic tone and renal ultrafiltration, implicate TRPC6 as an MS channel, and recent evidence indicates that TRPC6 may be directly activated by stretch applied to the patch.

7.6.7 TRPC7

Since TRPC7 belongs to the same subfamily as TRPC6, and also forms a ROC activated by DAG/OAG, it might be expected to display the same direct stretch sensitivity to Ca^{2+} block as reported for TRPC6. Immunoprecipitation and electro-physiological experiments indicate that TRPC6 and TRPC7 can co-assemble to form channel complexes in A7r5 vascular smooth muscle cells (Maruyama et al. 2006). However, the same study also demonstrated that the co-assembly of TRPC7 (or TRPC73) with TRPC6 can change specific channel properties compared with the homomeric TRPC6 channel. For example, whereas increasing external Ca^{2+} from 0.05 to 1 mM suppresses currents in HEK cells transfected with TRPC7 (or TRPC3) alone, or with TRPC6/7 (or TRPC3/6) in combination, it fails to suppress currents in TRPC6-transfected cells. Therefore, apart from the constitutive open-ing seen with TRPC3 but not TRPC6 (see Sect. 7.5.3), TRPC3/6/7 subfamily members differ in their sensitivity to Ca^{2+} block. Other studies indicate even more profound differences between TRPC7 and TRPC6 functions. For example, based on overexpression in HEK cells, it was concluded that mouse TRPC7 forms a ROC, whereas human TRPC7 forms a SOC (Okada et al. 1999; Riccio et al. 2002a). In this case, the initial explanation was that a proline at position 111 in mTRPC7 was replaced by leucine in the hTRPC7. However, hTRPC7 suppres-sion/knockout experiments indicate that TRPC7 is required for both the endog-enous SOC and ROC in HEK293 cells (Lièvremont et al. 2004; Zagranichnaya et al. 2005). Furthermore, when hTRPC7 (with leucine at position 111) was tran-siently expressed in HEK293 cells it enhanced ROC, but when it was stably expressed it enhanced both ROC and SOC (Lièvremont et al. 2004). In this case, the explanation was that stable transfection allowed for a time-dependent up-regulation of other ancillary components that were required to couple TRPC7 to store depletion (Lièvremont et al. 2004). On the other hand, although hTRPC7 suppression in DT40 B-cells also reduced receptor/DAG-activated and store-oper-ated Ca^{2+} entry, the latter effect appeared to arise because of increased Ca^{2+} stores and the greater difficulty in depleting them to activate SOC (Lièvremont et al. 2005). Indeed, when Ca^{2+} stores were more effectively depleted (i.e., with a combination of IP_3 and calcium chelator) there was no difference in SOC activa-tion between wild type and TRPC7$^{-/-}$ cells (Lièvremont et al. 2005). Similar findings have been reported for TRPC7 suppression in human keratinocytes (Beck et al. 2006). A still further complication is that, in cells lacking the IP_3R, the OAG-activated current is absent but can be restored by transient IP_3R expres-sion or by overexpression of TRPC7 (Vazquez et al. 2006). This was taken to indicate that the endogenous TRPC7 needs to interact with endogenous proteins including regulatory IP_3R but when TRPC7 is overexpressed the other proteins are not required for OAG activation.

The above review of the TRPC literature indicates the importance of measuring directly the stretch sensitivity of different TRP channels under conditions in which the stoichiometry and molecular nature of the TRPCs forming the channel complex are well defined.

7.7   Conclusions

At least three basic mechanisms, referred to as "bilayer", "conformational coupling" and "enzymatic", may confer mechanosensitivity on TRPCs. The bilayer mechanism should operate if the TRPC channel, in shifting between closed and open states, undergoes a change in its membrane occupied area, thickness and/or cross-sectional shape. Any one of these changes would confer mechanosensitivity on the channel. A bilayer mechanism may also underlie the ability of lipidic second messengers (e.g., DAG/OAG, LPL, AA and 5'6'-EET) to directly activate TRPC channels by inserting in the bilayer to alter its local bilayer packing, curvature and/or the lateral pressure profile. The only unequivocal way to demonstrate that a bilayer mechanism operates is to show that stretch sensitivity is retained when the purified channel protein is reconstituted in liposomes. After this stage, one can go on to measure channel activity as a function of changing bilayer thickness (i.e., by using phospholipids with different acyl length chains) and local curvature/pressure profile (i.e., by using lysophospholipids with different shapes) (Perozo et al. 2002).

The second mechanism involves conformational coupling (CC), which has been evoked to account for TRPC sensitivity to depletion of internal Ca^{2+} stores. CC was originally used to explain excitation–contraction (E–C) coupling involving the physical coupling between L-type Ca^{2+} channels (i.e., dihydropyridine receptors, DHPR) in the plasma membrane and ryanodine receptors (RyR1) that release Ca^{2+} from the sarcoplasmic reticulum (SR) (Protasi 2002). Subsequently, a retrograde form of CC was discovered between the same two proteins that regulate the organization of the DHPR into tetrads and the magnitude of the Ca^{2+} current carried by DHPR (Wang et al. 2001; Paolini et al. 2004; Yin et al. 2005). Another form of CC was associated with physiological stimuli that do not deplete Ca^{2+} stores yet activate Ca^{2+} entry through channels referred to as excitation-coupled Ca^{2+} entry channels to distinguish them from SOC (Cherednichenko et al. 2004). Interestingly, RyR1 is functionally coupled to both TRPC1-dependent SOC and TRPC3-dependent SR Ca^{2+} release (Sampieri et al. 2005; Lee et al. 2006).

A key issue for all forms of CC is whether the direct physical link that conveys mechanical conformational energy from one protein to another can also act as a pathway to either focus applied mechanical forces on the channel or alternatively constrain the channel from responding to mechanical forces generated within the bilayer. Another possibility is that reorganization or clustering of the resident ER protein (i.e., STIM) that senses Ca^{2+} stores may alter channel mechanosensitivity by increasing the strength of CC (Kwan et al. 2003).

Some insights into these possibilities can be provided by the process of "membrane blebbing", which involves decoupling of the plasma membrane from the underlying CSK, and has been shown to either increase or decrease the mechanosensitivity of MS channels depending upon the channel (Hamill and McBride 1997; Hamill 2006). Since membrane blebbing would also be expected to disrupt any dynamic interactions between TRPC channels and scaffolding proteins it should alter TRPC function. In one case it has been reported that Ca^{2+} store depletion after, but not before, formation of a tight seal is effective in blocking the activation of SOC channels in frog oocyte patches (Yao et al. 1999). Presumably, this occurs because the sealing process physically decouples the channels from ER proteins that sense internal Ca^{2+} stores. Tight seal formation using strong suction can also reduce MscCa mechanosensitivity and gating kinetics, possibly by a related mechanism (Hamill and McBride 1992). On the other hand, it has been reported that I_{CRAC} is retained following cell "ballooning" (i.e., a form of reversible membrane blebbing) indicating that the coupling between the channel and the Ca^{2+} sensor STIM may be relatively resistant to decoupling (Bakowski et al. 2001). In any case, in order to directly demonstrate a role for CC in mechanosensitivity, one needs to show that stretch sensitivity can be altered in mutants in which TRPC–ancillary protein interactions are disrupted (see Sect. 7.5.4).

The third mechanism of mechanosensitivity relates to functional coupling between TRPCs and putative MS enzymes. Evidence indicates that the PLA_2 and Src kinase may be MS, and both enzymes have been implicated in conferring mechanosensitivity on TRPV4 (Xu et al. 2003; Vriens et al. 2004; Cohen 2005a, 2005b). PLA_2 and Src kinase have also been implicated in the activation of TRPC-mediated SOC and ROC activities (Hisatsune et al. 2004; Bolotina and Csutora 2005; Vazquez et al. 2004b). There is also evidence that indicates PLC may be MS (Brophy et al. 1993), with some reports indicating that the mechanosensitivity depends upon Ca^{2+} influx (Basavappa et al. 1988; Matsumoto et al. 1995; Ryan et al. 2000; Ruwhof et al. 2001; Alexander et al. 2004) and others indicating independence of external Ca^{2+} and Ca^{2+} influx (Mitchell et al. 1997; Rosales et al. 1997; Moore et al. 2002). In either case, the combined evidence indicates that mechanical forces transduced by MscCa and/or by MS enzymes may modulate the gating of all TRP channels. The physiological and/or pathological effects of this MS modulation remain to be determined. The methods discussed in this chapter, including the application of pressure steps to measure the kinetics of MS enzyme–channel coupling and the use of membrane protein liposome reconstitution for identifying specific protein–lipid interactions should play an increasing role in understanding the importance of the different MS mechanisms underlying TRPC function.

Dietrich, A., Kalwa, H., Storch, U., Mederos y Schnitzler, M., Slananova, B., Pinkenburg, O., Dubrovska, G., Essin, K., Gollasch, M., Birnbaumer, L. and Guderman, T. (*Pflügers Archives* published on-line July 2007) have generated a TRPC1-deficient mouse that appears viable and fertile. The mice also appear to develop normally except for a ~20% increase in average body weight compared with WT mice. Furthermore, the pressure-induced constriction of their cerebral arteries are not impaired, and smooth muscle cells isolated from TRPC1$^{-/-}$ cerebral arteries show similar currents activated by osmotic/hydrostatic pressure, and similar Ca^{2+} influx induced by Ca^{2+} store depletion compared with those seen in WT smooth muscle cells. Based on these observations, it was concluded that TRPC1 is not an obligatory component of either the stretch-activated or the store-operated channel in vascular smooth muscle. The results of this study also indicate that there must be redundant mechanisms that compensate for deletion of the TRPC1 subunit which is the most widely if not ubiquitously expressed subunit in mammalian cells. For example, TRPC6 has been implicated as the myogenic channel in vascular smooth muscle. However, the literature indicates there are several classes of MscCa that differ in their single channel conductance (20-80 pS) and rectification (outward vs inward) properties. In this case, MscCa may be formed by different TRP subunit combinations in different cell types. Finally, the apparent normal phenotype of the TRPC1$^{-/-}$ mouse is somewhat reminiscent of an earlier study in which pharmacological knockout of the MscCa expressed in *Xenopus* oocytes had no effect on oocyte growth, maturation, fertilization or early embryogenesis and development of the tadpole (Wilkinson, N.C., Gao, F., and Hamill, O.P. (*Journal of Membrane Biology*, 165, 161-174, 1998). In this case there must be redundant mechanisms that can compensate for the absence of this MscCa activity.

Acknowledgments We thank the United States Department of Defense Prostate Cancer Research Program and the National Cancer Institute for funding support.

References

Ackerman MJ, Wickman KD, Clapham DE (1994) Hypotonicity activates a native chloride current in *Xenopus* oocyte. J Gen Physiol 103:153–179

Alexander D, Alagarsamy S, Douglas JG (2004) Cyclic stretch-induced cPLA$_2$ mediates ERK 1/2 signaling in rabbit proximal tubule cells. Kidney Int 65:551–563

Allen DG, Whitehead NP, Yeung EW (2005) Mechanisms of stretch-induced muscle damage in normal and dystrophic muscle: role of ionic changes. J Physiol 567 3:723–735

Ahmmed GU, Mehta D, Vogel S, Holinstat M, Paria BC, Tiruppathi C, Malik AB (2004) Protein kinase C$_\alpha$ phosphorylates the TRPC1 channels and regulates store-operated Ca^{2+} entry in endothelial cells. J Biol Chem 79:20941–20949

Ambudkar IS (2006) Ca^{2+} signaling microdomains: platforms for the assembly and regulation of the assembly and regulation of TRPC channels. Trends Pharmacol Sci 27:25–32

Andersen L, Seilhamer JA (1997) Comparison of selected mRNA and protein abundance in human liver. Electrophoresis 18:533–537

Antoniotti S, Pla F, Barrel S, Scalabrino L, Vovisolo, D (2006) Interaction between TRPC subunits in endothelial cells. J Recep Signal Transduc 26:225–240

Bakker EN, Kerkhof CJM, Sipkema P (1999) Signal transduction in spontaneous myogenic tone insolated arterioles from rat skeletal muscle. Cardiovasc Res 41:229–236

Bakowski D, Glitsch MD, Parekh AB (2001) An examination of the secretion-like coupling model for the activation of the Ca^{2+} release-activated Ca^{2+} current I_{crac} in RBL-1 cells. J Physiol 532 1:55–71

Barsanti C, Pellegrini M, Pellegrino M (2006a) Regulation of the mechanosensitive cation channels by ATP and cAMP in leech neurons. Biochim Biophys Acta 1758:666–672

Barsanti C, Pellegrini M, Ricci D, Pellegrino M (2006b) Effects of intracellular pH and Ca^{2+} on the activity of stretch-sensitive cation channels in leech neurons Pfluegers Arch 452:435–443

Basora N, Boulay G, Biloddeau L, Rousseau E, Marcel DP (2003) 20-Hydroxyeicosatetraenocic acid (20-HETE) activates mouse TRPC6 channels expressed in HEK293 cells. J Biol Chem 278:31709–31716

Basavappa S, Pedersen SF, Jorgensen NK, Ellory JC, Hoffmann EK (1988) Swelling-induced arachidonic acid release via a 85 kDa $cPLA_2$ in human neuroblastoma cells. J Neurophysiol 79:1441–1449

Bear CE (1990) A nonselective cation channel in rat liver cells is activated by membrane stretch. Am J Physiol 258:C421–C428

Beck B, Zholos A, Sydorenko V, Roudbaraki M, Lehenkyi V, Bordat P, Prevarskaya N, Skryma M (2006) TRPC7 is a receptor-operated DAG-activated channel in human keratinocytes. J Invest Dermatol 126:1982–1993

Beech DJ (2005) TRPC1: store-operated channel and more. Pfuegers Arch 451:53–60

Beech DJ (2006) Bipolar phospholipid sensing by TRPC5 calcium channel. Biochem Soc Trans 35:101–104

Beech DJ, Xu SZ, Flemming MR (2003) TRPC1 store operated cationic channel subunit. Cell Calcium 33:433–440

Besch SR, Suchyna T, Sachs F (2002) High speed pressure clamp. Pfluegers Arch 445:161–166

Bielfeld-Ackermann A, Range C, Korbmacher C (1998) Maitotoxin (MTX) activates a nonselective cation channel in *Xenopus laevis* oocytes. Pfluegers Arch 436:329–337

Bobanović LK, Laine M, Petersen CCH, Bennett DL, Berridge MJ, Lipp P, Ripley SJ, Bootman MD (1999) Molecular cloning and immunolocalization of a novel vertebrate trp homologue from *Xenopus*. Biochemistry 340:593–599

Boittin FX, Pettermann O, Hirn C, Mittaud P, Dorchies OM, Roulet E, Ruegg UT (2006) Ca^{2+}-independent phospholipase A_2 enhances store-operated Ca^{2+} entry in dystrophic skeletal muscle fibres. J Cell Sci 119:3733–3742

Bolotina VM, Csutora P (2005) CIF and other mysteries of the store-operated Ca^{2+}-entry pathway. Trends Neurosci 30:378–387

Boulay G, Zhu X, Peyton M, Jiang M, Hurst R, Stefani E, Birnbaumer L (1997) Cloning and expression of a novel mammalian homolog of Drosophila transient receptor potential (Trp) involved in calcium entry secondary to activation of receptors coupled by the G_q class of G protein. J Biol Chem 272:29672–29680

Brazier SW, Singh BB, Liu X, Swaim W, Ambudkar IS (2003) Caveolin-1 contributes to assembly of store-operated Ca^{2+} influx channels by regulating plasma membrane localization of TRPC1. J Biol Chem 29:27208–27215

Brereton HM, Harland ML, Auld AM, Barritt GJ (2000) Evidence that the TRP-1 protein is unlikely to account for store-operated Ca^{2+} inflow in *Xenopus laevis* oocytes. Mol Cell Biochem 214:63–74

Brereton HM, Chen J, Rychkov G, Harland ML, Barritt GJ (2001) Maitotoxin activates an endogenous non-selective cation channel and is an effective initiator of the activation of the heterologously expressed hTRPC-1 (transient receptor potential) non-selective cation channel in H4-IIE liver cells. Biochim Biophys Acta 1540:107–126

Brophy CM, Mills I, Rosales O, Isales C, Sumpio BE (1993) Phospholipase C: a putative mechanotransducer for endothelial cell response to acute hemodynamic changes, Biochem Biophys Res Commun 190:576–581

Bugaj V, Alexeenko V, Zubov A, Glushankova L, Nikalaev A, Wang Z, Kaznaceyeva I, Mozhayeva GN (2005) Functional properties of endogenous receptor and store operated calcium influx channels in HEK293 cells. J Biol Chem 280:16790–16797

Calabrese B, Manzi S, Pellegrini M, Pelligrino M (1999) Stretch activated cation channels of leech neurons: characterization and role in neurite outgrowth. Eur J Neurosci 11:2275–2284

Callamaras N, Sun XP, Ivorra I, Parker I (1998) Hemispheric asymmetry of macroscopic and elementary calcium signals mediated by InsP$_3$ in Xenopus oocytes. J Physiol 511:395–405

Cantiello HF, Montalbetti, N, Li Q, Chen XZ (2007) The cytoskeleton connection as a potential mechanosensory mechanism: lessons from polycystin 2 (TRPP2). Curr Top Membr 59:233–296

Casado M, Ascher P (1998) Opposite modulation of NMDA receptors by lysophospholipids and arachidonic acid: common features with mechanosensitivity. J Physiol 513:317–330

Cemerikic D, Sackin H (1993) Substrate activation of mechanosensitive, whole cell currents in renal proximal tubule. Am J Physiol 264:F697–F714

Chemin J, Patel AJ, Duprat F, Lauritzen I, Lazdunski M, Honoré E (2005) A phospholipid sensor controls mechanogating of the K$^+$ channel TREK-1. EMBO J 24:44–53

Chen J, Barritt GJ (2003) Evidence that TRPC1 (transient receptor potential canonical 1) forms a Ca^{2+}-permeable channels linked to the regulation of cell volume in liver cells obtaining using small interfering RNA targeted against TRPC1. Biochem J 373:327–336

Cherednichenko G, Hurne AM, Fessenden JD, Lee EH, Allen PD, Beam KG, Pessah IN (2004) Conformational activation of Ca^{2+} entry by depolarization of skeletal myotubes. Proc Natl Acad Sci USA 101:15793–15798

Christensen O (1987) Mediation of cell volume regulation by Ca^{2+} influx through stretch activated cation channels. Nature 330:66–68

Chu X, Cheung JY, Barber DL, Birnbaumer L, Rothblum LI, Conrad K, Abrason V, Chan Y, Stahl R, Carey DJ, Miller BA (2002) Erythropoietin modulates calcium influx through TRPC2. J Biol Chem 277:34375–34382

Chu X, Tong Q, Cheung JY, Wozney J, Conrad K, Maznack V, Zhang W, Stahl R, Barber DL, Miller BA (2004) Interaction of TRPC2 and TRPC6 in erythropoietin modulation of calcium influx. J Biol Chem 279:10514–10522

Cioffi DL, Wu S, Alexeyev M, Goodman SR, Zhu MX, Stevens T (2005) Activation of the endothelial store-operated ISOC Ca^{2+} channel requires interaction of protein 4.1 with TRPC4. Circ Res 97:1164–1172

Clapham DE (2003) TRP channels as cellular sensors. Nature 426:517–524

Cohen DM (2005a) SRC family kinases in cell volume regulation. Am J Physiol 288: C483–C493

Cohen DM (2005b) TRPV4 and the mammalian kidney. Pfluegers Arch 451:168–175

Cosens DJ, Manning A (1969) Abnormal electroretinogram from a Drosophila mutant. Nature 224:285–287

Delmas P (2004) Assembly and gating of TRPC channels in signaling microdomains. Novartis Found Symp 258:75–97

Delmas P, Wanaverbecq N, Abogadie FC, Mistry M, Brown DA (2002) Signaling microdomains define the specificity of receptor-mediated INsP$_3$ pathways in neurons. Neuron 14:209–220

Diakov A, Koch JP, Ducoudret O, Mueler-Berger S, Frömter E (2001) The disulfoic stilbene DIDS and the marine toxin maitotoxin activated the same two types of endogenous cation conductance in the cell membrane of Xenopus laevis oocytes. Pfluegers Arch 442:700–708

Dietrich AM, Schnitzler MM, Emmel J, Kallwa H, Hofmann T, Gundermann T (2003) N-linked protein glycosylation is a major determinant for basal TRPC3 and TRPC6 channel activity. J Biol Chem 278:47842–47852

Dietrich A, Schnitzker MM, Gollasch M, Gross V, Storch U, Dubrovska G, Obst M, Yildirim E, Salanova B, Kalwa H, Essin K, Pinkenburg O, Luft FC, Gudermann T, Birnbaumer L (2005) Increased vascular smooth muscle contractility in TRPC6$^{-/-}$ mice. Mol Cell Biol 25: 6980–6989

Drummond HA, Price MP, Welsh MJ, Abboud FM (1998) A molecular component of the arterial baroreceptor mechanotransducer. Neuron 21:1435–1441

Ducret T, Vandebrouck C, Cao ML, Lebacq J, Gailly P (2006) Functional role of store-operated and stretch-activated channels in murine adult skeletal muscle fibers. J Physiol 575 3:913–924

Durvasula RV, Shankland SJ (2006) Podocyte injury and targeting therapy: an update. Curr Opin Nephrol Hypertens 15:1–7

Escobar LI, Salvador C, Martinez M, Vaca L (1998) Maitotoxin, a cationic channel activator. Neurobiol 6:59–74

Estacion M, Li S, Sinkins WG, Gosling M, Bahra P, Poll C, Westwick J, Schilling WP (2004) Activation of human TRPC6 channels by receptor stimulation. J Biol Chem 279:22047–22056

Fensome-Green A, Stannard N, Li M, Bolsover S, Cockcroft S (2007) Bromoenol lactone, an inhibitor of group VIA calcium-independent phospholipase A2 inhibits antigen-stimulated mast cell exocytosis without blocking Ca^{2+} influx. Cell Calcium 41:145–153

Flemming PK, Dedman AM, Xu SZ, Li J, Zeng F, Naylor J, Benham CD, Bateson AN, Muraki K, Beech DJ (2006) Sensing of lysophospholipids by TRPC5 calcium channel. J Biol Chem 281:4977–4982

Flemming R, Xu SZ, Beech DJ (2003) Pharmacological profile of store-operated channels in cerebral arteriolar smooth muscle cells. Br J Pharm 139:955–965

Fong P, Turner PR, Denetclaw WF, Steinhardt RA (1990) Increased activity of calcium leak channels in myotubes of Duchenne human and mdx mouse origin. Science 250:673–676

Franco A, Lansman JB (1990) Calcium entry through stretch-inactivated channels in mdx myotubes. Nature 344:670–673

Franco-Obregon A, Lansman JB (2002) Changes in mechanosensitive channel gating following mechanical stimulation in skeletal muscle myotubes from the mdx mouse. J Physiol 539 2:391–407

Freichel M, Suh SH, Pfeifer A, Schweig U, Trost C, Weissgerber P, Biel M, Phillip S, Freise D, Droogmans G, Hofmann F, Flocerzi V, Nilius B (2003) Lack of an endothelial store-operated Ca^{2+} current impairs agonist-dependent vasorelaxation in TRPC4$^{-/-}$ mice. Nat Cell Biol 3:121–127

Freichel M, Vennekens R, Olausson J, Hoffmann M, Müller C, Stolz S, Scheunemann J, Weissgerber P, Flockerzi V (2004) Functional role of TRPC proteins in vivo: lessons from TRPC-deficient mouse models. Biochem Biophys Res Commun 322:1352–1358

Gailly P, Colson-Van Schoor M (2001) Involvement of TRP2 protein in store-operated influx of calcium in fibroblasts. Cell Calcium 30:157–165

Giamarchi A, Padilla F, Coste B, Raoux M, Crest M, Honoré E, Delmas P (2006) The versatile nature of the calcium-permeable cation channel TRPP2. EMBO Rep 7:787–793

Glazebrook PA, Schilling WP, Kunze DL (2005) TRPC channels as signal transducers. Pfluegers Arch 451:125–130

Goel M, Sinkins WG, Zuo CD, Estacion M, Schilling WP (2006) Identification and localization of TRPC channels in the rat kidney. Am J Physiol 290:F1241–F1252

Golovina VA, Platoshyn O, Bailey CL, Wang J, Limsuwan A, Sweeney M, Rubin LJ, Yuan JX (2001) Upregulated TRP and enhanced capacitative Ca^{2+} entry in human pulmonary artery myocytes during proliferation. Am J Physiol 280:H746–H755

Gottlieb PA, Suchyna TM, Ostrow LW, Sachs F (2004) Mechanosensitive ion channels as drug targets. Curr Drug Targets 3:287–295

Greka A, Navarro B, Duggan A, Clapham DE (2003) TRPC5 is a regulator of hippocampal neurite length and growth cone morphology. Nat Neurosci 6:837–845

Groschner K, Rosker C (2005) TRPC3: a versatile transducer molecule that serves integration and diversification of cellular signals. Naunyn Schmiedebergs Arch Pharmacol 371:251–256

Gu CX, Juranka PF, Morris CE (2001) Stretch-activation and stretch-inactivation of Shaker-IR, a voltage-gated K^+ channel. Biophys J 80:2678–2693

Guharay F, Sachs F (1984) Stretch-activated single ion channel currents in tissue cultured embryonic chick skeletal muscle. J Physiol 352:685–701

Hamill OP (1983) Potassium and chloride channels in red blood cells. In: Sakmann B, Neher E (eds) Single channel recording. Plenum, New York, pp 451–471

Hamill OP (2006) Twenty odd years of stretch-sensitive channels. Pfluegers Arch Eur J Phys 453.3:333–351

Hamill OP, Maroto R (2007) TRPCs as MS channels. Curr Top Membr 59:191–231

Hamill OP, Martinac B (2001) Molecular basis of mechanotransduction in living cells. Physiol Rev 81:685–740

Hamill OP, McBride DW Jr (1992) Rapid adaptation of the mechanosensitive channel in Xenopus oocytes. Proc Natl Acad Sci USA 89:7462–7466

Hamill OP, McBride DW Jr (1996) The pharmacology of mechanogated membrane ion channels. Pharmacol Rev 48:231–252

Hamill OP, McBride DW Jr (1997) Induced membrane hypo-/hyper-mechanosensitivity: a limitation of patch clamp recording. Annu Rev Physiol 59:621–631

Hamill OP, Marty A, Neher E, Sakmann B, Sigworth F (1981) Improved patch clamp techniques for high current resolution from cells and cell-free membrane patches. Pfluegers Arch 391:85–100

Hill AJ, Hinton JM, Cheng H, Gao Z, Bates DO, Hancox JC, Langton PD, James AF (2006) A TRPC-like non-selective cation current activated by α-adrenoceptors in rat mesenteric artery smooth muscle cells. Cell Calcium 40:29–40

Hille B (2001) Ion channels of excitable membranes, 3rd edn. Sinauer, Sunderland, MA

Hisatsune C, Kuroda Y, Nakamura K, Inoue T, Nakamura T, Michikawa T, Mizuntani A, Mikoshiba K (2004) Regulation of TRPC6 channel activity by tyrosine phosphorylation. J Biol Chem 279:18887–18894

Hofmann T, Obukhov AG, Schaefer M, Harteneck C, Gudermann T, Schultz G (1999) Direct activation of human TRPC6 and TRPC3 channels by diacylglycerol. Nature 397:259–263

Hofmann T, Schaeffer M, Schultz G, Gudermann T (2000) Cloning, expression and subcellular localization of two novel splice variants of mouse transient receptor potential 2. Biochem J 351:115–122

Hofmann T, Schaeffer M, Schultz G, Gudermann T (2002) Subunit composition of mammalian transient receptor potential channels in living cells. Proc Natl Acad Sci USA 99:7461–7466

Honoré E, Maingret F, Lazdunski M, Patel AJ (2002) An intracellular proton sensor commands lipid and mechano-gating of the K^+ channel TRE-1. EMBO J 21:2968–2976

Honoré E, Patel AJ, Chemin J, Suchyna T, Sachs F (2006) Desensitization of mechano-gated K2p channels. Proc Natl Acad Sci USA 103:6859–6864

Hopf FW, Reddy P, Hong J, Steinhardt RA (1996) A capacitive calcium current in cultured skeletal muscle cells is medicated by the Ca^{2+}-specific leak channel and inhibited by dihyropyridine compounds. J Biol Chem 271:22358–22367

Howard J, Bechstedt S (2004) Hypothesis: a helix of ankyrin repeats of the NOMPC-TRP ion channel is the gating spring of mechanoreceptors. Curr Biol 14:224–226

Howard J, Roberts WM, Hudspeth AJ (1988) Mechanoelectrical transduction by hair cells. Annu Rev Biophys Chem 17:99–124

Huang GN, Zeng W, Kim JY, Yuan JP, Han L, Muallem S, Worley PF (2006) STIM1 carboxl-terminus activates native SOC, ICRAC and TRPC1 channels. Nat Cell Biol 8:1003–1010

Huang M, Gu G, Ferguson E, Chalfie M (1995) A stomatin-like protein necessary for mechano-sensation in C. elegans. Nature 378:292–295

Huber TB, Scherner B, Müller RU, Höhne M, Bartram M, Calixto A, Hagmann H, Reinhardt C, Koos F, Kunzelmann K, Shirokova E, Krautwurst D, Harteneck C, Simons M, Pavenstadt H, Kerjaschki D, Thiele C, Walz G, Chalfie M, Benzing T (2006) Podocin and MEC-2 bind cholesterol to regulate the activity of associated proteins. Proc Natl Acad Sci USA 103:17079–17086

Hui H, McHugh D, Hannan M, Zeng F, Xu SZ, Khan SUH, Levenson R, Beech DJ, Weiss JL (2006) Calcium-sensing mechanism in TRPC5 channels contributing to retardation of neurite outgrowth. J Physiol 572 1:165–172

Hurst RS, Zhu X, Boulay G, Birnbaumer L, Stefani E (1998) Ionic currents underlying HTRP3 mediated agonist-dependent Ca^{2+} influx in stably transfected HEK293 cells. FEBS Lett 422:333–338

Inoue R, Okada T, Onoue H, Hara Y, Shimizu S, Naitoh S, Ito Y, Mori Y (2001) The transient receptor potential protein homologue TRP6 is the essential component of vascular α_1-adrenceptor-activated Ca^{2+}-permeable cation channel. Circ Res 88:325–332

Inoue R, Jensen LJ, Shi J, Morita H, Nishida M, Honda A, Ito Y (2006) Transient receptor potential channels in cardiovascular function and disease. Circ Res 99:119–131

Iwata Y, Katanosaka Y, Arai Y, Komanura K, Miyatake K, Shigekawa M (2003) A novel mechanism of myocyte degeneration involving the Ca^{2+}-permeable growth factor-regulated channel. J Cell Biol 161:957–967

Jaconi M, Pyle J, Bortolon R, Ou J, Clapham D (1999) Calcium release and influx colocalize to the endoplasmic reticulum. Curr Biol 7:599–602

Jacques-Fricke BT, Seow Y, Gottlieb PC, Sachs F, Gomez TM (2006) Ca^{2+} Influx through mechanosensitive channels inhibits neurite outgrowth in opposition to other influx pathways and release of intracellular stores. J Neurosci 26:5656–5664

Janssen LJ, Kwan CY (2007) ROCs and SOCs: what's in a name? Cell Calcium 41:245–247

Jung S, Strotmann R, Schultz G, Plant TD (2002) TRPC6 is a candidate channel involved in receptor-stimulated cation currents in A7r5 smooth muscle cells. Am J Physiol 282:C347–C359

Jung S, Mühle A, Shaefer M, Strotmann R, Schultz G, Plant TD (2003) Lanthanides potentiate TRPC5 currents by an action at the extracellular sites close to the pore mouth. J Biol Chem 278:3562–3571

Jungnickel MK, Marrero H, Birnbaumer L, Lemos JR, Florman HM, (2001) Trp2 regulates entry of Ca^{2+} into mouse sperm triggered by ZP3. Nature Cell Biol 3:499–502

Kamouchi M, Philipp S, Flockerzi V, Wissenbach U, Mamin A, Raemaekers L, Eggermont J, Droogmans G, Nilius B (1999) Properties of heterologously expressed hTRP3 channels in bovine pulmonary artery endothelial cells. J Physiol 518 2:345–359

Kanzaki M, Zhang YQ, Mashima H, Li L, Shibata H, Kojima I (1999) Translocation of a calcium-permeable cation channel induced in insulin-like growth factor-1. Nature Cell Biol 1:165–170

Kawasaki BT, Liao Y, Birnbaumer L (2006) Role of Src in C3 transient receptor potential channel function and evidence for a heterogeneous makeup of receptor- and store-operated Ca^{2+} entry channels. Proc Natl Acad Sci USA 103:335–340

Kim D (1992) A mechanosensitive K^+ channel in heart-cells: activation by arachidonic acid. J Gen Physiol 100:1021–1040

Kiselyov K, Xu X, Mozayeva G, Kuo T, Pessah I, Mignery G, Zhu X, Birnbaumer L, Muallen S (1998) Functional interaction between $InsP_3$ receptors and store-operated $Htrp_3$ channels. Nature 396:478–482

Kloda A, Martinac B (2001a) Structural and functional differences between two homologous mechanosensitive channels of *Methanocccus jannaschii*. EMBO J 20:1888–1896

Kloda A, Martinac B (2001b) Mechanosensitive channel of Thermoplasma, the cell wall-less Archae. Cell Biochem Biophys 34:321–347

Kriz W (2005) TRPC6 – a new podocyte gene involved in focal segmental glomerulosclerosis. Trends Mol Med 11:527–530

Kumar B, Dreja K, Shah SS, Cheong A, Xu SZ, Sukumar P, Naylor J, Forte A, Cipollaro M, McHugh D, Kingston PA, Heagerty AM, Munsch C, Bergdahl M, Hultgardh-Nilsson A, Gomez MF, Porter KE, Hellstrand P, Beech DJ (2006) Upregulated TRPC1 channel in vascular injury in vivo and its role in human neoitimal hyperplasia. Circ Res 98:557–563

Kung C (2005) A possible unifying principle for mechanosensation. Nature 436:647–654

Kunichika N, Yu Y, Remillard CV, Platoshyn O, Zhang S, Yuan JXL (2004) Overexpression of TRPC1 enhances pulmonary vasoconstriction induced by capacitative Ca^{2+} entry. Am J Physiol 287:L962–L969

Kwan HY, Leung PC, Huang Y, Yao X (2003) Depletion of intracellular Ca^{2+} stores sensitizes the flow-induced Ca^{2+} influx in rat endothelial cells. Circ Res 92:286–292

Lane JW, McBride DW Jr, Hamill OP (1991) Amiloride block of the mechanosensitive cation channel in Xenopus oocytes. J Physiol 441:347–366

Lane JW, McBride DW Jr, Hamill OP (1992) Structure–activity relations of amiloride and some of its analogues in blocking the mechanosensitive channel in Xenopus oocytes. Br J Pharmacol 106:283–286

Lee EH, Cherednichenko G, Pessah IN, Allen PD (2006) Functional coupling between TRPC3 and RyR1 regulates the expressions of key triadic proteins. J Biol Chem 281:10042–10048

Lee J, Ishihara A, Oxford G, Johnson B, Jacobson K (1999) Regulation of cell movement is mediated by stretch-activated calcium channels. Nature 400:382–386

Lehtonen JY, Kinnunen PK (1995) Phospholipase A_2 as a mechanosensor. Biophys J 68 1888–1894

Levina, N, Tötemeyer S, Stokes NR, Louis P, Jones MA, Booth IR (1999) Protection of *Escherichia coli* cells against extreme turgor pressure by activation of MscS and MscL mechanosensitive channels: identification of genes for MscS activity. EMBO J 18:1730–1737

Lièvremont JP, Bird GS, Putney JW Jr (2004) Canonical transient receptor potential TRPC7 can function as both a receptor- and store-operated channel in HEK-293 cells. Am J Physiol 287: C1709–C1716

Lièvremont JP, Numaga T, Vazquez G, Lemonnier L, Hara Y, Mori E, Trebak M, Moss SE, Bird GS, Mori Y, Putney JW Jr (2005) The role of canonical transient receptor potential 7 in B-cell receptor-activated channels. J Biol Chem 280:35346–35351

Lindahl M, Backman E, Henriksson KG, Gorospe JR, Hoffman EP (1995) Phospholipase A2 activity in dystrophinopathies. Neuromusc Disord 5:193–199

Lintschinger B, Balzer-Geldsetzer M, Baskaran T, Graier WF, Romanin C, Zhu MX, Groschner K (2000) Coassembly of Trp1 and Trp3 proteins generates diacylglycerol- and Ca^{2+}-sensitive cation channels. J Biol Chem 275:27799–27805

Liman E, Corey DP, Dulac C (1999) TRP2: a candidate transduction channel for mammalian pheromone sensory signaling. Proc Natl Aacd Sci USA 96:5791–5796

Liou J, Kim ML, Heo WD, Jones JT, Myers JW, Ferell JE, Meyer T (2005) STIM is a Ca^{2+} sensor essential for Ca^{2+}-store-depletion-triggered Ca^{2+} influx. Curr Biol 15:1235–1241

Liu X, Wang W, Singh BB, Lockwich T, Jadlowiec J, O'Connell B, Wellner R, Zhu MX, Ambudkar IS (2000) TRP1, a candidate protein for the store-operated Ca^{2+} influx mechanism in salivary gland cells. J Biol Chem 275:3403–3411

Liu X, Singh BB, Ambudkar IS (2003) TRPC1 is required for functional store-operated channels. J Biol Chem 278:11337–11343

Liu X, Bandyopadhyay BC, Singh BB, Groschner K, Ambudkar IS (2005) Molecular analysis of a store-operated and 2-acetyl-sn-glycerol-sensitive non-selective cation channel. J Biol Chem 280:21600–21606

Lockwich TP, Liu X, Singh BB, Jadlowiec J, Weiland S, Ambudkar IS (2000) Assembly of Trp1 in a signaling complex associated with caveolin-scaffolding lipid raft domains. J Biol Chem 275:11934–11942

Ma R, Rundle D, Jacks J, Koch M, Downs T, Tsiokas L (2003) Inhibitor of myogenic family, a novel suppressor of store-operated currents through an interaction with TRPC1. J Biol Chem 278:52763–52772

Markin VS, Sachs F (2007) Thermodynamics of mechanosensitivity. Curr Top Membr 58:87–119

Maroto R, Hamill OP (2007) MscCa regulation of tumor cell migration and metastasis. Curr Top Membr 59:485–509

Maroto R, Raso A, Wood TG, Kurosky A, Martinac B, Hamill OP (2005) TRPC1 forms the stretch-activated cation channel in vertebrate cells. Nat Cell Biol 7:179–185

Martinac B (2007) 3.5 Billion years of mechanosensory transduction: structure and function of mechanosensitive channels in prokaryotes. Curr Top Membr 58:25–57

Martinac B, Hamill OP (2002) Gramicidin A channels switch between stretch-activation and stretch-inactivation depending upon bilayer thickness. Proc Natl Acad Sci USA 99:4308–4312

Martinac B, Adler J, Kung C (1990) Mechanosensitive channels of *E. coli* activated by amphipaths. Nature 348:261–263

Maruyama Y, Nakanishi Y, Walsh EJ, Wilson DP, Welsh DG, Cole WC (2006) Heteromultimeric TRPC6-TRPC7 channels contribute to arginine vasopressin-induced cation current of A7r5 vascular smooth muscle. Circ Res 98:1520–1527

Matsumoto H, Baron CB, Coburn RF (1995) Smooth muscle stretch-activated phospholipase C activity. Am J Physiol 268:C458–C465

Matthews BD, Thodeki CK, Ingber DE (2007) Activation of mechanosensitive ion channels by forces transmitted through integrins and the cytoskeleton. Curr Top Membr 58:59–85

McBride DW Jr, Hamill OP (1992) Pressure-clamp: a method for rapid step perturbation of mechanosensitive channels. Pfluegers Arch 421:606–612

McBride DW Jr, Hamill OP (1993) Pressure-clamp techniques for measurement of the relaxation kinetics of mechanosensitive channels. Trends Neurosci 16:341–345

McBride DW Jr, Hamill OP (1995) A fast pressure clamp technique for studying mechano-gated channels. In: Sakmann B, Neher E (eds) Single channel recording, 2nd edn. Plenum, New York, pp 329–340

McBride DW Jr, Hamill OP (1999) A simplified fast pressure-clamp technique for studying mechanically-gated channels. Methods Enzymol 294:482–489

McKay RR, Szymeczek-Seay CL, Lièvremont JP, Bird GS, Zitt C, Jüngling E, Lückhoff A, Putney JW Jr (2000) Cloning and expression of the human transient receptor potential 4 (TRP4) gene: localization and functional expression of human TRP4 and TRP3. Biochem J 351:735–746

Mery L, Strauss B, Dufour JF, Krause KH, Hoth M (2002) The PDZ-interacting domain of TRPC4 controls its localization and surface expression in HEK293 cells. J Cell Sci 15:3497–3508

Minke B, Cook B (2002) TRP channel proteins and signal transduction. Physiol Revs 82 429–472

Minke B, Wu C, Pak WL (1975) Induction of photoreceptor voltage noise in the dark in Drosophila mutant. Nature 258:84–87

Mitchell CH, Zhang JJ, Wang L, Jacob TJC (1997) Volume-sensitive chloride current in pigmented ciliary epithelial cells: role of phospholipases. Am J Physiol 272:C212–C222.

Montell C (2005) The TRP superfamily of cation channels. Science STKE re3:1–24

Montell C, Rubin GM (1989) Molecular characterization of the *Drosophila* trp locus: a putative integral membrane protein required for phototransduction. Neuron 2:1313–1323

Moore AL, Roe MW, Melnick RF, Lidofsky SD (2002) Calcium mobilization evoked by hepatocellular swelling is linked to activation of phospholipase Cγ. J Biol Chem 277:34030–34035

Moore TM, Brough GH, Babal P, Kelly JJ, Li M, Stevens T (1998) Store-operated calcium entry promotes shape changes in pulmonary endothelial cells expressing TRP1. Am J Physiol 275: L574–L582

Mori Y, Wakamori M, Miyakaw T, Hermosura M, Hara Y, Nishida M, Hirose K, Mizushima A, Kurosaki M, Mori E, Gotoh K, Okada T (2002) Transient receptor potential regulates capacitative Ca^{2+} entry and Ca^{2+} release from endoplasmic reticulum in B lymphocytes. J Exp Med 195:673–681

Morris CE, Horn R (1991) Failure to elicit neuronal macroscopic mechanosensitive currents anticipated by single-channel studies. Science 251:1246–1249

Morris CE, Juranka PF (2007) Lipid stress at play: mechanosensitivity of voltage-gated channels. Curr Top Membr 59:297–338

Muraki K, Iwata Y, Katanosaka Y, Ito T, Ohya S, Shigekawa M, Imaizumi Y (2003) TRPV2 is a component of osmotically sensitive cation channels in murine aortic myocytes. Circ Res 93:829–838

Nauli SM, Zhou J (2004) Polycystins and mechanosensation in renal and nodal cilia. Bioessays 26:844–856

Nauli SM, Alenghat FJ, Luo Y, Williams E, Vassilev P, Elia A, Lu W, Brown EM, Quinn SJ, Ingber DE, Zhou J (2003) Polycystins 1 and 2 mediate mechanosensation in primary cilium of kidney cells. Nat Genet 33:129–137

Nilius B, Voets T (2005) TRP channels: a TR(I)P through a world of multifunctional cation channels. Pfluegers Arch 451:1–10

Nilius B, Owsianik G, Voets T, Peters JA (2007) Transient receptors potential cation channels in disease. Physiol Revs 87:165–217

Numata T, Shimizu T, Okada Y (2007) TRPM7 is a stretch- and swelling-activated cation channel involved in volume regulation in human epithelial cells. Am J Physiol 292:C460–C467

Obukhov AG, Nowycky MC (2004) TRPC5 activation kinetics are modulated by the scaffolding protein ezrin/radixin/moesin-binding phosphoprotein-50 (EBP50). J Cell Physiol 201:227–235

Obukhov AG, Nowycky MC (2005) A cytosolic residue mediates Mg^{2+} block and regulates inward current amplitude of a transient receptor potential channel. J Neurosci 25:1234–1239

Ohata H, Tanaka K, Maeyama N, Ikeuchi T, Kamada A, Yamamoto M, Momose K (2001) Physiological and pharmacological role of lysophosphatidic acid as modulator in mechanotransduction. Jpn J Physiol 87:171–176

Okada T, Inoue R, Yamazaki K, Maeda A, Kurosaki T, Yamakuni T, Tanaka I, Shimizu S, Ikenaka K, Imoto K, Mori Y (1999) Molecular and functional characterization of a novel mouse transient receptor potential homologue TRP7. J Biol Chem 274:27359–27370

O'Neil RG, Heller S (2005) The mechanosensitive nature of TRPV channels. Pfluegers Arch 451:193–203

Opsahl LR, Webb WW (1994) Transduction of membrane tension by the ion channel alamethicin. Biophys J 66:71–74

Owsianik G, D'Hoedt D, Voets T, Nilius B (2006) Structure–function relationship of the TRP channel superfamily. Rev Physiol Biochem Pharmacol 156:61–90

Palmer CP, Zhou XL, Lin J, Loukin SH, Kung C, Saimi Y (2001) A TRP homolog in Saccharomyces cerevisiae forms an intracellular Ca permeable channel in the yeast vacuolar membrane. Proc Natl Acad Sci USA 98:7801–7805

Paoletti P, Ascher P (1994) Mechanosensitivity of NMDA receptors in cultured mouse central neurons. Neuron 13:645–655

Paolini C, Fessenden JD, Pessah IN, Franzini-Armstrong C (2004) Evidence for conformational coupling between two calcium channels. Proc Natl Acad Sci USA 101:12748–12752

Parekh AB, Putney JW Jr (2005) Store-operated calcium channels. Physiol Rev 85:757–810

Paria PC, Malik AB, Kwiatek AM, Rahman A, May MJ, Ghosh S, Tiruppathi C (2003) Tumor necrosis factor-α induces nuclear factor-κB-dependent TRPC1 expression in endothelial cells. J Biol Chem 278 37195–37203

Park KS, Kim Y, Lee Y, Earm YE, Ho WK (2003) Mechanosensitive cation channels in arterial smooth muscle cells are activated by diacylglycerol and inhibited by phospholipase C inhibitor. Circ Res 93:557–564

Patel AJ, Honoré E (2001) Properties and modulation of mammalian 2P domain K⁺ channels. Trends Neurosci 24:339–346

Peinelt C, Vig M, Koomoa DL, Beck A, Nadler MJS, Koblan-Huberson M, Lis A, Fleig A, Penner R, Kinet JP (2006) Amplification of CRA current by STIM1 and CRACM1 (Orai1). Nat Cell Biol 8:771–773

Pellegrino M, Pellegrini M (2007) Mechanosensitive channels in neurite outgrowth. Curr Top Membr 59:111–125

Perbal G, Driss-Ecole D (2003) Mechanotransduction in gravisensing cells. Trends Plant Sci 8:498–504

Perozo E, Kloda A, Cortes DM, Martinac B (2002) Physical principles underlying the transduction of bilayer deformation forces during mechanosensitive channel gating. Nat Struct Biol 9:696–703

Perraud AL, Fleig A, Dunn CA, Bagley LA, Launay P, Schmitz C, Stokes AJ, Zhu Q, Bessman MJ, Penner R, Kinet JP, Scharenberg AW (2001) ADP-ribose gating of the calcium-permeable LTRPC2 channel revealed by Nudix motif homology. Nature 411:595–599

Philipp S, Hambrecht J, Braslavski L, Schroth G, Freichel M, Murakami M, Cavalié A, Flockerzi V (1998) A novel capacitive calcium entry channels expressed in excitable cells. EMBO J 17:4274–4282

Phillip S, Torst C, Warnat J, Rautmann J, Himmerkus N, Schroth G, Kretz O, Nastainczyk W, Cacalie A, Hoth M, Flockerzi V (2000) TRPC4 (CCE1) is part of native Ca^{2+} release-activated Ca2+ like channels in adrenal cells. J Biol Chem 275:23965–23972

Pickard BG (2007) Delivering force and amplifying signals in plant mechanosensing. Curr Top Membr 58:361–392

Plant TD, Schaefer M (2005) Receptor-operated cation channels formed by TRPC4 and TRPC5. Naunyn Schmiedebergs Arch Pharmacol 371:266–276

Powl AM, Lee AG (2007) Lipid effects on mechanosensitive channels. Curr Top Membr 58:151–178

Praetorius HA, Spring KR (2005) A physiological view of the primary cilium. Annu Rev Physiol 67:515–529

Protasi F (2002) Structural interactions between RYRs and DHPs in calcium release units of cardiac and skeletal muscle cells. Front Biosci 7:650–658

Putney JW Jr, Trebak M, Vazquez G, Wedel B, Bird GS (2004) Signaling mechanisms for TRPC3 channels. Novartis Found Symp 258:123–139

Ramsey IS, Delling M, Clapham DE (2006) An introduction to TRP channels. Annu Rev Physiol 68:619–647

Rao JN, Platoshyn O, Golovina VA, Liu L, Zou T, Marasa BS, Turner DJ, Yuan JXJ, Wang JY (2006) TRPC1 functions as a store-operated Ca^{2+} channel in intestinal epithelial cells and regulates mucosal restitution after wounding. Am J Physiol 290:G782–G792

Reading SA, Earley S, Waldron BJ, Welsh DJ, Brayden JE (2005) TRPC3 mediates pyridine receptor-induced depolarization of cerebral arteries. Am J Physiol 288 H2055–H2061

Reifarth FW, Clauss W, Weber WM (1999) Stretch-independent activation of the mechanosensitive cation channel in oocytes of *Xenopus laevis*. Biochim Biophys Acta 1417:63–76

Reiser J, Polu KR, Möller CC, Kemlan P, Altinas MM, Wei C, Faul C, Herbert S, Villegas I, Avila-Casado C, McGee M, Sugmoto H, (2005) TRPC6 is a glomerular slit diaphragm-associated channel required for normal renal function. Nat Genet 37:739–744

Riccio A, Mattei C, Kelsell RE, Medhurst AD, Calver AR, Randall AD, Davis JB, Benham CD, Pangalos MN (2002a) Cloning and functional expression of human short TRP7, a candidate protein for store operated Ca^{2+} influx. J Biol Chem 277:12302–12309

Riccio A, Medhurst AD, Mattei C, Kelsell RE, Calver AR, Randall AD, Benham CD, Pangalos MN (2002b) mRNA distribution analysis of human TRPC family in CNS and peripheral tissues. Mol Brain Res 109:95–104

Rohács T, Lopez Cm, Michailidis I, Logothetis DE (2005) PI(4,5)P2 regulates the activation and desensitization of TRPM6 channels through the TRP domain. Nat Neurosci 8:626–634

Roos J, DiGreorio PJ, Yeramin AV, Ohlsen K, Liodyno M, Zhang S, Safrina O, Kazak JA, Wagner SL, Cahalan MD, Velicelebi G, Stauderman KA (2005) STIM1, an essential and conserved component of store-operated Ca^{2+} channel function. J Cell Biol 169:435–445

Rosales OR, Isales CM, Barrett PQ, Brophy C, Sumpio BE (1997) Exposures of endothelial cells to cyclic strain induces elevations of cytosolic Ca^{2+} concentration through mobilization of intracellular and extracellular pools. Biochem J 326:385–392

Ruknudin A, Song MJ, Sachs F (1991) The ultrastructure of patch-clamped membranes: a study using high voltage electron microscopy. J Cell Biol 112:125–134

Rüsch A, Kros CJ, Richardson GP (1994) Block by amiloride and its derivatives of mechano-electrical transduction in outer hair cells of mouse cochlear cultures. J Physiol 474:75–86

Ruwhof C, Van Wamel JET, Noordzij LAW, Aydin S, Harper JCR, Van Der Laarse A (2001) Mechanical stress stimulates phospholipase C activity and intracellular calcium ion levels in neonatal rat cardiomyocytes. Cell Calcium 29:73–83

Ryan MJ, Gross KW, Hajduczok G (2000) Calcium-dependent activation of phospholipase C by mechanical distension in renin-expressing As4.1 cells. Am J Physiol 279:E823–E829

Sachs F (1988) Mechanical transduction in biological systems. CRC Crit Revs Biomed Eng 16:141–169

Sackin H (1989) A stretch-activated K^+ channel sensitive to cell volume. Proc Natl Acad Sci USA 86:1731–1735

Saimi Y, Zhou X, Loukin SH, Haynes WJ, Kung C (2007) Microbial TRP channels and their mechanosensitivity. Curr Top Membr 58:311–327

Saleh SN, Albert AP, Pepeiatt CM, Large WA (2006) Angiotensin II activates two cation conductances with distinct TRPC1 and TRPC6 channel properties in rabbit mesenteric artery myocytes. J Physiol 577 2:479–495

Sampieri A, Diaz-Munoz M, Antaramian A, Vaca L (2005) The foot structure form the type 1 ryanodine receptor is required for functional coupling to store-operated channels. J Biol Chem 280:24804–24815

Schaefer M, Plant TD, Obukhov AG, Hofmann T, Gudermann T, Shultz G (2000) Receptor-mediated regulation of the nonselective cation channels TRPC4 and TRPC5. J Biol Chem 275:17517–17526

Shuttleworth TJ, Thompson JL, Mignen O (2004) ARC channels: a novel pathway for receptor-activated Ca^{2+} entry. Physiology 19:355–361

Sinkins WG, Estacion M, Schilling WP (1998) Functional expression of TRPC1: a human homologue of the Drosophila TRP channel. Biochem J 331:331–339

Sinkins WG, Goel M, Estacion M, Schilling WP (2004) Association of immunophilins with mammalian TRPC channels. J Biol Chem 279:34521–34529

Slish DF, Welsh DG, Brayden JE (2002) Diacylglycerol and protein kinase C activate cation channels in myogenic tone. Am J Physiol 283:H2196–H2201

Soboloff J, Spassova M, Xu W, He LP, Cuesta N, Gill DL (2005) Role of endogenous TRPC6 channels in Ca^{2+} signal generation in A7r5 Smooth muscle cells. J Biol Chem 280:39786–39794

Sossey-Alaoui K, Lyon JA, Jones L, Abidi FE, Hartung AJ, Hane B, Schwarz CE, Stevenson RE, Srivastava AK (1999) Molecular cloning and characterization of TRPC5 (HTRPC5), the human homolog of a mouse brain receptor-activated capacitative Ca^{2+} entry channel. Genomics 60:330–340

Small DL, Morris CE (1994) Delayed activation of single mechanosensitive channels in Lymnaea neurons. Am J Physiol 267:C598–C606

Smani T, Zakharov SI, Leno E, Csutoras P, Trepakova ES, Bolotina VM (2003) Ca^{2+}-independent phospholipase A_2 is a novel determinant of store-operated Ca^{2+} entry. J Biol Chem 278:11909–11915

Smani T, Zakharov SI, Csutoras P, Leno E, Trepakova ES, Bolotina VM (2004) A novel mechanism for the store-operated calcium influx pathway. Nat Cell Biol 6:113–120

Spassova MA, Hewavitharana T, Xu W, Soboloff J, Gill DL (2006) A common mechanism underlies stretch activation and receptor activation of TRPC6 channels. Proc Natl Acad USA 103:16586–16591

Spehr M, Hatt H, Wetzel CH (2002). Arachidonic acid plays a role in rat vomeronasal signal transduction. J Neurosci 22:8429–8437

Stamboulian S, Moutin MJ, Treves S, Pochon N, Grunwald D, Zorzato F, Waard MD, Ronjat M, Arnoult C (2005) Junctate, an inositol 1,4,5-triphosphate receptor associated protein is present in sperm and binds TRPC2 and TRPC5 but not TRPC1 channels. Dev Biol 286:326–337

Strotmann R, Harteneck C, Nunnemacher K, Schultz G, Plant TD (2000) OTRPC4, a nonselective cation channel that confers sensitivity to extracellular osmolarity. Nat Cell Biol 2:695–702

Strübing C, Krapivinsky G, Krapivinsky L, Clapham DE (2001) TRPC1 and TRPC5 from a novel cation channel in mammalian brain. Neuron 29:645–655

Strübing C, Krapivinsky G, Krapivinsky L, Clapham DE (2003) Formation of novel TRPC channels by complex subunit interactions in embryonic brain. J Biol Chem 278:39014–39019

Suchyna TM, Sachs F (2007) Mechanosensitive channel properties and membrane mechanics in dystrophic myotubes. J Physiol 581:369–387

Suchyna TM, Johnson JH, Hamer K, Leykam JF, Hag DA, Clemo HF, Baumgarten CM, Sachs F (1998) Identification of a peptide toxin from Grammostola spatula spider venom that blocks cation selective stretch-activated channels. J Gen Physiol 115:583–598

Suchyna TM, Tape SE, Koeppe RE III, Anderson OS, Sachs F, Gottlieb PA (2004) Bilayer-dependent inhibition of mechanosensitive channels by neuroactive peptide enatiomers. Nature 430:235–240

Sukharev S (2002) Purification of the small mechanosensitive channel in *Escherichia coli* (MScS): the subunit structure, conduction and gating characteristics. Biophys J 83:290–298

Sukharev SI, Martinac B, Arshavsky VY, Kung C (1993) Two types of mechanosensitive channels in the *E. coli* cell envelope: solubilization and functional reconstitution. Biophys J 65:177–183

Sukharev SI, Blount P, Martinac B, Blattner FR, Kung C (1994) A large-conductance mechano-sensitive channel in *E. coli* encoded by MscL alone. Nature 368:265–268

Tang Y, Tang J, Chen Z, Torst C, Flockerzi V, Li M, Ramesh V, Zhu MX (2000) Association of mammalian trp4 and phospholipase C isozymes with a PDZ domain-containing protein, NHERF. J Biol Chem 275:37559–37564

Tomita Y, Kaneko S, Funayama M, Kondo H, Satoh M, Akaike A (1998) Intracellular Ca^{2+} store operated influx of Ca^{2+} through TRP-R a rat homolog of TRP, expressed in Xenopus oocyte. Neurosci Lett 248:195–198

Torihashi S, Fujimoto T, Trost C, Nakayama S (2002) Calcium oscillation linked to pacemaking of intestinal cells of Cajal: requirement of calcium influx and localization of TRPC4 in caveolae. J Biol Chem 277:19191–19197

Tsiokas L, Arnould T, Zhu C, Kim E, Walz G, Sukhatme VP (1999) Specific association of the gene product of pkD2 with the TRPC1 channel. Proc Natl Acad Sci USA 96:3934–3939

Trebak M, Bird GS, Mckay RR, Putney JW (2002) Comparison of human TRPC3 channels in receptor-activated and store-operated modes. Differential sensitivity to channel blockers suggests fundamental differences in channel composition. J Biol Chem 277:21617–21623

Treves S, Franzini-Armstrong C, Moccagatta L, Arnoult C, Grasso C, Schrum A, Ronjat M, Zorzato F (2004) Junctate is a key element in calcium entry induced by activation of InsP$_3$ receptors and/or calcium store depletion. J Cell Biol 166:537–548

Tryggvason K, Wartiovaara J (2005) How does the kidney filter plasma? Physiology 20:96–101

Vaca L, Sampieri A (2002) Calmodulin modulates the delay period between the release of calcium from internal stores and activation of calcium influx via endogenous TRP1 channels. J Biol Chem 277:42178–42187

Vandebrouck C, Martin D, Colson-Van Schoor M, Debaix H, Gailly P (2002) Involvement of TRPC in the abnormal calcium influx observed in dystrophic (mdx) mouse skeletal muscle fibers. J Cell Biol 158:1089–1096

Vandorpe DH, Morris CE (1992) Stretch activation of the Aplysia S-channel. J Membr Biol 127:205–214

Vannier B, Peyton M, Boulay G, Brown D, Qin N, Jiang M, Zhu X, Birnbaumer L (1999) Mouse trp2, the homologue of the human trpc2 pseudogene encodes mTrp2, a store depletion-activated capacitive Ca^{2+} entry channel. Proc Natl Acad USA 96:2060–2064

Vanoye CG, Reuss L (1999) Stretch-activated single K$^+$ channels account for whole-cell currents elicited by swelling. Proc Natl Acad Sci USA 96:6511–6516

Vazquez G, Lièvremont PP, Bird GS, Putney JW Jr (2001) Human Trp3 forms both inositol tri-sphosphate receptor-dependent and receptor-independent store-operated cation channels in DT40 avian B lymphocytes. Proc Natl Acad Sci USA 98:11777–11782

Vazquez G, Wedel BJ, Trebak M, Bird GS, Putney JW Jr (2003) Expression levels of the canonical transient receptor potential 3 (TRPC3) channels determine its mechanism of activation. J Biol Chem 278:21649–21654

Vazquez G, Wedel BJ, Aziz O, Trebak M, Putney JW Jr (2004a) The mammalian TRPC cation channels. Biochim Biophys 1742:21–36

Vazquez G, Wedel BJ, Kawasaki BT, Bird GS, Putney JW (2004b) Obligatory role of src kinase in the signaling mechanism for TRPC3 cation channels. J Biol Chem 279:40521–40528

Vazquez G, Bird GSJ, Mori Y, Putney JW (2006) Native TRPC7 channel activation by an inositol trisphosphate receptor-dependent mechanism. J Biol Chem 281:25250–25258

Venkatachalam K, Zheng F, Gill DL (2003) Regulation of canonical transient receptor potential (TRPC) channel function by diacylglycerol and protein kinase C. J Biol Chem 278:29031–29040

Villereal ML (2006) Mechanism and functional significance of TRPC channel multimerization. Semin Cell Devel Biol 17:618–629

Voets T, Talavera K, Owsiannik G, Nilius B (2005) Sensing with TRP channels. Nat Chem Biol 1:85–92

Vriens J, Watanabe H, Janssens A, Droogmans G, Voets T, Nilius B (2004) Cell swelling, heat, and chemical agonists use distinct pathways for the activation of the cation channel TRPV4. Proc Natl Acad Sci USA 101:396–401

Walker RG, Willingham AT, Zucker CS (2000) A Drosophilia mechanosensory transduction channel. Science 287:2229–2234

Wang JHC, Thampatty BP (2006) An introductory review of cell mechanobiology. Biomech Model Mechanobiol 5:1–16

Wang SQ, Song LS, Lakatta EG, Cheng H (2001) Ca^{2+} signaling between single L-type Ca^{2+} channels and ryanodine receptors in heart cells. Nature 410:592–596

Watanabe H, Vriens J, Prenen J, Droogmans G, Voets T, Nilius B (2003) Anandamide and arachidonic acid use epoxyeicosatrienoic acids to activate TRPV4 channels. Nature 424:434–438

Weber WM, Popp C, Clauss W, van Driessche W (2000) Maitotoxin induces insertion of different ion channels into the Xenopus oocyte plasma membrane via Ca^{2+}-stimulated exocytosis. Pfluegers Arch 439:363–369

Welsh DG, Nelson MT, Eckman DM, Brayden JE (2000) Swelling activated cation channels mediate depolarization of rat cerebrovascular smooth muscle by hypotonicity and intravascular pressure. J Physiol 527 1:139–148

Welsh DG, Morielli AD, Nelson MT, Brayden JE (2002) Transient receptor potential channels regulate myogenic tone of resistance arteries. Circ Res 90:248–250

Wes PD, Chevesich J, Jeromin A, Rosenberg C, Stetten G, Montell C (1995) TRPC1, a human homolog of a Drosophila store operated channel. Proc Natl Acad Sci USA 92:9652–9656

Winn MP, Conlon PJ, Lynn KL, Farrington MK, Creazzo T, Hawkins AF, Daskalakis N, Kwan SY, Ebersviller S, Burchette JL, Pericak-Vance MA, Howell DN, Vance JM, Rosenberg PB (2005) A mutation in the TRPC6 cation channel causes familial focal segmental Glomerulosclerosis. Science 308:1801–1804

Wu X, Babnigg G, Zagranichnaya T, Villereal ML (2002) The role of endogenous human TRP4 in regulating carbachol-induced calcium oscillations in HEK-293 cells. J Biol Chem 277:13597–13608

Xu H, Zhao H, Tian W, Yoshida K, Roullet JP, Cohen DM (2003) Regulation of a transient receptor potential (TRP) channel by tyrosine phosphorylation. J Biol Chem 278:11520–11527

Xu SZ, Beech DJ (2001) TRPC1 is a membrane-spanning subunit of store-operated Ca^{2+} channels in native vascular smooth muscle cells. Circ Res 88:84–87

Xu SZ, Zeng F, Boulay G, Grimm C, Harteneck C, Beech DJ (2005) Block of TRPC5 channels by 2-aminoethoxydiphenyl borate: a differential, extracellular and voltage-dependent effect. Br J Pharmacol 145:405–414

Xu SZ, Muraki K, Zeng F, Li J, Sukumar P, Shah, S, Dedman AM, Flemming PK, McHugh D, Naylor J, Gheong A, Bateson AN, Munsch CM, Porter KE, Beech DJ (2006) a sphingosine-1-phosphate-activated calcium channel controlling vascular smooth muscle cell motility. Circ Res 98:1381–1389

Yamada H, Wakamori M, Hara Y, Takahashi Y, Konishi K, Imoto K, Mori Y (2000) Spontaneous single channel activity of neuronal TRP5 channel recombinantly expressed in HEK293 cells. Neurosci Lett 285:111–114

Yang XC, Sachs F (1989) block of stretch activated ion channels in *Xenopus* oocytes by gadolinium and calcium ions. Science 243:1068–1071

Yao Y, Ferrer-Montiel AV, Montal M, Tsien RY (1999) Activation of store-operated Ca^{2+} current in Xenopus oocytes requires SNAP-25 but not a diffusible messenger. Cell 98:475–485

Yeung EW, Allen DG (2004) Stretch-activated channels in stretch-induced muscle damage: role in muscular dystrophy. Clin Exp Pharmacol Physiol 31:551–556

Yildrin E, Dietrich A, Birnbaumer L (2003) The mouse C-type transient receptor potential 2 (TRPC2) channel: alternative splicing and calmodulin binding to its N terminus. Proc Natl Acad Sci USA 100:2220–2225

Yu Y, Sweeney M, Zhang S, Platoshyn O, Landsberg, J, Rothman A, Yuan JX (2003) PDGF stimulates pulmonary vascular smooth muscle cells proliferation by upregulating TRPC6 expression. Am J Physiol 284:C316–C330

Yu Y, Fantozzi I, Remillard CV, Landsberg JW, Kunichika N, Platoshyn O, Tigno DD, Thistlethwaite PA, Rubin LJ, Yuan JX (2004) Enhanced expression of transient receptor potential channels in idiopathic pulmonary arterial hypertension. Proc Natl Acad USA 101:13861–13866

Yuan JP, Kislyoy K, Shin DM, Chen J, Shcheynikov N, Kang SH, Dehoff MH, Schwarz MK, Seeberg PH, Muallem S, Worley PF (2003) Homer binds TRPC family channels and is required for gating of TRPC1 by IP$_3$ receptors. Cell 114:777–789

Yin CC, Blayney LM, Lai FA (2005) Physical coupling between ryanodine receptor-calcium release channels. J Mol Biol 349:538–546

Zagranichnaya TK, Wu X, Villereal ML (2005) Endogenous TRPC1, TRPC3 and TRPC7 proteins combined to form native store-operated channels in HEK-293 cells. J Biol Chem 280:29559–29569

Zeng F, Xu SZ, Jackson PK, McHugh D, Kumar B, Fountain SJ, Beech DJ (2004) Human TRPC5 channel activated by a multiplicity of signals in a single cell. J Physiol 559 3:739–750

Zhang Y, Hamill OP (2000a) Calcium-, voltage and osmotic stress-sensitive currents in *Xenopus* oocytes and their relationship to single mechanically gated channels. J Physiol 523 1:83–90

Zhang Y, Hamill OP (2000b) On the discrepancy between membrane patch and whole cell mechanosensitivity in *Xenopus* oocytes. J Physiol 523 1:101–115

Zhang Y, Gao F, Popov V, Wan J, Hamill OP (2000) Mechanically-gated channel activity in cytoskeleton deficient blebs and vesicles from *Xenopus* oocytes. J Physiol 523 1:117–129

Zhang Y, Guo F, Kim JY, Saffen D (2006) Muscarinic acetylcholine receptors activated TRPC6 channels in PC12D cells via Ca^{2+} store-independent mechanisms. J Biochem 139:459–470

Zhou XL, Batiza AF, Loukin SH, Palmer CP, Kung C, Saimi Y (2003) The transient receptor potential channels on the yeast vacuole is mechanosensitive. Proc Natl Acad Sci USA 100:7105–7110

Zhu X, Chu PB, Peyton M, Birnbaumer L (1995) Molecular cloning of a widely expressed human homologue for the *Drosophila* trp gene. FEBS Lett 373:193–198

Zitt C, Zobei A, Obukhov AG, Harteneck C, Kalkbrenner F, Lückhoff A, Schultz G (1996) Cloning and functional expression of a human Ca^{2+}-permeable cation channel activated by calcium store depletion. Neuron 16:1189–1196

Zitt C, Obukhov AG, Strübing C, Zobel A, Kalkbrenner F, Lückhoff A, Schultz G (1997) Expression of TRPC3 in Chinese hamster ovary cells results in calcium-activated cation currents not related to store depletion. J Cell Biol 138:1333–1341

Zufall F, Ukhanov K, Lucas P, Leinders-Zufall T (2005) Neurobiology of TRPC2: from gene to behavior. Pfluegers Arch 451:61–71

Chapter 8
Mechano- and Chemo-Sensory Polycystins

Amanda Patel(✉), Patrick Delmas, and Eric Honoré

Abstract Polycystins belong to the superfamily of transient receptor potential (TRP) channels and comprise five PKD1-like and three PKD2-like (TRPP) subunits. In this chapter, we review the general properties of polycystins and discuss their specific role in both mechanotransduction and chemoreception. The heteromer PKD1/PKD2 expressed at the membrane of the primary cilium of kidney epithelial cells is proposed to form a mechano-sensitive calcium channel that is opened by physiological fluid flow. Dysfunction or loss of PKD1 or PKD2 polycystin genes may be responsible for the inability of epithelial cells to sense mechanical cues, thus provoking autosomal dominant polycystic kidney disease (ADPKD), one of the most prevalent genetic kidney disorders. *pkd1* and *pkd2* knock-out mice recapitulate the human disease. Similarly, PKD2 may function as a mechanosensory calcium channel in the immotile monocilia of the developing node transducing leftward flow into an increase in calcium and specifying the left–right axis. *pkd2*, unlike *pkd1* knock-out embryos are characterized by right lung isomerism (situs inversus). Mechanical stimuli also induce cleavage and nuclear translocation of the PKD1 C-terminal tail, which enters the nucleus and initiates signaling processes involving the AP-1, STAT6 and P100 pathways. This intraproteolytic mechanism is implicated in the transduction of a change in renal fluid flow to a transcriptional long-term response. The heteromer PKD1L3/PKD2L1 is the basis for acid sensing in specialised sensory cells including the taste bud cells responsible for sour taste. Moreover, PKD1L3/PKD2L1 may be implicated in the chemosensitivity of neurons surrounding the spinal cord canal, sensing protons in the cerebrospinal fluid. These recent results demonstrate that polycystins fulfill a major sensory role in a variety of cells including kidney epithelial cells, taste buds cells and spinal cord neurons. Such mechanisms are involved in short- and long-term physiological

IPMC-CNRS, 660 Route des Lucioles, 06560 Valbonne, France, patel@ipmc.cnrs.fr regulation. Alteration of these pathways culminates in severe human pathologies, including ADPKD.

8.1 Introduction

Autosomal dominant polycystic kidney disease (ADPKD) results from mutations in the *PKD1* or *PKD2* genes encoding polycystin 1 (PKD1) and polycystin 2 (PKD2), respectively (for recent reviews, see Boucher and Sandford 2004; Delmas 2005; Lakkis and Zhou 2003; Nauli and Zhou 2004; Sutters and Germino 2003). This disease is the most frequent genetically inherited renal disorder, affecting approximately 1:1,000. It is characterised by the formation of multiple fluid-filled cysts in the kidney that lead to early onset renal failure.

Polycystins belong to the superfamily of transient receptor potential (TRP) channels (Montell et al. 2002) – so named because Drosophila photoreceptors lacking TRP exhibit a transient voltage response to continuous light. Based on their structure (Fig. 8.1), the polycystin (PC) subfamily (TRPP) is divided into two groups, namely polycystin 1-like (PKD1-like) including *PKD1*, *PKD1L1*, *-2* and *-3* and *PKDREJ*, and polycystin 2-like (PKD2-like), including *PKD2*, *PKD2L1* and *PKD2L2* (Guo et al. 2000b; Yuasa et al. 2002, 2004).

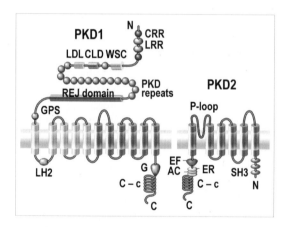

Fig. 8.1 Membrane topology of the polycystins. The predicted architecture of human PKD1 (TRPP1) (*left*) and PKD2 (TRPP2) (*right*) is illustrated. *AC* Acid cluster, *C-c* capacity for coiled-coil formation, *CLD* C-type lectin domain, *CRR* cysteine-rich region, *EF* EF-hand Ca²⁺-binding domain, *ER* endoplasmic reticulum retention signal, *G* G-protein-binding-/activating site, *GPS* G-protein-coupled receptor proteolytic site, *LDL* low-density lipoprotein-like domain, *LH2* lipoxygenase homology, *LRR* leucine-rich repeats, *PKD* repeats, *REJ* receptor for egg jelly, *SH3* src homology 3, *WSC* cell wall integrity and stress response component. The last six transmembrane segments of PKD1 show homology to PKD2. Eight different PKD subunits have been identified, including five PKD1-like and three PKD2-like proteins. The PKD1/PKD2 complex plays a functional role in kidney epithelial cell mechanotransduction. The PKD1L3/PKD2L1 heteromer is involved in proton and sour taste sensing

PKD1-like proteins are large (~460 kDa) integral membrane glycoproteins with an N-terminal extracellular region, 11 predicted transmembrane-spanning segments and a short intracellular C-terminal domain (Hughes et al. 1995) (Fig. 8.1). The extracellular region comprises up to ~3,000 amino acids (in the case of PC1) and contains a number of recognisable protein motifs including ligand binding sites and adhesive domains (Ibraghimov-Beskrovnaya et al. 2000; Moy et al. 1996; Weston et al. 2003). The presence of these domains suggests that PKD1 is involved in interactions with proteins (homophilic and/or heterophilic interactions) and carbohydrates on the extracellular side of the membrane (Malhas et al. 2002). Cellular partners of polycystins are involved mainly in stabilizing the architecture of the cell, a process that seems to be affected in ADPKD. The intracellular domain of PKD1 is rather short (~200 amino acids) and, for most PKD1 proteins, contains a G-protein activation site that can mediate G-protein intracellular signaling (Delmas et al. 2002). The cytoplasmic C-terminal domain of PKD1 can interact with a variety of other proteins involved in cellular signaling (for recent and extensive reviews, see Boucher and Sandford 2004; Delmas 2005; Lakkis and Zhou 2003; Nauli and Zhou 2004; Sutters and Germino 2003).

PKD2-like proteins show moderate similarity to the last six transmembrane segments of PKD1. The PKD2-like proteins have a predicted topology of an integral membrane protein (~110 kDa) with six transmembrane-spanning segments with the N- and C-terminal domains located intracellularly (Mochizuki et al. 1996) (Fig. 8.1). PKD2 contains an endoplasmic reticulum (ER) retention signal within its C-terminal domain that prevents trafficking to the cell surface. The intracellular C-terminal region of PKD2 also contains a Ca^{2+}-binding EF-hand domain. The extracellular loop linking putative transmembrane segments 5 and 6 is thought to harbour the pore-forming sequence. PKD2 proteins form non-selective cationic channels that conduct both monovalent (Na^+, K^+) and divalent (Ca^{2+}) ions (Hanaoka et al. 2000).

Members of the PKD1 and PKD2 groups physically interact through a coiled-coil domain that links their intracellular C-termini regions together (Qian et al. 1997; Tsiokas et al. 1997). The PKD1/PKD2 complex functions as a "receptor-ion channel" complex, with PKD1 acting as the receptor that transduces signals to PKD2, which acts as an ion channel (Delmas et al. 2002, 2004; Hanaoka et al. 2000). PKD1 signaling also involves the activation of a G-protein pathway via the activation of $G\alpha$ proteins and the release of $G\beta\gamma$ proteins (Parnell et al. 1998). For this reason, PKD1 is likely compared to a G-protein coupled receptor (GPCR). However, the G-protein pathway seems no longer predominant when PKD1 is associated with PKD2 (Delmas et al. 2002). The same is true for the constitutive activity of PKD2 channels, which are inhibited by PKD1 interaction (Delmas et al. 2004). PKD1 also promotes the translocation of PKD2 to the membrane. Thus, PKD1 may be a component of the ion channel complex and/or may act as a chaperone partner (Delmas 2005).

In this chapter, we will review recent evidence that the heteromers PKD1/PKD2 and PKD1L3/PKD2L1 are involved in mechanotransduction and chemoreception, respectively.

8.2   Role of the Heteromer PKD1/PKD2
in Mechanotransduction

The PKD1/PKD2 complex is expressed at the plasma membrane of the primary cilium in renal epithelial cells (Nauli et al. 2003). Almost every cell (with the exception of the immune system) has one primary cilium during interphase or during the G0 phase of the cell cycle (Badano et al. 2006; Davenport and Yoder 2005; Eley et al. 2005; Praetorius and Spring 2005; Salisbury 2004; Singla and Reiter 2006). Primary cilia disassemble during the mitotic phase and are reassembled at early interphase. In differentiated kidney epithelial cells, this specialised structure projects into the fluid-filled tubular lumen of the epithelium and is thought to behave as a mechano-sensor that regulates tissue morphogenesis (Nauli et al. 2003; Nauli and Zhou 2004). The dimensions of the primary cilium are about 0.2 μm in diameter and 5 μm in length on average. Structurally, the cilium consists of a microtubule-based axoneme covered by a specialised plasma membrane (Badano et al. 2006; Davenport and Yoder 2005; Eley et al. 2005; Praetorius and Spring 2005; Salisbury 2004; Singla and Reiter 2006). The cilia axoneme emerges from the basal body – a centriole-derived, microtubule organising centre – and extends from the cell surface into the extracellular space (Fig. 8.2). The axoneme is composed of nine separate microtubule doublets at the periphery but lacks a central pair of microtubes. Except for nodal cilia, which exhibit an unusual twirling movement, primary cilia are thought to be non-motile because of the lack of axonemal dyneins (Badano et al. 2006; Davenport and Yoder 2005; Eley et al. 2005; Praetorius and Spring 2005; Salisbury 2004; Singla and Reiter 2006). Both PKD1 and PKD2 are co-localised with other ciliary proteins including cystin, polaris, and alpha and gamma tubulins (Nauli et al. 2003). Although the plasma membrane of the cilium is continuous with the remainder of the cell, protein transit into the cilium is highly regulated. Proteins are moved bidirectionally along the cilium by intraflagellar transport (Badano et al. 2006; Davenport and Yoder 2005; Eley et al. 2005; Praetorius and Spring 2005; Salisbury 2004; Singla and Reiter 2006). Membrane proteins are transported from the cytoplasm onto the ciliary membrane by vesicles targeted for exocytosis at a point adjacent to the ciliary basal body. The kinesin-II motor complex and the cytoplasmic dynein motor complex are responsible for anterograde and retrograde transport, respectively. The localisation of functional polycystins to the primary cilia may be one of the basic cellular requirements to maintain normal kidney structure and function (Nauli et al. 2003; Nauli and Zhou 2004). Mutations that disrupt the function of primary cilia result in a broad spectrum of disorders, including cystic kidneys, hepatic and pancreatic abnormalities, skeletal malformation, obesity, and severe developmental defects (Badano et al. 2006; Davenport and Yoder 2005; Eley et al. 2005; Praetorius and Spring 2005; Salisbury 2004; Singla and Reiter 2006). For instance, mutations in polaris, a structural protein that functions as a cargo molecule in the cilia, provoke PKD (Yoder et al. 2002). In the polaris mutants, cilia are absent or reduced in size, suggesting that defects in ciliogenesis are indeed linked to PKD.

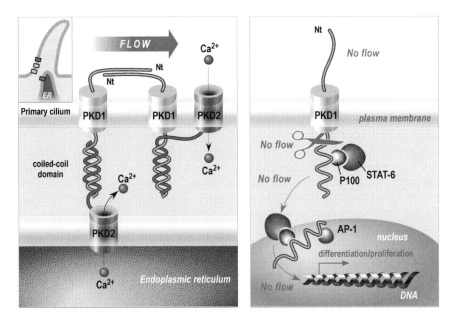

Fig. 8.2 Polycystins and mechanotransduction. PKD1 and PKD2 are expressed at the plasma membrane of the primary cilium of renal epithelial cells. Within this complex, PKD1 acts as a cell surface receptor, while PKD2 is the ion-translocating pore. At rest, the PKD1/PKD2 complex has little background activity and the channel is closed. Bending of the primary cilium by mechanical stimulation (shear stress) and activation of the PKD1/PKD2 complex results in a calcium influx though the open pore of PKD2. In addition, PKD1 may form cis-dimers via PKD domain interaction and may also form bonds between adjacent cells via trans-interactions. Moreover, PKD1 has been shown to be coupled to G protein pathways. Finally, PKD2 also forms a calcium release channel in the endoplasmic reticulum (ER) and may physically interact with PKD1 located in the plasma membrane (*left panel*). Polycystin-2 cation channel function is under the control of microtubular structures in primary cilia of renal epithelial cells. Mechanical stimuli induce cleavage and nuclear translocation of the polycystin-1 C terminus (*right panel*). Polycystin-1, AP-1, STAT6, and P100 function in pathways that transduce the ciliary mechanosensation that is activated in polycystic kidney disease

Cilia are also the sites at which the TRPP ion channels function in the nematode *Caenorhabditis elegans* (Barr and Sternberg 1999). LOV-1 is the closest *C. elegans* homologue of PKD1. LOV-1 is expressed in the sensory cilia of adult *C. elegans* male-specific sensory neurons of the rays, hook and head, which mediate response, vulva location and, potentially, chemotaxis to hermaphrodites, respectively (Barr and Sternberg 1999). PKD-2, the *C. elegans* homologue of PKD2, is expressed in the same neurons as LOV-1, suggesting that they function in the same pathway (Barr and Sternberg 1999).

In renal epithelial cells, the primary cilium is able to bend when the cells are superfused to flow. Increase in intracellular Ca^{2+} concentration is induced by bending the primary cilium with a micropipette or by flow (Praetorius and Spring 2001, 2003a, 2003b, 2005). In the kidney cell line MDCK, this flow response is critically dependent on the primary cilium as immature cells that do not present a primary cilium, or cells from which the cilium has been removed by chloral hydrate treatment, do not respond to flow by increasing intracellular calcium (Praetorius and Spring 2003a). The cilium-dependent calcium response depends on extracellular calcium for initiation of the signal (Nauli et al. 2003). This calcium influx is inhibited by gadolinium and amiloride although it is resistant to blockers of voltage-gated calcium channels. Recently, the polycystin complex has been elegantly demonstrated to be part of the mechano-transduction pathway that senses fluid flow in renal epithelial cells (Fig. 8.2). Cultured epithelial cells lacking PKD1 fail to induce a Ca^{2+} signaling response when exposed to fluid shear stress (Nauli et al. 2003). Similarly, when PKD2 channels are inactivated by antibodies, the Ca^{2+} signal induced by mechanical stimulation is impaired as by inhibitors of the ryanodine receptor, whereas inhibitors of G proteins, phospholipase C and inositol 1,4,5-trisphosphate (InsP3) receptors have no effect (Nauli et al. 2003). Moreover, cells from *pkd2* knockout mice do not respond to flow stimulation. However, when these cells were transfected with WT *pkd2*, the mechanosensory function is recovered. It has recently been shown that the extracellular N-terminal region of PKD1 presents characteristic mechanical properties of stability, strength and elasticity, apparently provided by their PKD domains, supporting a possible mechanosensory function (Forman et al. 2005; Qian et al. 2005). These results suggest that the primary cilium from renal epithelial cells is a functional site for the PKD1/PKD2 complex mediating mechanotransduction, i.e. sensitivity to shear stress. The conformational change of the large extracellular domain of PKD1 upon mechanical stimulation leads to the opening of the PKD2 ionic channel through its interaction by its C terminus. The signal is then amplified through the release of intracellular calcium via a calcium-induced calcium release mechanism by the ryanodine receptors.

In the embryonic node, a group of motile primary cilia generates a leftward flow of extraembryonic fluid that is thought to generate the first cues for the left–right (LR) axis of symmetry (Eley et al. 2005; Nauli and Zhou 2004). A group of non-motile cilia senses the flow (McGrath et al. 2003). Disruption of the nodal flow or its sensing is responsible for situs inversus. *pkd2* knock-out embryos are characterised by right lung isomerism, i.e. mirror-image duplication of the right lung on the left side, randomisation of heart looping and placement of stomach and spleen (Pennekamp et al. 2002). The mechanosensory cilia at the periphery of the node transform the directional extracellular fluid flow into an asymmetric intracellular calcium signal. Mouse gastrula have higher intracellular calcium concentrations on the left margin of the node that are proposed to activate the Nodal cascade (McGrath et al. 2003; Pennekamp et al. 2002). The absence of PKD1 localisation to cilia and a lack of laterality defects in *pkd1* knock-out embryos demonstrates a PKD1-independent function of PKD2 in LR axis formation (Karcher et al. 2005). Furthermore, PKD2 has recently been shown to traffic to cilia independently of PKD1 by using an N-terminal RVxP motif (Geng et al. 2006). PKD2 may thus function as a mechanosensory calcium channel, independently of PKD1, in the immotile monocilia of the developing node (Karcher et al. 2005). The exact PKD2 mechanism in LR axis formation remains to be elucidated (Pennekamp et al. 2002).

A recent study provided evidence for the presence of single cation channels in isolated kidney primary cilia (Raychowdhury et al. 2005). Ciliary membrane channels were also observed by reconstitution in a lipid bilayer system (Raychowdhury et al. 2005). The most frequent cation-selective channel had a single channel conductance of 156 pS (Na^+ gradient with 150 mM versus 15 mM, in cis- and trans-compartment, respectively), which was inhibited by an anti-PKD2 antibody. A comparison of reconstituted ciliary versus plasma membranes indicated as much as 400-fold higher channel activity in ciliary membranes as averaged mean currents were divided by protein content (Li et al. 2006). At least three channel proteins, PKD-2, TRPC1 and, interestingly, the epithelial sodium channel, were immunodetected in this organelle (Raychowdhury et al. 2005). The microtubular disrupter colchicine abolished, while addition of the microtubule stabilizer taxol increased, ciliary PKD2 channel activity (Li et al. 2006). These data suggest that PKD2 cation channel function is under the control of microtubular structures in primary cilia of renal epithelial cells. It is of interest to note that PKD2 associates with TRPC1, another TRP cationic channel that has recently been shown to be stretch-activated (Maroto et al. 2005; Tsiokas et al. 1999). Moreover, TRPV4 and PKD2 interact and also co-localise in the kidney primary cilium (Giamarchi et al. 2006). The physiological significance of the TRPC1/PKD2 and TRPV4/PKD2 complexes remains however, to be determined.

PKD1 undergoes a proteolytic cleavage that releases part of its C-terminal tail that enters the nucleus and initiates signaling processes (Chauvet et al. 2004; Low et al. 2006) (Fig. 8.2). The cleavage occurs in vivo in association with alterations in mechanical stimuli. The putative cleavage site has not yet been exactly defined. After ureteral ligation, which reduces the tubular flow and elevates intratubular pressure, the PKD1 C-terminal tail accumulates in the nucleus (Chauvet et al. 2004). The nuclear accumulation of the C-terminal fragment of PKD1 also occurs in polycystic kidney Kif3A mutant mice lacking primary cilium (Chauvet et al. 2004). Interestingly, PKD2 impairs the nuclear localisation of the PKD1 C-terminal tail (Chauvet et al. 2004). These results suggest that PKD2 may act as a buffer regulating the nuclear translocation of the released PKD1 C terminus. The C-terminal domain of PKD1 has been shown to stimulate the AP-1 pathway, although these results have recently been challenged (Chauvet et al. 2004; Low et al. 2006). The PKD1 tail also interacts with the transcription factor STAT6 and the co-activator P100, and it stimulates STAT6-dependent gene expression (Low et al. 2006). Under normal conditions, STAT6 localises to primary cilia of renal epithelial cells (Low et al. 2006). Cessation of apical fluid flow results in nuclear translocation of STAT6. Cyst-lining cells in ADPKD exhibit elevated levels of nuclear STAT6, P100, and the PKD1 C-terminal tail. These results suggest that this mechanism may play an important role in the pathogenesis of human kidney disease. Indeed, exogenous expression of the human PKD1 tail results in renal cyst formation in zebrafish embryos (Low et al. 2006). The finding that STAT6 translocates from cilia to nuclei in the absence of apical fluid flow makes it highly likely that these PKD1/AP-1/STAT6/P100 pathways are involved in the transduction of a mechanical signal into a transcriptional response and may play a role in a long-term response to changes in renal fluid flow (Chauvet et al. 2004; Low et al. 2006) (Fig. 8.2).

Renal injury may create conditions of ceased lumenal fluid flow, to which epithelial cells need to respond by initiating a repair program that typically involves dedifferentiation and proliferation and that is dependent on AP-1, STAT6 and P100. Such a proteolytic system bypasses adaptor proteins and kinase/phosphatase cascades. If PKD1 is lost (knock-out and mutations), STAT6 can no longer be sequestered at the cilia and may become constitutively active in the nucleus and contribute to cyst formation (Low et al. 2006). Flow- and/or pressure-dependent regulated intramembrane proteolysis of PKD1 may thus play a key role in ADPKD (Chauvet et al. 2004; Low et al. 2006). It should also be noted that another internal proteolysis occurs at the G protein coupled receptor proteolytic site (GPS) in the amino terminal domain of PKD1. This cleavage requires the receptor for egg jelly domain and is disrupted in ADPKD.

In ER membrane stores, PKD2 acts as a Ca^{2+} release channel that amplifies transient Ca^{2+} changes initiated by $InsP_3$-generating membrane receptors (Koulen et al. 2002) (Fig. 8.2). PKD2 opening is induced by an increase in cytosolic Ca^{2+} (Ca^{2+}-induced Ca^{2+} release) (Koulen et al. 2002). PKD2 has also been shown to interact with type I $InsP_3$ receptor to modulate intracellular calcium signaling (Li et al. 2005). Stretch-sensitive intracellular calcium stores distinct from the $InsP_3$-, ryanodine- and NAADP-sensitive stores have been described (Ji et al. 2002; Mohanty and Li 2002). It remains to be determined whether PKD2 contributes to the stretch-sensitive intracellular calcium release channels. As in the case of the nodal primary cilia, PKD2 may by itself contribute to the release of calcium or, as described in the kidney primary cilia, PKD1 expressed at the plasma membrane may interact via its C-terminal coiled-coil domain with the C terminus of PKD2 inserted in the ER membrane (Bichet et al. 2006; Delmas 2004) (Fig. 8.2). A conformational change of PKD1 induced by mechanical stimulation could lead to the opening of PKD2 and release of intracellular calcium.

8.3   Role of the Heteromer PKD1L3/PKD2L1 in Chemoreception

Recent reports based on subtracted cDNA libraries enriched in sequences expressed in taste buds of the circumvallate papillae, in situ hybridisation using probes of the 33 TRP mouse genes as well as multi-step bioinformatics and expression screening strategies have revealed that the PKD1L3/PKD2L1 complex may be the basis for acid sensing in specialised sensory cells (Huang et al. 2006; Ishimaru et al. 2006; LopezJimenez et al. 2006).

Taste reception occurs at the apical tip of taste cells that form taste buds (Fig. 8.3). Each taste bud is composed of about 50–100 cells that possess microvilli. PKD1L3 and PKD2L1 are expressed in a subset of taste cells that are different from those that express components of sweet, bitter and umami signal transduction pathways (Huang et al. 2006; Ishimaru et al. 2006; LopezJimenez et al. 2006). PKD1L3 shares the membrane topology of PKD1 with 11 transmembrane segments, a large

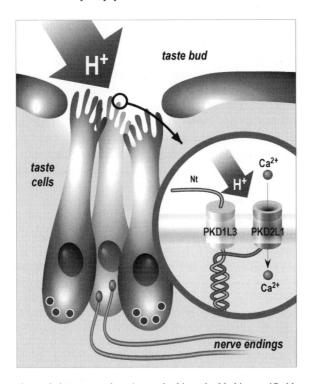

Fig. 8.3 Polycystins and chemoreception. A taste bud is embedded in stratified layers of epithelial cells (not depicted). Cells coupled by gap junctions are the taste cells responding to specific stimuli including sour, sweet, salt and bitter stimuli. The stratum corneum of the epithelium opens to form a taste pore through which microvilli of taste cells protude. Sour taste cells respond to acidic stimuli including citric acid. Taste cells terminate at the basement membrane that separates the epithelium from the papillary layer. A taste cell is shown to synapse with chorda tympani neurons. Expression of the PKD1L3/PKD2L1 complex in specific cells is responsible for sour taste

extracellular N-terminal domain and a cytosolic C-terminal domain (Li et al. 2003). PKD2L1 (Polycystin-L) is homologous to PKD2 (71% similarity in protein sequence) and shares the same topology, with six transmembrane segments and both N- and C-termini facing the cytosol (Basora et al. 2002; Guo et al. 2000a; Nomura et al. 1998; Wu et al. 1998). PKD2L1 acts as a calcium-regulated non-selective cation channel permeable to mono- and di-valent cations with a unitary conductance of 137 pS (Chen et al. 1999). Channel activity is increased when either the extracellular or intracellular calcium ion concentration is raised, indicating that PKD2L1 may act as a transducer of calcium-mediated signaling in vivo (Chen et al. 1999). The EF-hand and other parts of the carboxyl tail of PKD2L1 are not determinants of channel activation/inactivation (Li et al. 2002). Depending on the expression system (i.e. Xenopus oocyte versus mammalian cells) the interaction with PKD1 is necessary for functional expression of PKD2L1 (Chen et al. 1999; Murakami et al. 2005). Co-expression of PKD2L1 together with PKD1 resulted in the expression of PKD2L1 channels on the cell surface of mammalian cells, whereas PKD2L1 expressed alone was retained within the ER (Murakami et al. 2005). The coiled-coil domain is responsible for retaining PKD2L1 within the ER. Co-expression of PKD1 and PKD2L1 resulted in the formation of functional cation channels that were opened by hypo-osmotic stimulation, whereas neither molecule formed functional channels when expressed alone (Murakami et al. 2005).

PKD1L3 is co-expressed with PKD2L1 in about 20% of taste receptor cells of the circumvallate and foliate papillae (Huang et al. 2006; Ishimaru et al. 2006; LopezJimenez et al. 2006) (Fig. 8.3). In contrast, PKD1L3 was absent from the fungiform papillae or the palate although PKD2L1 was expressed in these taste cells. PKD1L3 is not co-expressed with either gustducin, a G protein alpha subunit that mediates bitter, and perhaps sweet and umami, taste transduction, or T1R3, a GPCR involved in the detection of sweet-tasting components and amino acids (Ishimaru et al. 2006; LopezJimenez et al. 2006). Moreover, PKD1L3 and PKD2L1 do not co-express with TRPM5, a TRP cationic channel responsible for normal bitter, sweet and umami sensation (Huang et al. 2006). PKD1L3 and PKD2L1 co-immunoprecipitate from co-transfected cells (Ishimaru et al. 2006). The PKD2L1 proteins are accumulated in the taste pore region, where taste chemicals are detected (Huang et al. 2006). When PKD1L3 was expressed alone in HEK cells, little surface expression was observed (Ishimaru et al. 2006). However, in the presence of PKD2L1, robust cell surface expression is observed, suggesting that interaction between both subunits is required for their cell surface expression (Ishimaru et al. 2006).

PKD1L3/PKD2L1 are activated by various acids when co-expressed in heterologous cells but not by other classes of tastants (Huang et al. 2006; Ishimaru et al. 2006). Cells expressing both subunits respond by an increase in intracellular calcium upon stimulation by acidic solutions, including citric acid, HCl and malic acid (Ishimaru et al. 2006) (Fig. 8.3). When stimulated by citric acid, cells co-expressing PKD1L3 and PKD2L1 show large inward non-selective cationic currents (reversing at 0 mV). A delay was observed between acid stimulation and current activation. HCl was less potent than citric acid at the same pH (Ishimaru et al. 2006). This is consistent with the fact that weak acids taste more sour than strong acids. In cells expressing PKD1L3 or PKD2L1 alone no signal is detected (Ishimaru et al. 2006). No activation was obtained with other taste stimuli including NaCl, bitter chemicals, sucrose, saccharin or umami. By analogy to PKD1 and PKD2, PKD1L3 might function as a sour-sensing receptor and PKD2L1 as an associated ion channel (Huang et al. 2006; Ishimaru et al. 2006; LopezJimenez et al. 2006). The exact mechanism of acidic activation, including calcium dependency, remains to be defined.

These results suggest that PKD1L3 and PKD2L1 heteromers may function as sour taste receptors (Huang et al. 2006; Ishimaru et al. 2006; LopezJimenez et al. 2006). To dissect the function of the PKD2L1-expressing cells in the tongue, cellular ablation was performed in mice by targeted expression of attenuated diphtheria toxin (Huang et al. 2006). These mice have a complete loss of PKD2L1-expressing taste receptor cells. Remarkably, these animals lost their ability to recognise sour taste, with a lack of response to citric acid, HCl, tartaric acid and acetic acid (Huang et al. 2006). However, responses to sweet, umami, bitter or salty tastants are preserved.

These results elegantly confirm that PKD2L1-expressing cells are indeed the sour taste sensors (Huang et al. 2006).

In the mouse, PKD1L3 and PKD2L1 are also expressed in the testis as well as in a subset of neurons surrounding the central canal of the spinal cord (Huang et al. 2006; Veldhuisen et al. 1999). These neurons may function as chemoreceptors sensing protons in the cerebrospinal fluid. PKD2L1-expressing neurons are sensitive to pH stimulation. Acidification of the extracellular medium from pH 7.4 to 6.0 evokes a reversible increase in the firing rate of these neurons, although control neurons that do not express PKD2L1 are not responsive to acidosis (Huang et al. 2006). It is therefore suggested that the PKD2L1-expressing neurons of the spinal cord are sensors of cerebrospinal and ventricular pH.

Taken together, these recent findings indicate that the heteromer PKD1L3/PKD2L1 plays a key role in proton sensing for both taste cells and spinal cord neurons (Huang et al. 2006; Ishimaru et al. 2006; LopezJimenez et al. 2006). In contrast to PKD1 and PKD2, both PKD2L1 and PKD1L3 are excluded as candidate genes for autosomal recessive polycystic kidney disease, autosomal dominant polycystic liver disease, and the third form of ADPKD (Veldhuisen et al. 1999).

In conclusion, the discussion of recent breakthroughs in polycystin (TRPP) research presented in this chapter indicates that specific PKD heteromultimers fulfill key sensory functions in specialised cells. A key feature is the expression of these molecular complexes in the primary cilium of various cells, including kidney epithelial cells and embryonic node cells. However, it should be remembered that the PKD1/PKD2 complex has also been visualised in other cellular locations such as lateral cell junctions, where it may be involved in different functions such as cell–cell interactions and cell–matrix adhesion. Future work will be needed to understand how sensory stimuli gate polycystin receptor/ion channel complexes.

Acknowledgements This work was supported by the Agence Nationale de la Recherche (ANR - 2005) Cardio-vasculaire-Obésité-Diabète (E.H., A.P. and P.D.), the Fondation del Duca (E.H.), The Association for Information and Research on Genetic Kidney Diseases (AIRG-France) (E.H.), the Fondation de France (A.P.) and the Fondation Schlumberger (P.D.).

References

Badano JL, Mitsuma N, Beales PL, Katsanis N (2006) The ciliopathies: an emerging class of human genetic disorders. Annu Rev Genomics Hum Genet 7:125–148

Barr MM, Sternberg PW (1999) A polycystic kidney-disease gene homologue required for male mating behaviour in *C. elegans*. Nature 401:386–389

Basora N, Nomura H, Berger UV, Stayner C, Guo L, Shen X, Zhou J (2002) Tissue and cellular localization of a novel polycystic kidney disease-like gene product, polycystin-L. J Am Soc Nephrol 13:293–301

Bichet D, Peters D, Patel A, Delmas P, Honoré E (2006) The cardiovascular polycystins: insights from autosomal dominant polycystic kidney disease and transgenic animal models. Trends Cardiovasc Res 16:292–298

Boucher C, Sandford R (2004) Autosomal dominant polycystic kidney disease (ADPKD, MIM 173900, PKD1 and PKD2 genes, protein products known as polycystin-1 and polycystin-2). Eur J Hum Genet 12:347–354

Chauvet V, Tian X, Husson H, Grimm DH, Wang T, Hiesberger T, Igarashi P, Bennett AM, Ibraghimov-Beskrovnaya O, Somlo S, Caplan MJ (2004) Mechanical stimuli induce cleavage and nuclear translocation of the polycystin-1 C terminus. J Clin Invest 114:1433–1443

Chen XZ, Vassilev PM, Basora N, Peng JB, Nomura H, Segal Y, Brown EM, Reeders ST, Hediger MA, Zhou J (1999) Polycystin-L is a calcium-regulated cation channel permeable to calcium ions. Nature 401:383–386

Davenport JR, Yoder BK (2005) An incredible decade for the primary cilium: a look at a once-forgotten organelle. Am J Physiol Renal Physiol 289:F1159–F1169

Delmas P (2004) Polycystins: from mechanosensation to gene regulation. Cell 118:145–148

Delmas P (2005) Polycystins: polymodal receptor/ion-channel cellular sensors. Pfluegers Arch 451:264–276

Delmas P, Nomura H, Li X, Lakkis M, Luo Y, Segal Y, Fernandez-Fernandez JM, Harris P, Frischauf AM, Brown DA, Zhou J (2002) Constitutive activation of G-proteins by polycystin-1 is antagonized by polycystin-2. J Biol Chem 277:11276–11283

Delmas P, Nauli SM, Li X, Coste B, Osorio N, Crest M, Brown DA, Zhou J (2004) Gating of the polycystin ion channel signaling complex in neurons and kidney cells. FASEB J 18:740–742

Eley L, Yates LM, Goodship JA (2005) Cilia and disease. Curr Opin Genet Dev 15:308–314

Forman JR, Qamar S, Paci E, Sandford RN, Clarke J (2005) The remarkable mechanical strength of polycystin-1 supports a direct role in mechanotransduction. J Mol Biol 349:861–871

Geng L, Okuhara D, Yu Z, Tian X, Cai Y, Shibazaki S, Somlo S (2006) Polycystin-2 traffics to cilia independently of polycystin-1 by using an N-terminal RVxP motif. J Cell Sci 119:1383–1395

Giamarchi A, Padilla F, Coste B, Raoux M, Crest M, Honore E, Delmas P (2006) The versatile nature of the calcium-permeable cation channel TRPP2. EMBO Rep 7:787–793

Guo L, Chen M, Basora N, Zhou J (2000a) The human polycystic kidney disease 2-like (PKDL) gene: exon/intron structure and evidence for a novel splicing mechanism. Mamm Genome 11:46–50

Guo L, Schreiber TH, Weremowicz S, Morton CC, Lee C, Zhou J (2000b) Identification and characterization of a novel polycystin family member, polycystin-L2, in mouse and human: sequence, expression, alternative splicing, and chromosomal localization. Genomics 64:241–251

Hanaoka K, Qian F, Boletta A, Bhunia AK, Piontek K, Tsiokas L, Sukhatme VP, Guggino WB, Germino GG (2000) Co-assembly of polycystin-1 and -2 produces unique cation-permeable currents. Nature 408:990–994

Huang AL, Chen X, Hoon MA, Chandrashekar J, Guo W, Trankner D, Ryba NJ, Zuker CS (2006) The cells and logic for mammalian sour taste detection. Nature 442:934–938

Hughes J, Ward CJ, Peral B, Aspinwall R, Clark K, San Millan JL, Gamble V, Harris PC (1995) The polycystic kidney disease 1 (PKD1) gene encodes a novel protein with multiple cell recognition domains. Nat Genet 10:151–160

Ibraghimov-Beskrovnaya O, Bukanov NO, Donohue LC, Dackowski WR, Klinger KW, Landes GM (2000) Strong homophilic interactions of the Ig-like domains of polycystin-1, the protein product of an autosomal dominant polycystic kidney disease gene, PKD1. Hum Mol Genet 9:1641–1649

Ishimaru Y, Inada H, Kubota M, Zhuang H, Tominaga M, Matsunami H (2006) Transient receptor potential family members PKD1L3 and PKD2L1 form a candidate sour taste receptor. Proc Natl Acad Sci USA 103:12569–12574

Ji G, Barsotti RJ, Feldman ME, Kotlikoff MI (2002) Stretch-induced calcium release in smooth muscle. J Gen Physiol 119:533–544

Karcher C, Fischer A, Schweickert A, Bitzer E, Horie S, Witzgall R, Blum M (2005) Lack of a laterality phenotype in Pkd1 knock-out embryos correlates with absence of polycystin-1 in nodal cilia. Differentiation 73:425–432

Koulen P, Cai Y, Geng L, Maeda Y, Nishimura S, Witzgall R, Ehrlich BE, Somlo S (2002) Polycystin-2 is an intracellular calcium release channel. Nat Cell Biol 4:191–197

Lakkis M, Zhou J (2003) Molecular complexes formed with polycystins. Nephron Exp Nephrol 93:e3–e8

Li A, Tian X, Sung SW, Somlo S (2003) Identification of two novel polycystic kidney disease-1-like genes in human and mouse genomes. Genomics 81:596–608

Li Q, Liu Y, Zhao W, Chen XZ (2002) The calcium-binding EF-hand in polycystin-L is not a domain for channel activation and ensuing inactivation. FEBS Lett 516:270–278

Li Q, Montalbetti N, Wu Y, Ramos AJ, Raychowdhury MK, Chen XZ, Cantiello HF (2006) Polycystin-2 cation channel function is under the control of microtubular structures in primary cilia of renal epithelial cells. J Biol Chem 281:37566–37575

Li Y, Wright JM, Qian F, Germino GG, Guggino WB (2005) Polycystin 2 interacts with type I inositol 1,4,5-trisphosphate receptor to modulate intracellular Ca^{2+} signaling. J Biol Chem 280:41298–41306

LopezJimenez ND, Cavenagh MM, Sainz E, Cruz-Ithier MA, Battey JF, Sullivan SL (2006) Two members of the TRPP family of ion channels, Pkd1l3 and Pkd2l1, are co-expressed in a subset of taste receptor cells. J Neurochem 98:68–77

Low SH, Vasanth S, Larson CH, Mukherjee S, Sharma N, Kinter MT, Kane ME, Obara T, Weimbs T (2006) Polycystin-1, STAT6, and P100 function in a pathway that transduces ciliary mechanosensation and is activated in polycystic kidney disease. Dev Cell 10:57–69

Malhas AN, Abuknesha RA, Price RG (2002) Interaction of the leucine-rich repeats of polycystin-1 with extracellular matrix proteins: possible role in cell proliferation. J Am Soc Nephrol 13:19–26

Maroto R, Raso A, Wood TG, Kurosky A, Martinac B, Hamill OP (2005) TRPC1 forms the stretch-activated cation channel in vertebrate cells. Nat Cell Biol 7:179–185

McGrath J, Somlo S, Makova S, Tian X, Brueckner M (2003) Two populations of node monocilia initiate left-right asymmetry in the mouse. Cell 114:61–73

Mochizuki T, Wu G, Hayashi T, Xenophontos SL, Veldhuisen B, Saris JJ, Reynolds DM, Cai Y, Gabow PA, Pierides A, Kimberling WJ, Breuning MH, Deltas CC, Peters DJ, Somlo S (1996) PKD2, a gene for polycystic kidney disease that encodes an integral membrane protein. Science 272:1339–1342

Mohanty MJ, Li X (2002) Stretch-induced Ca(2+) release via an IP(3)-insensitive Ca(2+) channel. Am J Physiol Cell Physiol 283:C456–462

Montell C, Birnbaumer L, Flockerzi V, Bindels RJ, Bruford EA, Caterina MJ, Clapham DE, Harteneck C, Heller S, Julius D, Kojima I, Mori Y, Penner R, Prawitt D, Scharenberg AM, Schultz G, Shimizu N, Zhu MX (2002) A unified nomenclature for the superfamily of TRP cation channels. Mol Cell 9:229–231

Moy GW, Mendoza LM, Schulz JR, Swanson WJ, Glabe CG, Vacquier VD (1996) The sea urchin sperm receptor for egg jelly is a modular protein with extensive homology to the human polycystic kidney disease protein, PKD1. J Cell Biol 133:809–817

Murakami M, Ohba T, Xu F, Shida S, Satoh E, Ono K, Miyoshi I, Watanabe H, Ito H, Iijima T (2005) Genomic organization and functional analysis of murine PKD2L1. J Biol Chem 280:5626–5635

Nauli SM, Zhou J (2004) Polycystins and mechanosensation in renal and nodal cilia. Bioessays 26:844–856

Nauli SM, Alenghat FJ, Luo Y, Williams E, Vassilev P, Li X, Elia AE, Lu W, Brown EM, Quinn SJ, Ingber DE, Zhou J (2003) Polycystins 1 and 2 mediate mechanosensation in the primary cilium of kidney cells. Nat Genet 33:129–137

Nomura H, Turco AE, Pei Y, Kalaydjieva L, Schiavello T, Weremowicz S, Ji W, Morton CC, Meisler M, Reeders ST, Zhou J (1998) Identification of PKDL, a novel polycystic kidney disease 2-like gene whose murine homologue is deleted in mice with kidney and retinal defects. J Biol Chem 273:25967–25973

Parnell SC, Magenheimer BS, Maser RL, Rankin CA, Smine A, Okamoto T, Calvet JP (1998) The polycystic kidney disease-1 protein, polycystin-1, binds and activates heterotrimeric G-proteins in vitro. Biochem Biophys Res Commun 251:625–631

Pennekamp P, Karcher C, Fischer A, Schweickert A, Skryabin B, Horst J, Blum M, Dworniczak B (2002) The ion channel polycystin-2 is required for left-right axis determination in mice. Curr Biol 12:938–943

Praetorius HA, Spring KR (2001) Bending the MDCK cell primary cilium increases intracellular calcium. J Membr Biol 184:71–79

Praetorius HA, Spring KR (2003a) Removal of the MDCK cell primary cilium abolishes flow sensing. J Membr Biol 191:69–76

Praetorius HA, Spring KR (2003b) The renal cell primary cilium functions as a flow sensor. Curr Opin Nephrol Hypertens 12:517–520

Praetorius HA, Spring KR (2005) A physiological view of the primary cilium. Annu Rev Physiol 67:515–529

Qian F, Germino FJ, Cai Y, Zhang X, Somlo S, Germino GG (1997) PKD1 interacts with PKD2 through a probable coiled-coil domain. Nat Genet 16:179–183

Qian F, Wei W, Germino G, Oberhauser A (2005) The nanomechanics of polycystin-1 extracellular region. J Biol Chem 280:40723–40730

Raychowdhury MK, McLaughlin M, Ramos AJ, Montalbetti N, Bouley R, Ausiello DA, Cantiello HF (2005) Characterization of single channel currents from primary cilia of renal epithelial cells. J Biol Chem 280:34718–34722

Salisbury JL (2004) Primary cilia: putting sensors together. Curr Biol 14:R765–767

Singla V, Reiter JF (2006) The primary cilium as the cell's antenna: signaling at a sensory organelle. Science 313:629–633

Sutters M, Germino GG (2003) Autosomal dominant polycystic kidney disease: molecular genetics and pathophysiology. J Lab Clin Med 141:91–101

Tsiokas L, Kim E, Arnould T, Sukhatme VP, Walz G (1997) Homo- and heterodimeric interactions between the gene products of PKD1 PKD2. Proc Natl Acad Sci USA 94:6965–6970

Tsiokas L, Arnould T, Zhu C, Kim E, Walz G, Sukhatme VP (1999) Specific association of the gene product of PKD2 with the TRPC1 channel. Proc Natl Acad Sci USA 96:3934–3939

Veldhuisen B, Spruit L, Dauwerse HG, Breuning MH, Peters DJ (1999) Genes homologous to the autosomal dominant polycystic kidney disease genes (PKD1 and PKD2). Eur J Hum Genet 7:860–872

Weston BS, Malhas AN, Price RG (2003) Structure–function relationships of the extracellular domain of the autosomal dominant polycystic kidney disease-associated protein, polycystin-1. FEBS Lett 538:8–13

Wu G, Hayashi T, Park JH, Dixit M, Reynolds DM, Li L, Maeda Y, Cai Y, Coca-Prados M, Somlo S (1998) Identification of PKD2L, a human PKD2-related gene: tissue-specific expression and mapping to chromosome 10q25. Genomics 54:564–568

Yoder BK, Tousson A, Millican L, Wu JH, Bugg CE Jr, Schafer JA, Balkovetz DF (2002) Polaris, a protein disrupted in orpk mutant mice, is required for assembly of renal cilium. Am J Physiol Renal Physiol 282:F541–552

Yuasa T, Venugopal B, Weremowicz S, Morton CC, Guo L, Zhou J (2002) The sequence, expression, and chromosomal localization of a novel polycystic kidney disease 1-like gene, PKD1L1, in human. Genomics 79:376–386

Yuasa T, Takakura A, Denker BM, Venugopal B, Zhou J (2004) Polycystin-1L2 is a novel G-protein-binding protein. Genomics 84:126–138

Chapter 9
Biophysics of CNG Ion Channels

Peter H. Barry(✉), Wei Qu, and Andrew J. Moorhouse

Abstract Cyclic nucleotide-gated (CNG) ion channels are cation-selective, opened by intracellular cyclic nucleotides like cAMP and cGMP, and present in many different neurons and non-neuronal cells. This chapter will concentrate primarily on the biophysical aspects of retinal and olfactory CNG channels, with special reference to ion permeation and selectivity and their underlying molecular basis, and will include a brief overview of the physiological function of CNG channels in both olfaction and phototransduction. We will review the subunit composition and molecular structure of the CNG channel and its similarity to the closely related potassium channels, and will also briefly outline the currently accepted molecular basis underlying activation of the channel and the location of the channel 'gate'. We will then outline some general methodologies for investigating ion permeation and selectivity, before reviewing the ion permeation and selectivity properties of native and recombinant CNG channels. We will discuss divalent ion permeation through the channel and the mechanism of channel block by divalent ions. The chapter will conclude by discussing the results

Department of Physiology and Pharmacology, School of Medical Sciences, The University of New South Wales, UNSW Sydney 2052, Australia, P.Barry@unsw.edu.au of recent experiments to investigate the molecular determinants of cation-anion selectivity in the channel.

9.1 Introduction

Cyclic nucleotide-gated (CNG) ion channels are "non-selective" cation channels that link cellular excitability to the intracellular concentration of cyclic nucleotides such as guanosine 3′, 5′-cyclic monophosphate (cGMP) or adenosine 3′, 5′-cyclic monophosphate (cAMP). CNG channels have been identified in many neurons and non-neuronal cells, although the physiological function and properties are most clearly established for CNG channels of photoreceptors and olfactory sensory neurons. Binding of cyclic nucleotides to the CNG channel results in opening of the integral membrane pore, which is readily permeable to Na^+, K^+ and Ca^{2+} ions, and a depolarisation of the membrane potential.

The activation, modulation and permeability properties differ somewhat amongst different CNG channel subtypes and under different physiological conditions, making the elucidation of the molecular and biophysical principles of channel gating and ion permeation important for a complete understanding of their physiological role. The CNG channels belong to the large family of "P-loop"-containing cation channels and show some structural and functional similarities to K^+, Na^+ and Ca^{2+} channels. Recent advances in elucidating the molecular and structural determinants of function in these related channels therefore have implications for the understanding of CNG channel function and, conversely, knowledge of the biophysics of the CNG channel can help elucidate the properties of those other channels. This chapter will concentrate primarily on the biophysical aspects of retinal and olfactory CNG channels, with special reference to ion permeation and selectivity and their underlying molecular basis. We will also give a brief overview of the physiological function of CNG channels in olfaction and phototransduction, and of the molecular structure of the pore and its role in activation of the channel. For a series of general reviews on CNG channels, see Yau and Baylor (1989), Menini (1995), Finn et al. (1996), Zagotta and Siegelbaum (1996), Kaupp and Seifert (2002), Matulef and Zagotta (2003) and Pifferi et al. (2006).

9.2 Physiological Function of Retinal and Olfactory CNG Channels

9.2.1 Visual Transduction

CNG channels were first identified in the plasma membrane of frog retinal rod outer segments (Fesenko et al. 1985) and in the outer segments of catfish cones (Haynes and Yau 1985), and their role in photoreceptor transduction has been subsequently extensively studied (e.g. see reviews of Yau and Baylor 1989; Burns and Baylor 2001; Kaupp and Seifert 2002; Matulef and Zagotta 2003). We will describe their role in rod phototransduction (Fig. 9.1a), responsible for vision in low ambient light levels, though the basic transduction principles apply also to the cones, responsible for colour vision and vision in bright light.

The initial steps of phototransduction occur in the outer segment of the rod, where there is a stack of free-floating disk-shaped membranes, with a special set of functional proteins for responding to a light signal (e.g., Fig. 9.1a, c). In the dark, there is a high concentration of cGMP, formed from GTP by guanylate cyclase, which results in the CNG channels being open and an inward movement of cations down their electrochemical gradient; Na^+, Ca^{2+} (Fig. 9.1b) and a very small proportion being Mg^{2+} ions. There will also be a small efflux of K^+ ions down its electrochemical gradient, but since this driving force is small compared to the inward driving force for cations, its contribution to phototransduction is minimal. In retinal rods under physiological conditions, this steady-state "dark" current is predominantly carried by Na^+ influx into the rod outer segment, with about 15% of this current being carried by Ca^{2+} and about 5% by Mg^{2+} (Yau and Baylor 1989). An electrogenic Na/Ca–K exchanger extrudes most of the incoming Ca^{2+} ions along with some K^+, exchanging them for Na^+ ions (at a ratio of 1:1:4; see Fig. 9.1b). The α-subunit of the CNG channel can bind directly to this exchanger (Schwarzer et al. 2000), suggesting a very close physical and functional relationship between channel and exchanger. The net inward cation current in the rod outer segment will depolarise the photoreceptor. In the rod inner segment, there is a different distribution of channels and transporters, with a high density of the Na^+/K^+ ATPase pumps and K^+ channels. The inner segment outward K^+ current completes the circuit for the dark current (Fig. 9.1a). In the dark, this steady-state dark current results in a depolarised membrane potential (V_m about $-40\,mV$) being in between E_{Na} and E_K, the equilibrium potentials for Na^+ and K^+, respectively (Fig. 9.1d).

In the light, the chromophore retinal (a derivative of vitamin A that is embedded within rhodopsin in the outer segment) absorbs light and undergoes isomerisation, resulting in a conformational change of rhodopsin to a photo-activated state (R^* in Fig. 9.1c), which results in the activation of a specific rod G protein (transducin; G in Fig. 9.1c). Transducin binds GTP in exchange for GDP and activates a phosphodiesterase (P in Fig. 9.1c), which hydrolyses cGMP into 5'-GMP. In response to the resulting decrease in cGMP concentration, the CNG channels close, the dark current is radically reduced and the membrane potential hyperpolarises towards E_K due to the now more dominant influence of the K^+ channels in the rod inner segment (Fig. 9.1e). The reduced cation influx through the CNG channels, coupled with the continued activity of the Na/Ca–K exchanger results in a decrease in the local Ca^{2+} concentration in the rod outer segment. The membrane hyperpolarisation causes a decrease in the release of glutamate from the rod presynaptic terminals and ultimately the sensation of light.

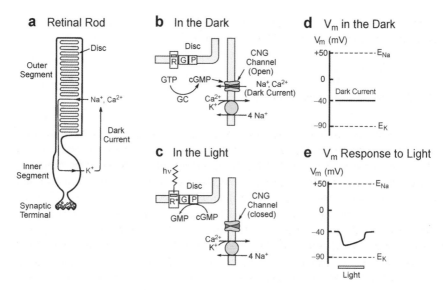

Fig. 9.1 The role of cyclic nucleotide-gated (CNG) channels in visual transduction. **a** Schematic diagram of a retinal rod. **b** Illustration of the situation in the dark with the presence of a high concentration of cGMP, generated from GTP in the presence of guanylate cyclase (*GC*), which binds to the CNG channel and opens it, allowing an inflow of Na^+ and Ca^{2+} to generate the dark current. Ca^{2+} is then extruded by means of a Na/Ca–K exchanger. **c** In response to light, rhodopsin (*R*) adopts a photo-activated state (*R**), which causes a G protein (*G*) to be activated and a phosphodiesterase (*P*) to hydrolyse cGMP to 5'-GMP. The resultant reduced cGMP concentration causes the CNG channel to close. The internal $[Ca^{2+}]$ is lowered as it continues to be extruded by the Na/Ca–K exchanger. **d** The depolarisation of the membrane potential (V_m) photoreceptor to about –40 mV is caused by the inflow of cations resulting from the dark current. **e** The hyperpolarisation of the photoreceptor membrane potential results in response to a turning off of the dark current and the resultant decrease in Ca^{2+} within the rod (see text). Based on Fig. 16.7 in Shepherd (1994), Fig. 2 in Kaupp and Seifert (2002) and Fig. 1 in Matulef and Zagotta (2003)

The Ca^{2+} permeability of rod CNG channels is not only important for contributing to the dark current, but the local intracellular Ca^{2+} concentration has additional implications for adaptation in phototransduction. The higher relative level of intracellular Ca^{2+} in the dark inhibits guanylate cyclase activity and enhances phosphodiesterase activity, both contributing to reductions in cGMP and dark adaptation. The relative decrease in the local intracellular Ca^{2+} concentration in the light has the opposite effect on these enzymes, increasing cGMP and contributing to light adaptation (see Fig. 9.1e).

Another feature of rod CNG channel permeation properties with relevance for phototransduction is that the typical physiological concentration of extracellular Ca^{2+} partially blocks the channel, reducing the single channel currents at –40 mV from about 1 pA in a calcium-free solution to about 4 fA in a physiological solution. Having a large number of channels in the outer segment, each with a very small conductance, is of great advantage for reducing the signal-to-noise ratio in the dark current (Yau and Baylor 1989; Shepherd 1994).

9.2.2 Olfactory Transduction

Shortly after the demonstration of CNG channels in retinal rods, they were demonstrated to be present also in the plasma membrane of the cilia of toad olfactory sensory neurons (OSNs, also called olfactory receptor neurons) (Nakamura and Gold 1987) and their critical role in olfaction was demonstrated. Unlike in photoreceptors, the CNG channels are closed in the absence of sensory stimulation, and it is the opening of the CNG channels and the resulting depolarisation of the olfactory sensory neurons that signals the response to odours (for reviews see e.g., Schild and Restrepo 1998; Menini 1999; Nakamura 2000; Frings 2001; Kaupp and Seifert 2002; Pifferi et al. 2006). In OSNs, in the absence of odorant activation, the cytoplasmic cAMP and cGMP concentrations are low and the CNG channels remain closed. Binding of an odorant to its receptor, located with their highest density in the hair-like cilia protruding from a dendritic knob at the end of a bipolar OSN (Fig. 9.2a), activates a specific olfactory G-protein (G_{olf}), which stimulates an adenylate cyclase (A_c; Fig. 9.2b) resulting in the generation of cAMP from ATP. The cAMP binds to and activates the olfactory CNG channel, with a resultant influx of both Na^+ and Ca^{2+} ions.

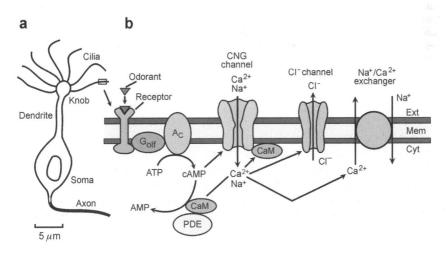

Fig. 9.2 The role of CNG channels in olfactory transduction. **a** Schematic diagram of an olfactory sensory neuron (OSN; previously also known as an olfactory receptor neuron) of about the size of a rat OSN. **b** Schematic diagram of a cAMP-mediated transduction cascade for the cilial membrane, where most of the odorant receptors are located. Based on Fig. 3 in Schild and Restrepo (1998), Fig. 1 in Menini (1999) and Fig. 1 in Pifferi et al. (2006)

The influx of Ca^{2+} ions also activates a Ca^{2+}-activated Cl^- channel, which results in the efflux of Cl^- ions (which are unusually highly concentrated inside OSNs). The influx of both Na^+ and Ca^{2+} cations together with the efflux of Cl^- anions result in membrane depolarisation, which is conducted passively along the dendrite to the OSN soma where an action potential is triggered and conducted along an axon projecting to the olfactory bulb.

As with phototransduction, the influx of Ca^{2+} plays additional roles in adaptation of the sensory signal. Ca^{2+} ions entering the OSN combine with calmodulin (CaM) to activate a phosphodiesterase (PDE), that hydrolyses cAMP and reduces CNG channel activity. In addition, Ca^{2+} ions can also directly reduce the sensitivity of the CNG channel for cAMP via both a Ca^{2+}-calmodulin-dependent process and an unidentified membrane-bound endogenous factor (Balasubramanian et al. 1996).

9.3   Subunit Composition of CNG Channels

CNG channels are tetrameric proteins comprised of four principle and modulatory subunits, with the subunit nomenclature now being based on gene sequence and classified into two major subfamilies, A and B, each of which is further divided according to amino acid homology and functional context (Table 9.1; see also Bradley et al. 2001; Kaupp and Seifert 2002). The 'A' subfamily of CNG channel genes (equivalent to the old 'α' subunits) is comprised of the CNGA1 (rods), CNGA2 (OSN), CNGA3 (cones) and CNGA4 (OSN) subunits. Each can be expressed as functional homomers in HEK 293 cells or oocytes activated by cAMP or cGMP (although CNGA4 has closer homology to the B subunits and homomeric CNGA4 channels can be activated only by nitric oxide). The A subunits play a critical role in determining the channel's permeation properties, and homomeric CNGA1 and CNGA2 channels provide a valuable model for structure–function

Table 9.1 Current nomenclature for subunits of rod, cone, and olfactory cyclic nucleotide-gated (CNG) channels (See Table 2 of Kaupp and Seifert 2002 for original cloning references). *OSN* Olfactory sensory neuron

Type of channel	Old notation	New notation	Function as homomeric channels
Rods	CNG1, CNGα1, RCNC1	CNGA1	Yes
OSNs	CNG2, CNGα3, and OCNC1	CNGA2	Yes
Cones	CNG3, CNGα2, CCNC1	CNGA3	Yes
OSNs	CNG5, CNGB2, CNGα4 and OCNC2	CNGA4	Yes[a]
Rods and OSNs	CNG4, CNGβ1;	CNGB1	No
Cones	CNG6, CNGβ3	CNGB3	No

[a] CNGA4 can form NO-activated homomeric channels, but cannot be activated by cAMP

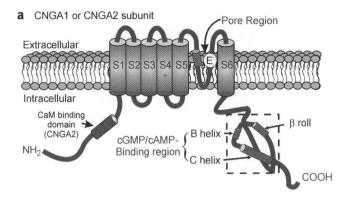

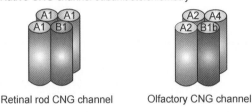

Fig. 9.3 The overall structure of the CNG channel. **a** Schematic diagram of the membrane topology and functional domains of one of the CNG subunits that form the tetrameric CNG channel, in this case either the retinal rod CNGA1 or olfactory CNGA2 subunit [based on Fig. 16 in Kaupp and Seifert (2002) for the CNGA1 subunit, and Fig. 4 in Frings (2001) for the CNGA2 subunit]. The cGMP/cAMP binding region in the CNGA2 subunit is considered to be very similar to that of CNGA1, and the calmodulin (*CaM*) binding domain is intended to relate only to CNGA2. **b** The accepted subunit stochiometry of both the retinal rod and olfactory CNG channels (Weitz et al. 2002; Zheng and Zagotta 2004)

studies of permeation and other basic properties of rod and olfactory CNG channels. The B family consists of CNGB1 and CNGB3 (the original CNGB2 being later redesignated as CNGA4) with two major short splice variants formed from the CNGB1 gene: CNGB1a in rods and CNGB1b in OSNs (reviewed in Kaupp and Siefert 2002). The B subunits do not form functional homomeric channels, but can modify channel properties when co-expressed with A subunits. The tetrameric stochiometry of the native rod CNG channel is 3:1 CNGA1:CNGB1 (Weitz et al. 2002) and that of the native olfactory CNG channel is 2:1:1 CNGA2:CNGA4:CNGB1b (Zheng and Zagotta 2004; see Fig. 9.3b).

The overall structure of the CNGA1 and CNGA2 subunits is shown in Fig. 9.3a, with each subunit having six transmembrane segments (S1–S6). They also have a pore region between S5 and S6 and an intracellular amino and carboxy terminus, with the former harbouring a CaM-Ca^{2+} binding domain and the latter containing the cyclic nucleotide binding domain.

9.4   Structure of the CNG Channel Pore

As in other "P-loop" cation channels, the pore region plays a major role in determining the ion selectivity of the channel (Heginbotham et al. 1992). Figure 9.4a schematically illustrates the pore-forming region based on the structure of the closely related KcsA K⁺ channel (Doyle et al. 1998). The inner wall of the channel pore is lined by the S6 α-helix, tilted (like an "inverted tepee") to form a constriction at the inner end. The centre of the pore contains a cavity and the outer end contains the selectivity filter, formed by residues extending from the S6 helix to the pore-helix. The pore-helix is angled so as to project its carboxy terminus towards the central cavity. These general features of the pore have recently also been confirmed for the crystal structure of a non-selective cation channel from *Bacillus cereus* (the "NaK channel"; Shi et al. 2006), which is functionally more closely related to CNG channels than are the crystallised K⁺ channels. Figure 9.4b shows a suggested model of the CNG channel (Qu et al. 2006).

In KcsA, K⁺ selectivity is achieved by the backbone carbonyl oxygens of residues T75–Y78 (see Fig. 9.5) forming four K⁺ co-ordination/binding sites within the selectivity filter. The carbonyl oxygens from G79 (and possibly from D80) form an electronegative entrance to the selectivity filter, and the negative ends of the pore–helix dipole create an electronegative environment in the central cavity that is critical for cation selectivity (Zhou et al. 2001; Roux and MacKinnon 1999). This critical sequence of residues, the K⁺ channel signature sequence, is incomplete in CNG channels, with the Y78 and G79 residues missing (Fig. 9.5). Substituting the CNG P-loop sequence into K⁺ channels renders K⁺ channels non-selective between Na⁺ and K⁺ (Heginbotham et al. 1992). Only the inner two of these ion co-ordination sites are present in the *Bacillus* NaK channel, and they can accommodate a range of monovalent cations (Shi et al. 2006). Both divalent and monovalent cations could be stabilised at the extracellular vestibule of the NaK channel.

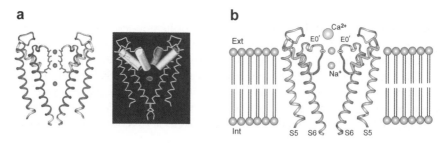

Fig. 9.4 A comparison of the structure of the P-loops of the K⁺ channel from *Streptomyces lividans* (KcsA) and CNG channels. **a** The P-loop of KcsA channels. [*Left panel* Reprinted with permission from Fig. 5 in Roux et al. (2000), © 2000 American Chemical Society. *Right panel* From Fig. 1 in Roux and MacKinnon (1999), reprinted with permission from the American Association for the Advancement of Science]. **b** The P-loop of wild-type CNGA2 channels (modified from Fig. 11A in Qu et al. 2006)

Relative residue position		-4'	-3'	-2'	-1'	–	–	0'	1'	2'	3'	
Channel Type	PL											PR
Bovine Rod CNGA1	359	T	T	I	G	–	–	[E]	T	P	P	366
Bovine Rod CNGB1	958	I	T	I	G	–	–	G	L	P	[D]	965
Rat Olfactory CNGA2	338	T	T	I	G	–	–	[E]	T	P	P	345
Rat Olfactory CNGA4	230	T	T	V	G	–	–	[D]	T	P	L	237
NaK Bacillus cereus	62	T	T	V	G	[D]	G	N	F	S	P	71
KcsA	74	T	T	V	G	Y	G	[D]	L	Y	–	82
Shaker K+	441	T	T	V	G	Y	G	[D]	M	T	–	449

Fig. 9.5 The relative amino acid sequence alignment in the pore-loop region of bovine retinal rod and rat olfactory CNG channels compared with the KcsA and Shaker K+ channels. Residue positions are relative to the negatively charged glutamate or aspartate in the CNGA sequences. *PL* and *PR* refer to the absolute residue number at the left and right of each pore sequence, respectively. The charged residues in the P-loop are *boxed*. The reference for NaK *Bacillus cereus* is in the supplementary material of Shi et al. (2006) and the references for the other sequences may be found in Fig. 1A in Qu et al. (2006)

Also indicated in Fig. 9.5 is a negatively charged glutamate residue (E0') at the extracellular end of the selectivity filter (Figs. 9.3a, 9.4b), which is a critical determinant of ion permeation in CNG channels (see Sect. 9.6.3). In KcsA, the analogous aspartate residue (D80) projects its side-chain towards the interior of the protein (behind the selectivity filter), where it forms a carboxy-carboxylate interaction with a glutamate residue (E71) within the pore helix (Zhou et al. 2001). In the *Bacillus* NaK channel, a potentially analogous aspartate (D66) is found within the extracellular end of the selectivity filter. Its side-chain projects upwards and is exposed to the extracellular solution, contributing to a vestibule that can weakly (relative to the inner coordination sites) interact with cations (Shi et al. 2006).

9.5 Activation of CNG Channels

The increase in structural information and data from mutagenesis experiments are beginning to result in plausible models for the structural basis of the conformational change that occurs during activation (Fig. 9.6; see also Flynn et al. 2001; Matulef and Zagotta 2003). Binding of cyclic nucleotides results in a re-arrangement of the cyclic nucleotide-binding domain (CNBD) in the cytoplasmic carboxy terminus, allowing the C-helix to move closer to the β-roll (Fig. 9.3). This conformational change is transmitted to the transmembrane helices via the C-linker, connecting S6 to the CNBD. Specifically, a series of mutagenesis and accessibility studies (using Ni2+ and histidine substitutions) suggest a clockwise movement of a helical segment within this C-linker region, which results in the subsequent movement of the S6 transmembrane helices. The S6 helices come together to form a helical bundle at the inner end of the "inverted tepee" forming a "smokehole" (Fig. 9.7). In K+ channels, this smokehole is the actual channel "gate" and an increase in its diameter

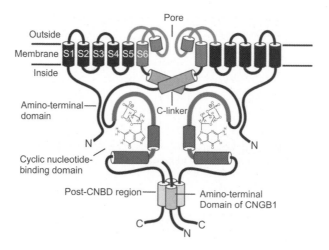

Fig. 9.6 A simplified schematic diagram of the structure of CNGA1 channels to show the critical elements involved in channel activation and gating. For easier visualisation, only two of the four subunits comprising the CNG channel are shown. The cylinders represent proposed α-helices. Reprinted from Fig. 3 in the supplementary material of Matulef and Zagotta (2003), with permission from the Annual Review of Cell and Developmental Biology, vol 19, © 2003 by Annual Reviews (www.annualreviews.org)

results in ion permeation (Liu et al. 1997). In CNG channels, cysteine mutagenesis studies have demonstrated that while the smokehole also widens in response to cyclic nucleotide binding, somewhat surprisingly, the smokehole does not appear to be the actual channel gate since small cations can readily permeate into the channel interior when the channel is closed (Flynn and Zagotta 2001).

If the smokehole is not the actual gate, then where is the gate? This critical question has not yet been definitively resolved, but there are four different pieces of evidence suggesting that the gate is localised to the selectivity filter (Matulef and Zagotta 2003): (1) mutations to E0′ (E363 in CNGA1) could reduce channel open probability and confer desensitisation-like properties. For example, in the presence of 500 μM cGMP, the open probability for the glycine E0′G mutant CNGA1 channel was about 0.1, compared to about 0.8 for the WT channel (Bucossi et al. 1996). Gavazzo et al. (2000) similarly found reductions in open probability for the olfactory CNGA2 channel mutations E0′G and asparagine E0′N. Furthermore, the CNGA1 mutations, alanine E0′A, serine E0′S and E0′N, displayed a rapid desensitisation-like decline in channel open probability (Bucossi et al. 1996), although this does not seem to occur in the CNGA2 E0′A mutant (W. Qu et al. personnel communication). (2) The pore-blockers tetracaine and dequalinium display state-dependent block of CNGA1 and CNGA2 channels, respectively (binding more tightly in the closed state), which is eliminated by mutations to neutralise E0′ (Fodor et al. 1997a, 1997b; Qu et al. 2005). (3) The permeation properties of CNG channels can vary

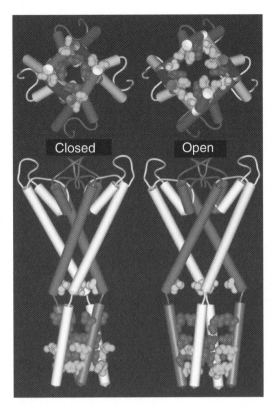

Fig. 9.7 A model of the conformational changes in the helix bundle of post-transmembrane C-linker segments of CNGA1 channels during gating. Bottom and side views in the closed and open states of the channel. Reprinted from Fig. 7 in Flynn et al. (2001) with permission from Macmillan: Nature Reviews Neuroscience, vol 2, © 2001

with the level of activation. For example, the relative P_{Ca}/P_{Na} value has been shown to increase continuously as the channel open probability rises in both CNGA1 (rod) and CNGA3 (cone) channels (Hackos and Korenbrot 1999), although the selectivity between monovalent inorganic and some organic cations did not change. (4) Cysteine-scanning mutagenesis studies with MTS reagents of the pore helix residues indicated that the helix underwent a change in conformation, such as a rotation, during channel opening (Liu and Sigelbaum 2000).

A recent computational model to further extend the structural basis of CNG channel gating has been developed by Giorgetti et al. (2005), incorporating homology models of the ligand-bound C linker region of mouse HCN channels and the KcsA channel with an impressive array of experimentally derived spatial constraints based on electrophysiological measurements in CNGA1 channels. They suggest a bending and anticlockwise rotation by about 60° (around the helix axis as

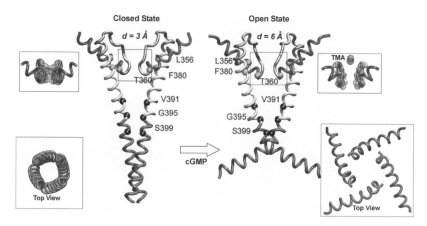

Fig. 9.8 Structural models of CNG gating, showing the closed and open states of the P-loop of the CNG channel based on homology modelling and other data in the protein databank (PDB) database (www.pdb.org). Only two subunits are shown for clarity; *insets* cross-sections of the selectivity filter region and the "smokehole". Reprinted from Fig. 2 in Giorgetti et al. (2005), with permission from the Federation of the European Biochemical Societies, FEBS Lett, vol 579 © 2005, with permission from Elsevier

seen from this extracellular end) of the C-linker N-terminal section (Fig. 9.8), the motion transmitted upwards to cause the upper part of S6 to rotate about 30° anti-clockwise. Due to the interaction of the S6 segments with the pore helix, this S6 rotation is proposed to re-arrange the pore helix so as to widen the lower end of the selectivity filter and allow ions to permeate (Fig. 9.8; Giorgetti et al. 2005). On the basis of this model, and using cysteine-scanning mutagenesis, they were also able to show that in the presence of a mild oxidising agent, copper phenanthroline (CuP), certain cysteine mutations were able to lock the channel in either the closed or open state, depending on whatever state they happened to be in at the time of CuP application (Nair et al. 2006). It will be interesting to examine how this gating model may incorporate the structure of the *Bacillus* NaK channel, particularly as its structure reveals an N-terminal helix adjacent to the membrane and perpendicular to the pore, seemingly located in an ideal position to pull open the smokehole (Shi et al. 2006).

9.6   Permeation and Selectivity of CNG Channels

Before specifically reviewing the ion selectivity of CNG channels, we will briefly consider the general methodologies for determining permeation properties of ion channels: relative permeabilities between ions of the same valency, relative perme-abilities between anions and cations, the minimum pore diameter of the selectivity filter and anomalous mole fraction effects.

9.6.1   General Methodologies for Permeation Measurements

To measure relative permeabilities for ions of the same valency, measurements should be taken of reversal potentials (V_{rev}) in bi-ionic solutions, known as bi-ionic potentials. For monovalent ions and purely cation-selective channels, the solutions should be of the form NaCl:XCl, where X is a test cation, and both NaCl and XCl are at the same concentration. V_{rev} should be very close to 0 mV in symmetrical NaCl solutions. The bath solution can then be changed to XCl and the change in V_{rev} measured. As with all permeation measurements, V_{rev} should be corrected for liquid junction potentials (e.g. Barry and Lynch 1991; Barry 1994) and concentrations converted to activities. The relative permeability can then be determined using the zero-current Goldman-Hodgkin-Katz (GHK) equation:

$$\Delta V_{rev} = \frac{RT}{F} \ln\left(\frac{a_{Na}^{o} + (P_X / P_{Na})a_X^{o}}{a_{Na}^{i} + (P_X / P_{Na})a_X^{i}} \right) \tag{9.1}$$

in which R, T and F are the gas constant, temperature in K and Faraday constant, respectively; a_{Na} and a_X are the activities of Na$^+$ and X$^+$ ions respectively; superscripts 'o' and 'i' refer to external and internal solutions, respectively, and P_{Na} and P_X refer to the relative permeability of the Na$^+$ and X$^+$ ions, respectively. If the channel is not perfectly cation selective, then the appropriate P_{Cl}/P_{Na} also needs to be incorporated into the GHK equation (e.g. Keramidas et al. 2000). Although the GHK equation assumes ions permeate independently of each other and that the electric field across the membrane is constant, Eq. 9.1 can be derived from a range of different permeation models (see discussion and references in Barry 2006) and hence the resultant relative permeabilities obtained from reversal potential measurements are somewhat model independent and are reliable.

The most straightforward and unequivocal approach to measure anion-to-cation permeability ratios (ion charge selectivities) is to use dilution potential measurements where shifts in V_{rev} are measured during dilutions of the solution bathing the cell or membrane patch, for example to about 50% and then 25% of the control salt solution. As with bionic measurements, the composition of the control salt solution should be as simple as possible without adversely affecting cell viability, ideally just a single salt (e.g. 150 mM NaCl) together with pH buffer. The diluted solutions are then osmotically balanced by the addition of an appropriate concentration of a non-electrolyte osmolyte like sucrose (see Barry 2006 for further discussion). In addition to again correcting V_{rev} measurements for liquid junction potentials, it is now absolutely essential to convert ion concentrations to activities because of the substantial changes in ionic strength of the solutions. The relative permeability ratios can be again determined by fitting the V_{rev} measurements to the zero-current GHK equation, but now in the following form for a mixture of permeant cations and anions:

$$\Delta V_{\text{rev}} = \frac{RT}{F} \ln \left(\frac{a_{\text{Na}}^{o} + (P_{\text{Cl}}/P_{\text{Na}})a_{\text{Cl}}^{i}}{a_{\text{Na}}^{i} + (P_{\text{Cl}}/P_{\text{Na}})a_{\text{Cl}}^{o}} \right) \tag{9.2}$$

with a_{Cl} and P_{Cl} referring to the activity and relative permeability of the Cl⁻ ions, respectively, and the rest of the parameters being as previously defined. This form of the GHK equation can again be derived without making any assumption about a constant electric field across the membrane and, like the bi-ionic situation, is essentially model-independent (Keramidas et al. 2004; Barry 2006) so that derived permeabilities are reliable. In practice, shifts in V_{rev} in approximately 50% and 25% dilution can be plotted against the log of the ionic activity and the data fitted to the GHK equation to determine the relative anion–cation permeability.

A critical parameter for permeation information is the minimum pore diameter of the channel – i.e. the mean diameter of the most constricted point of its selectivity filter region. The technique for determining this parameter was first used for the muscle end-plate channel by Hille and his co-workers (Dwyer et al. 1980), using bi-ionic V_{rev} measurements with large monovalent cationic organic test ions and assuming that the major factor determining permeation for these large ions was simply frictional interactions of the ion with the 'walls' of the selectivity filter region. The minimal cross-sectional area of the pore can be estimated by finding an organic ion of known dimensions, which would just permeate the channel, and a slightly larger one, which does not permeate, and/or by plotting the square root of the relative permeability against the mean dimension of a range of organic test ions, and extrapolating the fitted line to the abscissa (e.g. Dwyer et al. 1980; Cohen et al. 1992; Lee et al. 2003).

Anomalous mole fraction effects (AMFE) in ion channels are defined as those in which the permeation parameter, single channel conductance or reversal potential, in different mole fractions of mixtures of two permeant ions of the same sign, deviates from a simple monotonic dependence on the mole fraction of that mixture. Instead, the permeation parameter might go through a minimum or maximum at a certain intermediate mole fraction of the two ions. AMFE are considered to be an indication of the simultaneous presence of two different permeating ions in the channel pore and of the ions interacting with each other and with a membrane site in that channel (e.g. see discussion in Hille 2001).

9.6.2 Permeation Parameters in Native and Recombinant CNG Channels

This section will review the permeation properties of native rod and OSN CNG channels followed by the properties of their relevant recombinant channels. All the permeability and conductance measurements, unless indicated otherwise, were performed in the absence of divalent ions.

9.6.2.1 Monovalent Ion Permeability

Using bi-ionic potential measurements, the relative permeability sequence of alkali cations through both native rod and OSN CNG channels is shown in Table 9.2. For tiger salamander rod channels (Menini 1990) the values were corrected for liquid junction potentials and for activity coefficients and were the average of 15–30 patches. For rat OSN channels (Frings et al. 1992) values were again corrected for liquid junction potentials and represent the average of 5–10 patches. For newt OSN channels (Kurahashi 1990), the data was averaged from 3–5 patches and also corrected for liquid junction potentials. Although the range of permeability magnitudes is not great, the sequences correspond to Eisenman sequences between IX and XI, suggestive of the cations binding to a very high field strength site in the selectivity filter of the channel (Eisenman and Horn 1983). The relative permeability sequence in wild-type homomeric rat CNGA2 channels is P_{Na} (1) $\geq P_K$ (0.97) > P_{Li} (0.77) > P_{Cs} (0.62) $\geq P_{Rb}$ (0.57) (Qu et al. 2000), which is similar to that of native OSN channels (Frings et al. 1992) except for P_{Rb}, which is relatively more permeant in native OSN channels, being placed in between P_{Li} and P_{Cs}.

In principle, relative conductance measurements can give valuable permeability data but there are complications with such measurements conducted using bionic situations. At very large membrane potentials, the conductance of cation-selective channels would be expected to reflect predominantly the contribution of cations entering the channel from just one of the solutions. At more typical potentials, the channel would tend to be occupied by a mixture of cations from both intra- and extra-cellular solutions and the conductances should reflect this mixture. The situation can be simplified by using the same cation on both sides of the membrane. An example of a problem with the CNG channel is the linear current–voltage (I–V)

Table 9.2 Alkali cation permeability sequences of native CNG channels

Permeability (P) and conductance (g) sequences	Eisenman sequence	Channel type	Reference
P_{Li} (1.14) > P_{Na} (1) > P_K (0.98) > P_{Rb} (0.84) > P_{Cs} (0.58)	XI	Rod PR (salamander)	Menini 1990
P_{Na} (1) > P_K (0.81) > P_{Li} (0.74) > P_{Rb} (0.60) > P_{Cs} (0.52)	IX	OSN (rat)	Frings et al. 1992
P_{Na} (1) > P_K (0.93) $\approx P_{Li}$ (0.93) $\approx P_{Rb}$ (0.91) > P_{Cs} (0.72)	IX/X	OSN (newt)	Kurahashi 1990
g_{Na} (1) $\approx g_K$ (0.86) > g_{Rb} (0.55) > g_{Li} (0.42) > g_{Cs} (0.22)[a]	VIII	Rod PR (salamander) at + 100 mV [NaCl in pipette]	Menini 1990
g_{Na} (1) > g_{Li} (0.56) > g_K (0.52) > g_{Rb} (0.24) > g_{Cs} (0.23)	X	OSN (rat) [NaCl in pipette] at +50 mV	Frings et al. 1992

[a]The conductances were chord conductances estimated from Menini's relative outward currents at +100 mV and corrected for estimates of her reversal potential values in millivolts (obtained from Fig. 1C, Menini 1990) of approximately 0 (NaCl), −0.5 (KCl), −4 (RbCl) and −13 (CsCl)

curve in asymmetrical NaCl solutions (e.g. 50:100 mM), in contrast to a rectifying I–V curve predicted by the Goldman current equation (see Fig. 6 in Balasubramanian et al. 1997). This suggests that the conductances reflect the average of the Na^+ concentrations on both sides of the membrane, rather than the concentrations on the side from which the ions were mainly moving into the channel. In addition, conductance depends on both ion concentration and ion permeability. Furthermore, macroscopic conductance sequences depend on channel open probability being independent of the nature of the cation present, although this seems to hold for CNGA2 channels, since both single channel and macroscopic sequences were the same (Qu et al. 2000). Nevertheless, conductance sequences of both rat OSNs and salamander rods, which are also shown in Table 9.2, correspond to Eisenman sequences VIII and X. To interpret the different sequences, consider the results of Menini (1990) in salamander rods. Although Li^+ was the most permeant ion, it had a much lower relative conductance between Rb^+ and Cs^+, indicating that Li^+ binds more strongly than other cations to a binding site within the channel (with consequently a slower dissociation and rate of flux). In frog OSN channels, the macroscopic conductance sequence (at +50 mV) is similar in sequence order to the permeability sequence, except that the positions of K and Li have been reversed. More importantly, in comparison to the rod sequence, the permeability of Li^+ is much less than those of Na^+ and K^+, suggesting that Li^+ may bind less strongly to a binding site in the larger minimum pore diameter OSN channels than it does in the smaller diameter rod channels (see Sect. 9.6.2.3).

9.6.2.2 Anion–Cation Permeability

Menini (1990) investigated the presence of any Cl^- permeability in rod CNG channels both by measuring V_{rev} when the Cl^- in the internal NaCl solution was replaced by the much larger isethionate$^-$, and also by using dilution potential measurements in different activity gradients of NaCl, KCl and LiCl. In all cases, the data was best fitted by assuming zero Cl^- permeability (Menini 1990). Similarly, Balasubramanian et al. (1997), using dilution potential measurements in native rat OSN channels, reported a P_{Cl}/P_{Na} very close to zero when the external (pipette) NaCl concentration was constant at a value between 150 and 250 mM and the internal (cytoplasmic) NaCl concentration was varied. However, when the external NaCl concentration was reduced to either 100 mM or 50 mM, and the internal NaCl concentration varied, P_{Cl}/P_{Na} increased to 0.13 or 0.21, respectively. These results may reflect an increased minimum pore diameter of the channel in very low external (but not internal) ionic strength solutions. Also, recombinant CNGA2 channels are less cation-selective than their native counterparts, even with external 150 mM NaCl, with a P_{Cl}/P_{Na} of about 0.1 (Qu et al. 2000, 2006). This may simply reflect subtle differences in the selectivity pathway between homomeric and heteromeric CNG channels.

9.6.2.3 Minimum Pore Diameter

Using large organic cations (see Sect. 9.6.1), the minimum pore diameter of the native retinal rod CNG channel was shown to be about $3.8\,\text{Å} \times 5.0\,\text{Å}$; with tetramethylammonium (TMA), tetraethylammonium (TEA) and choline all impermeant (Picco and Menini 1993). Similar measurements by Balasubramanian et al. (1995) in the rat OSN channel revealed a significantly wider pore, at least as large as the acetylcholine receptor (AChR) channel ($6.5\,\text{Å} \times 6.5\,\text{Å}$; Adams et al. 1980) to permit the permeation of large ions like TMA and TEA. A similar set of bi-ionic measurements of permeabilities of large organic cations in rat CNGA2 channels produced a selectivity sequence of $P_{NH3OH} > P_{NH4} > P_{Na} > P_{Tris} > P_{choline} > P_{TEA}$, again indicating a corresponding minimum pore diameter of at least $6.5\,\text{Å} \times 6.5\,\text{Å}$ (Qu et al. 2000), as found for the native olfactory CNG channel (Balasubramanian et al. 1995).

9.6.2.4 Single Channel Conductance

The single channel conductance, γ, of the bovine rod CNG channel in the absence of divalent ions was found to be 6 pS with a linear I–V curve (Quandt et al. 1991), whereas in the native amphibian rod channel, the most prominent conductance level was much larger ($\gamma = 24$–$25\,\text{pS}$; Haynes et al. 1986; Zimmerman and Baylor 1986). This larger value was also measured in purified and reconstituted bovine rod channels inserted into lipid bilayers ($\gamma = 26\,\text{pS}$; e.g. Hanke et al. 1988). Overall, values of between 6 and 38 pS have been reported for photoreceptor CNG channels (cited in Frings et al. 1992). Some of this variability may relate to the very flickery nature of native channel openings. In comparison, homomeric bovine CNGA1 channels have both more stable openings and a single larger conductance state of about 28 pS ($32 \pm 2\,\text{pS}$ at $+80\,\text{mV}$ and $25 \pm 4\,\text{pS}$ at $-80\,\text{mV}$; Nizzari et al. 1993).

Native rat and frog OSN CNG channels have very rapid and flickery channel openings, with variable conductance levels, most commonly between 12 and 15 pS (e.g. Frings et al. 1992). Homomeric bovine CNGA2 channels again have more stable openings and a larger conductance of about 40 pS ($40 \pm 2\,\text{pS}$ at $+80\,\text{mV}$ and $38 \pm 2\,\text{pS}$ at $-80\,\text{mV}$; Gavazzo et al. 1997), similar to that of rat CNGA2 homomers (48 pS at $+60\,\text{mV}$; Bradley et al. 1994). Incorporation of the CNGA4 subunit, or CNGA4 and CNGB1b subunits, renders the channels flickery and reduces their conductance, and confers a rectification pattern more resembling that of native OSN channels (Bradley et al. 1994; Bönigk et al. 1999; see review by Kaupp and Siefert 2002). The increased rectification upon addition of CNGA4 to CNGA2 has been suggested to be due to the larger number of positive charges in S4 in the CNGA4 subunit (Bradley et al. 1994). For more information on conductances see also Table 1 in Zagotta and Siegelbaum (1996).

9.6.2.5 Multi- or Single-Ion Pore Permeation

Initial measurements of conductances in different mole-fraction mixtures of Li^+ and Na^+ seemed to indicate no AMFE in either rod (Menini 1990) or olfactory channels (Frings et al. 1992), suggesting that only a single ion was present in the pore or that these particular ionic species did not interact with each other in the pore. However, AMFE have been observed using other pairs of ions in rod channels; Na^+–NH_4^+ and Cs^+–Na^+ (Furman and Tanaka 1990), Ca^{2+}–Na^+ (Rispoli and Detwiler 1990) and Li^+–Cs^+ (Sesti et al. 1995b). Similar AMFE have also been reported for both native and CNGA2 OSN channels with both Na^+–NH_4^+ and Li^+–Cs^+ mixtures (Qu et al. 2001). In the case of the Na^+–NH_4^+ interaction, there was a conductance minimum when the inside $[Na^+]_i$ mole fraction ($[Na^+]_i$ / {$[Na^+]_i$ + $[NH_4^+]_i$}) = 0.8, inferring that the concurrent presence of both Na^+ and NH_4^+ ions stabilised their residence in the channel and hence reduced conductance. On the other hand, for Li^+–Cs^+ mixtures, there was a conductance maximum when the inside $[Cs^+]_i$ mole fraction ($[Cs^+]_i$ / {$[Cs^+]_i$ + $[Li^+]_i$}) = 0.6; presumably this combination of both ions together resulted in a more unstable, shallower energy well. Hence, multiple ions can reside and interact within the pores of native photoreceptor and OSN CNG channels.

9.6.2.6 Divalent Ion Permeability and Block

Although CNG channels are permeable to both monovalent and divalent ions, increasing the external concentration of divalent ions (e.g. Ca^{2+}, Mg^{2+}) reduces monovalent ion permeability and the overall magnitude of the channel conductance (e.g. Yau and Baylor 1989; Zufall et al. 1994). As described earlier (Sect. 9.2), the reduced channel conductance and the influx of Ca^{2+} has important implications for visual and olfactory transduction, improving the signal:noise ratio, amplifying the depolarisation and contributing to response adaptation.

The block by physiological concentrations of external Ca^{2+} is much greater at negative potentials, as the external Ca^{2+} ions are drawn into the channel causing a flickery block and outward rectification (e.g. Zufall et al. 1994). In addition, block of current by Ca^{2+} on the cytoplasmic side requires much greater concentrations of Ca^{2+} than from the extracellular side. The magnitude of the relative Ca^{2+} to monovalent cation permeability depends on the subunit composition of the CNG channel. For example, in inside-out patches under bi-ionic conditions (100 mM K^+ in pipette and 50 mM Ca^{2+} in cytoplasm) the relative divalent/monovalent ion permeability determined from reversal potential measurements was significantly greater than 1 (e.g. P_{Ca}/P_K = 2 for CNGA1, 5 for CNGA2 and 8 for CNGA3 channels; Frings et al. 1995). In a series of elegant experiments, using both electrophysiological and fluorescence measurements, Frings et al. (1995) and Dzeja et al. (1999) measured both Ca^{2+} block and fractional Ca^{2+} current in CNG channels. The amount of block similarly varies between different CNG channels (see Fig. 9.9), with the K_i for Ca^{2+} block varying from about 6 μM for rod CNGA1 channels, to 60 μM for

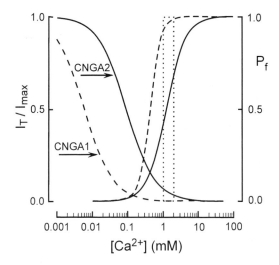

Fig. 9.9 Blockage of retinal CNGA1 (*dashed lines*) and olfactory CNGA2 (*solid lines*) channels by extracellular Ca^{2+} and the fraction of current carried by Ca^2, P_f, at $-70\,mV$. The Ca^{2+} blockage is defined by I_T/I_{max}, as the total fraction of current remaining unblocked in the presence of Ca^{2+} relative to the current in Ca^{2+}-free solution. The *dotted box* indicates that P_f has almost reached unity for CNGA1 channels in the range of $1-2\,mM$ [Ca^2]. It should be noted that the Ca^{2+} dependence of P_f in the native channel is much less steep than it is in the CNGA1 channel (see text). Redrawn and modified from Fig. 7B of Dzeja et al. (1999)

cone CNGA3 channels, to about $90\,\mu M$ for olfactory CNGA2 channels (Frings et al. 1995). The relationship between Ca^{2+} block, defined as the total current remaining unblocked relative to the total current in a Ca^{2+}-free solution (I_T/I_{max}), and the fraction of current carried by Ca^{2+}, P_f, is clearly demonstrated in Fig. 9.9 (Dzeja et al. 1999). As external [Ca^{2+}] increases, the relative fraction of current carried by Ca^{2+} increases, so that although I_{Ca}/I_{Na} can become very large, the total current also drops towards zero. It should be noted that native rods and cones have a very much shallower dependence of P_f on extracellular Ca^{2+} than do the recombinant channels, so that even when it is $10\,mM$, P_f is still only 0.5 (Kaupp and Seifert 2002). The ratio of P_{Ca}/P_{Na} also depends on the level of channel activation, increasing with increasing cGMP concentration (see Sect. 9.5).

It has been considered that the presence of a Ca^{2+} ion in or near the pore region of the CNG channel makes it energetically very difficult for a monovalent cation to pass through the channel (Frings et al. 1995; Seifert et al. 1999; Kaupp and Seifert 2002). This has recently been demonstrated using a Brownian dynamics modelling of ion permeation through a number of related P-loop channels with a structure very similar to that of the CNG channel (KcsA, sodium channel, L-type Ca^{2+} channel), and suggests the following mechanism for external divalent ion block and permeation (Corry et al. 2005). For both the sodium and calcium channels (which readily conduct sodium ions in the absence of external Ca^{2+}), either two Na^+ ions or one Ca^{2+} ion are readily bound in an energy well arising from negative glutamate residues at the external entrance to the pore. If an external Ca^{2+} ion enters the channel and is only weakly bound, it can be displaced by either an incoming Na^+ ion or Ca^{2+} ion and both ions will conduct. If it is more strongly bound (but not too strongly), it can be displaced only by an incoming Ca^{2+} ion. This model can also replicate the outward rectification of channels such as the Na^+ channel, where the divalent ion block is also much greater for inward than for outward currents (Corry et al. 2005).

9.6.3 Structural Basis of Ion Permeation and Selectivity in Recombinant CNG Channels

This section will review the effects of mutations on permeation in recombinant CNG channels, focussing on the effects of mutations to the negatively charged glutamate at 0′, as this residue has been thoroughly studied and seems to be the dominant pore residue for determining pore permeation properties (Eismann et al. 1994). Of relevance to this is the proposed aqueous accessibility of the P-loop residues, once mutated to cysteine, as shown in Fig. 9.10 [based on Fig. 13 in Becchetti et al. (1999) for the bovine rod CNGA1 channel and modified as per Fig. 11 in Qu et al. (2006) for rat CNGA2 residues]. The residues in blue represent those accessible to methanethiosulfonate (MTS) compounds from the internal solution and those in orange are those accessible from the external solution. The cysteine E0′C mutation was reported to be non-functional, so its accessibility to MTS compounds could not be tested (Becchetti et al. 1999). However, interactions between E0′ and external ions (see below) suggest it is also accessible to the external solution.

9.6.3.1 Effects of E0′ Pore Loop Glutamate Mutations on Inter-Cation Selectivity, Permeation, Ionic Block, AMFE and Conductances

A number of studies have demonstrated the dramatic effects of mutations to the E0′ residue on the above permeation properties (see review by Kaupp and Seifert 2002). For example, replacing the negative glutamate with a polar glutamine E0′Q (E363Q in bovine rod channels) caused the channels to be strongly outwardly rectifying, with conductances for all alkali cations radically reduced at negative voltages but with very similar magnitudes of conductance at positive voltages suggesting an increased energy barrier at the external entrance to the channel (Eismann et al. 1994). Conductance in both the E0′Q and alanine E0′A mutant CNGA1 channels was decreased from a wild-type value of about 26–27 pS (Hanke et al. 1988; Sesti et al. 1995a) to about 1–2 pS (Sesti et al. 1995a). In contrast, the charge-conserving aspartate mutation, E0′D, actually increased conductance up to 48 pS at large negative potentials (Sesti et al. 1995a). The cation selectivity sequence in the E0′Q mutation was basically similar, although the range of relative permeabilities was

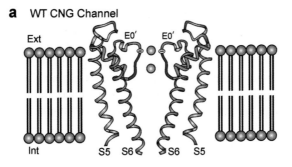

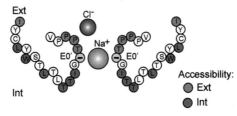

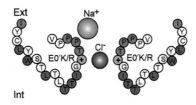

Fig. 9.10 Schematic diagram of the CNGA2 channel pore illustrating the role of the charges at the 0′ position in determining cation–anion selectivity. Reproduced from Fig. 11 in Qu et al. (2006), with permission from The Journal of General Physiology 2006, vol 127 © 2006 The Rockefeller University Press

much reduced (Eismann et al. 1994), suggesting that the 0′ residue may not directly contribute to the high affinity ion co-ordination site in the channel pore but rather may affect conductance by dictating the ion concentration within the pore (see also Laio and Torre 1999).

The E0′Q mutation also caused the CNGA1 channel to no longer exhibit AMFE between Li^+ and Cs^+ ions, suggesting that, instead of being a multi-ion channel, it now just had a single ion occupying it at any one time (Sesti et al. 1995b). The mutation to the polar asparagine (E0′N) also caused a loss of AMFE (Li^+–Cs^+), although the charge-conserving mutation, E0′D, resulted in channels that still displayed AMFE, but with a reduced magnitude (Sesti et al. 1995b).

Mutations to neutralise the charge at 0′ also markedly reduced the block of current by external divalent ions such as Ca^{2+} and Mg^{2+} (e.g. CNGA1, Eismann et al. 1994; CNGA2, Gavazzo et al. 2000), and abolished voltage-dependent block by external large mono- or poly-valent cations, such as spermine and dequalinium (CNGA2, Nevin et al. 2000; Qu et al. 2005). Concurrent with the reduced current block is an increase in the fractional Ca^{2+} current, indicating a reduced energy well for Ca^{2+} permeation. In bovine CNGA2 channels, for example, replacing the negative charge by a neutral glycine (E0′G, E340G) increased I_{Ca}/I_{Na} from about 0.05 in wild-type channels to about 0.16 in the mutant channels.

For at least one of the 0′ mutations, serine (E0′S), the minimum pore diameter seemed to increase, since the mutant channel, unlike the wild-type channel, became permeable to dimethylammonium ions (Bucossi et al. 1996).

9.6.3.2   Ion Charge Selectivity in Recombinant CNG Channels

Despite intensive mutagenesis experiments on E0′, until very recently there were no reports of effects on anion/cation selectivity (Qu et al. 2006). Inverting the charge at 0′ in CNGA2 by mutations to a positively charged lysine (E0′K) or arginine (E0′R) radically altered permeation, switching the selectivity of the channel from cations to anions (as depicted in Fig. 9.10), with P_{Cl}/P_{Na} changing from a wild-type value of 0.07 ± 0.01 to 14.4 ± 2.5 for E0′K and 10.8 ± 1.8 for E0′R (Qu et al. 2006). Relative anion selectivities for halide anions and NO_3^- for both anion-selective mutants were $P_{NO3} > P_I > P_{Br} > P_{Cl} > P_F > P_{acetate}$, which for the halide ions were increasing with the size of the dehydrated ion (or inversely with the size of the hydrated ion). This was the same selectivity sequence seen in the anion-selective $GABA_A$ and glycine receptor channels (Fatima-Shad and Barry 1993). Mutations to a neutral alanine (E0′A) or cysteine (E0′C) did not abolish selectivity, but did reduce cation selectivity, resulting in a change of P_{Cl}/P_{Na} from a wild-type value of 0.06 to values of about 0.2 and 0.4, respectively (W. Qu et al. manuscript in preparation).

The charge-inverting mutations also resulted in a decrease in γ (measured using noise analysis) from a wild-type value of $29 \pm 6 \, pS$ to $0.6 \pm 0.2 \, pS$ for E0′K and $2 \pm 1 \, pS$ for E0′R mutants (Qu et al. 2006). This indicates some other part of the pore remains averse to high anion flux. In addition to any possible changes in pore diameter, we suggest that this may arise due to the negative ends of the pore helix dipoles creating an energy barrier for anions at the intracellular end of the selectivity filter. The latter suggestion would be consistent with the clear outward rectification of currents in these mutant anion-selective channels, analogous to the rectification patterns in cation-selective mutant glycine receptor channels (Moorhouse et al. 2002). The residual cation selectivity in the charge-neutralising mutants could also result from the influence of the pore helix dipoles, although it cannot be discounted that the minimum pore diameter has increased to be large enough to slightly favour Na^+ over Cl^- (see discussion on glycine receptor selectivity and pore diameter in Keramidas et al. 2004).

9.7 Conclusion

The role of CNG channels in olfactory and visual transduction has demonstrated the importance of a solid biophysical understanding of the components of channel function, such as ion permeation, and the subtle differences in channel permeation properties that can result from different subunit combinations. Major advances in our understanding of the structure and function of CNG and other P-loop cation channels have resulted from a wealth of mutagenesis studies, coupled with patch-clamp recordings, biophysical analysis and, more recently, crystal structures. The recently resolved structure of the non-selective NaK channel from *Bacillus* highlights important structural differences between non-selective cation channels, such as CNG channels, and the highly selective K^+ channels. In CNG channels, the selectivity filter is somewhat different than it is in K^+ channels, and the charged glutamate residue (at $0'$) seems to dominate many properties of CNG channels, as has been recently demonstrated with mutations that radically invert ion-charge selectivity. It will now be important to further investigate the mechanisms underlying this selectivity conversion to enable the development of a complete biophysical model of ion permeation in CNG channels.

Acknowledgements This work was supported by the Australian Research Council and the National Health and Medical Research Council of Australia.

References

Adams DJ, Dwyer TM, Hille B (1980) The permeability of endplate channels to monovalent and divalent metal cations. J Gen Physiol 75:493–510

Balasubramanian S, Lynch JW, Barry PH (1995) The permeation of organic cations through cAMP-gated channels in mammalian olfactory receptor neurons. J Membr Biol 146:177–191

Balasubramanian S, Lynch JW, Barry PH (1996) Calcium-dependent modulation of the agonist affinity of the mammalian olfactory cyclic nucleotide-gated channel by calmodulin and a novel endogenous factor. J Membr Biol 152:13–23

Balasubramanian S, Lynch JW, Barry PH (1997) Concentration dependence of sodium permeation and sodium ion interactions in the cyclic AMP-gated channels of mammalian olfactory receptor neurons. J Membr Biol 159:41–52

Barry PH (1994) JPCalc: a software package for calculating liquid junction potential corrections in patch-clamp intracellular epithelial and bilayer measurements and for correcting liquid junction potential measurements. J Neurosci Methods 51:107–116

Barry PH (2006) The reliability of relative anion-cation permeabilities deduced from reversal (dilution) potential measurements in ion channel studies. Cell Biochem Biophys 46:143–154

Barry PH, Lynch JW (1991) Liquid junction potentials and small cell effects in patch-clamp analysis. J Membr Biol 121:101–117

Becchetti A, Gamel K, Torre V (1999) Cyclic nucleotide-gated channels. Pore topology studied through the accessibility of reporter cysteines. J Gen Physiol 114:377–392

Bönigk W, Bradley J, Müller F, Sesti F, Boekhoff I, Ronnett GV, Kaupp UB, Frings S (1999) The native rat olfactory cyclic nucleotide-gated channel is composed of three distinct subunits. J Neurosci 13:5332–5347

Bradley J, Li J, Davidson N, Lester HA, Zinn K (1994) Heteromeric olfactory cyclic nucleotide-gated channels: a subunit that confers increased sensitivity to cAMP. Proc Natl Acad Sci USA 91:8890–8894

Bradley J, Reuter D, Frings S (2001) Facilitation of calmodulin-mediated odor adaptation by cAMP-gated channel subunits. Science 294:2176–2178

Bucossi G, Eismann E, Sesti F, Nizzari M, Seri M, Kaupp UV, Torre V (1996) Time-dependent current decline in cyclic GMP-gated bovine channels caused by point mutations in the pore region expressed in Xenopus oocytes. J Physiol 493:409–418

Burns ME, Baylor DA (2001) Activation deactivation and adaptation in vertebrate photoreceptor cells. Annu Rev Neurosci 24:779–805

Cohen B, Labarca C, Davidson N, Lester H (1992) Mutations in M2 alter the selectivity of the mouse nicotinic acetylcholine receptor for organic and alkali metal cations. J Gen Physiol 100:373–400

Corry B, Vora T, Chung SH (2005) Electrostatic basis of valence selectivity in cationic channels. Biochim Biophys Acta 1711:72–86

Doyle DA, Cabral JM, Pfuetzzner RA, Kuo A, Gulbis JM, Cohen SL, Chait BT, MacKinnon R (1998) The structure of the potassium channel: molecular basis of K^+ conduction and selectivity. Science 280:69–77

Dwyer TM, Adams DJ, Hille B (1980) The permeability of the endplate channel to organic cations in frog muscle. J Gen Physiol 75:469–492

Dzeja C, Hagen V, Kaupp UB, Frings S (1999) Ca^{2+} permeation in cyclic nucleotide-gated channels. EMBO J 18:131–144

Eisenman G, Horn R (1983) Ionic selectivity revisited: the role of kinetic and equilibrium processes in ion permeation through channels. J Membr Biol 76:197–225

Eismann E, Müller F, Heinemann SH, Kaupp UB (1994) A single negative charge within the pore region of a cGMP-gated channel controls rectification Ca^{2+} blockage and ionic selectivity. Proc Natl Acad Sci USA 91:1109–1113

Fatima-Shad K, Barry PH (1993) Anion permeation in GABA- and glycine-gated channels in mammalian cultured hippocampal neurons. Proc R Soc London Ser B 253:69–75

Fesenko EE, Kolesnikov SS, Lyubarsky AL (1985) Induction by cyclic GMP of cationic conductance in plasma membrane of retinal rod outer segment. Nature 313:310–313

Finn JT, Grunwald ME, Yau KW (1996) Cyclic nucleotide-gated ion channels: an extended family with diverse functions. Annu Rev Physiol 58:395–426

Flynn GE, Zagotta WN (2001) Conformational changes in S6 coupled to the opening of cyclic nucleotide-gated channels. Neuron 30:689–698

Flynn GE, Johnson JP Jr, Zagotta WN (2001) Cyclic nucleotide-gated channels: shedding light on the opening of a channel pore. Nat Rev Neurosci 2:643–651

Fodor AA, Black KD, Zagotta WN (1997a) Tetracaine reports a conformational change in the pore of cyclic nucleotide-gated channels. J Gen Physiol 110:591–600

Fodor AA, Gordon SE, Zagotta WN (1997b) Mechanism of tetracaine block of cyclic nucleotide-gated channels. J Gen Physiol 109:3–14

Frings S (2001) Chemoelectrical signal transduction in olfactory sensory neurons of air-breathing vertebrates. Cell Mol Life Sci 58:510–519

Frings S, Lynch JW, Lindemann B (1992) Properties of cyclic nucleotide-gated channels mediating olfactory transduction: activation selectivity and blockage. J Gen Physiol 100:45–67

Frings S, Seifert R, Godde M, Kaupp UB (1995) Profoundly different calcium permeation and blockage determine the specific function of distinct cyclic nucleotide-gated channels. Neuron 15:169–179

Furman RE, Tanaka JC (1990) Monovalent selectivity of the cyclic guanosine monophosphate-activated ion channel. J Gen Physiol 96:57–82

Gavazzo P, Picco C, Menini A (1997) Mechanisms of modulation by internal protons of cyclic nucleotide-gated channels cloned from sensory receptor cells. Proc R Soc Lond B Biol Sci 264:1157–1165

Gavazzo P, Picco C, Eismann E, Kaupp UB, Menini A (2000) A point mutation in the pore region alters gating Ca^{2+} blockage and permeation of olfactory cyclic nucleotide-gated channels. J Gen Physiol 116:311–325

Giorgetti A, Nair AV, Codega P, Torre V, Carloni P (2005) Structural basis of gating of CNG channels. FEBS Lett 579:1968–1972

Hackos DH, Korenbrot JI (1999) Divalent cation selectivity is a function of gating in native and recombinant cyclic nucleotide-gated ion channels from retinal photoreceptors. J Gen Physiol 113:799–818

Hanke W, Cook NJ, Kaupp UB (1988) cGMP-dependent channel protein from photoreceptor membranes: single-channel activity of the purified and reconstituted protein. Proc Natl Acad Sci USA 85:94–98

Haynes LW, Yau KW (1985) Cyclic GMP-sensitive conductance in outer segment membrane of catfish cones. Nature 317:61–64

Haynes LW, Kay AR, Yau KW (1986) Single cyclic GMP-activated channel activity in excised patches of rod outer segment membrane. Nature 321:66–70

Heginbotham L, Abramson T, MacKinnon R (1992) A functional connection between the pores of distantly related ion channels as revealed by mutant K^+ channels. Science 258:1152–1155

Hille B (2001) Ion channels of excitable membranes, 3rd edn. Sinauer, Sunderland, MA

Kaupp UB, Seifert R (2002) Cyclic nucleotide-gated ion channels. Physiol Rev 82:769–824

Keramidas A, Moorhouse AJ, French CR, Schofield PR, Barry PH (2000) M2 pore mutations convert the glycine receptor channel from being anion to cation selective. Biophys J 78:247–259

Keramidas A, Moorhouse AJ, Schofield PR, Barry PH (2004) Ligand-gated ion channels: mechanisms underlying ion selectivity. Prog Biophys Mol Biol 86:161–204

Kurahashi T (1990) The response induced by intracellular cyclic AMP in isolated olfactory receptor cells of the newt. J Physiol 430:355–371

Laio A, Torre V (1999) Physical origin of selectivity in ionic channels of biological membranes. Biophys J 76:129–148

Lee DJ-S, Keramidas A, Moorhouse AJ, Schofield PR, Barry PH (2003) The contribution of proline 250 (P–2) to pore diameter and ion selectivity in the human glycine receptor channel. Neurosci Lett 351:96–200

Liu J, Siegelbaum SA (2000) Change of pore helix conformational state upon opening of cyclic nucleotide-gated channels. Neuron 28:899–909

Liu Y, Holmgren M, Jurman ME, Yellen G (1997) Gated access to the pore of a voltage-dependent K^+ channel. Neuron 19:175–184

Matulef K, Zagotta WN (2003) Cyclic nucleotide-gated ion channels. Annu Rev Cell Dev Biol 19:23–44

Menini A (1990) Currents carried by monovalent cations through cyclic GMP-activated channels in excised patches from salamander rods. J Physiol 424:167–185

Menini A (1995) Cyclic nucleotide-gated channels in visual and olfactory transduction. Biophys Chem 55:185–196

Menini A (1999) Calcium signalling and regulation in olfactory neurons. Curr Opin Neurobiol 9:419–426

Moorhouse AJ, Keramidas A, Zaykin A, Schofield PR, Barry PH (2002). Single channel analysis of conductance and rectification in cation-selective, mutant glycine receptor-channels. J Gen Physiol 119:411–425

Nair AV, Mazzolini M, Codega P, Giorgetti A, Torre V (2006) Locking CNGA1 channels in the open and closed state. Biophys J 90:3599–3607

Nakamura T (2000) Cellular and molecular constituents of olfactory sensation in vertebrates. Comp Biochem Physiol A 126:17–32

Nakamura T, Gold GH (1987) A cyclic nucleotide-gated conductance in olfactory receptor cilia. Nature 325:442–444

Nevin ST, Haddrill JL, Lynch JW (2000) A pore-lining glutamic acid in the rat olfactory cyclic nucleotide-gated channel controls external spermine block. Neurosci Lett 296:163–167

Nizzari M, Sesti F, Giraudo MT, Virginio C, Cattaneo A, Torre V (1993) Single-channel properties of cloned cGMP-activated channels from retinal rods. Proc R Soc Lond B Biol Sci 254:69–74

Picco C, Menini A (1993) The permeability of the cGMP-activated channel to organic cations in retinal rods of the tiger salamander. J Physiol 460:741–758

Pifferi S, Boccaccio A, Menini A (2006) Cyclic nucleotide-gated ion channels in sensory transduction. FEBS Lett 580:2853–2859

Qu W, Zhu XO, Moorhouse AJ, Bieri S, Cunningham AM, Barry PH (2000) Ion permeation and selectivity of wild-type recombinant rat CNG (rOCNC1) channels expressed in HEK293 cells. J Membr Biol 178:137–150

Qu W, Moorhouse AJ, Cunningham AM, Barry PH (2001) Anomalous mole fraction effects in recombinant and native cyclic nucleotide-gated channels in rat olfactory receptor neurons. Proc Roy Soc Lond Ser B 268:1395–1403

Qu W, Moorhouse AJ, Lewis TM, Pierce KD, Barry PH (2005) Mutation of the pore glutamate affects both cytoplasmic and external dequalinium block in the rat olfactory CNGA2 channel. Eur Biophys J 34:442–453

Qu W, Moorhouse AJ, Chandra M, Lewis TM, Pierce KD, Barry PH (2006) A single P-loop glutamate point mutation to either lysine or arginine switches the cation-anion selectivity of the CNGA2 channel. J Gen Physiol 127:375–389

Quandt FN, Nicol GD, Schnetkamp PPM (1991) Voltage-dependent gating and block of the cyclic-GMP-dependent current in bovine rod outer segments. Neuroscience 42:629–638

Rispoli G, Detwiler PB (1990) Nucleoside triphosphates modulate the light-regulated channel in detached rod outer segments. Biophys J 57:368A

Roux B, MacKinnon R (1999) The cavity and the pore helices in the KcsA K$^+$ channel: electrostatic stabilization of monovalent cations. Science 285:100–102

Roux B, Bernèche S, Im W (2000) Ion channels, permeation, and electrostatics: insight into the function of KcsA. Biochemistry 39:13295–13306

Schild D, Restrepo D (1998) Transduction mechanisms in vertebrate olfactory receptor cells. Physiol Rev 78:428–466

Schwarzer A, Schauf H, Bauer PJ (2000) Binding of the cGMP-gated channel to the Na/Ca-K exchanger in rod photoreceptors. J Biol Chem 275:13448–13454

Seifert R, Eismann E, Ludwig J, Baumann A, Kaupp UB (1999) Molecular determinants of a Ca^{2+}-binding site in the pore of cyclic nucleotide-gated channels: S5/S6 segments control affinity of intrapore glutamates. EMBO J 18:119–130

Sesti F, Kaupp UB, Eismann E, Nizzari M, Torre V (1995a) Glutamate 363 of the cyclic GMP-gated channel controls both the single conductance and the open probability. Biophys J 68:A243

Sesti F, Eismann E, Kaupp UB, Nizzari M, Torre V (1995b) The multi-ion nature of the cGMP-gated channel from vertebrate rods. J Physiol 487:17–36

Shepherd GM (1994) Neurobiology, 3rd edn. Oxford University Press, New York, pp 355–360

Shi N, Sheng Y, Alam A, Chen L, Jiang Y (2006) Atomic structure of a Na$^+$- and K$^+$- conducting channel. Nature 440:570–574

Weitz D, Ficek N, Kremmer E, Bauer PJ, Kaupp UB (2002) Subunit stoichiometry of the CNG channel of rod photoreceptors. Neuron 36:881–889

Yau KW, Baylor DA (1989) Cyclic GMP-activated conductance of retinal photoreceptor cells. Annu Rev Neurosci 12:289–327

Zagotta WN, Siegelbaum SA (1996) Structure and function of cyclic nucleotide-gated channels. Annu Rev Neurosci 19:235–263

Zheng J, Zagotta WN (2004) Stoichiometry and assembly of olfactory cyclic nucleotide-gated channels. Neuron 42:411–421

Zimmerman AL, Baylor DA (1986) Cyclic GMP-sensitive conductance of retinal rods consists of aqueous pores. Nature 321:70–72

Zufall F, Firestein S, Shepherd M (1994) Cyclic nucleotide-gated ion channels and sensory transduction in olfactory receptor neurons. Annu Rev Biophys Biomol Struct 23:577–607

Zhou Y, Morais-Cabral JH, Kaufman A, MacKinnon R (2001) Chemistry of ion coordination and hydration revealed by a K$^+$ channel-FAB complex at 2Å resolution. Nature 414:43–48

Chapter 10
Sensory Transduction in *Caenorhabditis elegans*

Austin L. Brown, Daniel Ramot, and Miriam B. Goodman(✉)

Abstract The roundworm *Caenorhabditis elegans* has a well-defined and comparatively simple repertoire of sensory-guided behaviors, all of which rely on its ability to detect chemical, mechanical or thermal stimuli. In this chapter, we review what is known about the ion channels that mediate sensation in this remarkable model organism. Genetic screens for mutants defective in sensory-guided behaviors have identified genes encoding channel proteins, which are likely transducers of

Department of Molecular & Cellular Physiology, Stanford University School of Medicine, B-111 Beckman Center, 279 Campus Dr., Stanford, CA 94305, USA, mbgoodman@stanford.edu

B. Martinac (ed.), *Sensing with Ion Channels. Springer Series in Biophysics 11*
© 2008 Springer-Verlag Berlin Heidelberg

chemical, thermal, and mechanical stimuli. Such classical genetic approaches are now being coupled with molecular genetics and in vivo cellular physiology to elucidate how these channels are activated in specific sensory neurons. The ion channel superfamilies implicated in sensory transduction in *C. elegans* – CNG, TRP, and DEG/ENaC – are conserved across phyla and also appear to contribute to sensory transduction in other organisms, including vertebrates. What we learn about the role of these ion channels in *C. elegans* sensation is likely to illuminate analogous processes in other animals, including humans.

10.1 Introduction

The roundworm *Caenorhabditis elegans* lives in a world that, at first glance, seems very different from our own. Through its brief life cycle, it crawls through a layer of moisture, grazing on bacteria. It has no sense of sight, and gravity is a gentle tug compared to the force of surface tension. At 1 mm in length and only slightly wider than a typical human hair, it would have been easy to overlook this tiny – but useful – organism. But for all the differences between this worm and human beings, it is the similarities that have made the study of *C. elegans* so fruitful.

This chapter focuses on how studies in *C. elegans* have advanced our understanding of the physiology and biophysics of sensory transduction. A significant advantage of studying *C. elegans* is the ability to conduct rapid genetic screens for mutations that disrupt sensory-guided behaviors. Critical to the success of this approach is the fact that (with the exception of male-specific mechanotransduction) sensation is not required for survival or reproduction in the laboratory.

As genetic screens can be conducted without prior knowledge of the molecules likely to be critical for sensation, they have the power to reveal the unexpected. Indeed, genetic studies of sensation in *C. elegans* led to the discovery of the first odorant receptor with a known natural ligand, the ODR-10 diacetyl receptor (Sengupta et al. 1996; Zhang et al. 1997) and the first transient receptor potential (TRP) channel subunit implicated in mechanosensation (Colbert et al. 1997). The first members of the degenerin/epithelial Na^+ (DEG/ENaC) family of ion channels, DEG-1 and MEC-4, were also discovered using genetic approaches in *C. elegans* (Chalfie and Wolinsky 1990; Driscoll and Chalfie 1991).

10.1.1   C. elegans *as a Simple Sensation-Action Machine*

With only 302 neurons and 56 glial and support cells, the nervous system of the *C. elegans* hermaphrodite is a compact biological machine capable of sensorimotor integration. Because the cell lineage of *C. elegans* is invariant, each neuron (indeed, each cell) has been given a unique identifier[1] and classified as a sensory neuron, interneuron or motor neuron based on anatomical criteria. Thanks to meticulous mapping of electrical and chemical synapses by John White and colleagues (White et al. 1986), the connectivity of the entire hermaphrodite nervous system is known. Individual neurons have been assigned functions based on the loss of behavioral responses induced by killing identified neurons with a laser. All these data set the stage for modeling the neural circuits that govern behavior.

Broadly speaking, *C. elegans* hermaphrodites execute two kinds of sensory-dependent behaviors: taxes and avoidances. They use taxis-like behaviors to locate chemical signals (chemotaxis) as well as favorable temperatures (thermotaxis) and oxygen concentrations (aerotaxis) in the environment. The behavioral strategy responsible for chemotaxis and thermotaxis is reminiscent of a biased random walk in which animals suppress turning when moving in a favorable direction (Clark et al. 2006b; Pierce-Shimomura et al. 1999; Ryu and Samuel 2002; Zariwala et al. 2003). Modulation of turning may also contribute to behavioral responses to O_2 (Cheung et al. 2005), suggesting that this navigation strategy is common to all *C. elegans* taxis behaviors. Diverse stimuli evoke avoidance behaviors, including aversive chemical stimuli applied to amphid sensilla in the head and phasmid sensilla in the tail (reviewed by Bargmann 2006), and touch applied along the body wall and to the nose (reviewed by Goodman 2006). Interested readers are advised to consult Wormbook (Girard et al. 2006) at http://www.wormbook.org.

C. elegans males have 87 sex-specific neurons, the vast majority of which are sensory neurons devoted to mating behaviors. Males locate hermaphrodites via chemical signals. Contact with a hermaphrodite triggers a series of stereotyped behaviors that culminate in sperm transfer, all of which involve the action of one or more putative sensory neurons. The contribution of individual male-specific neurons to mating has been investigated using laser ablation. Readers are directed to Barr and Garcia (2006) for a detailed review of male mating behavior.

10.1.2 The Senses of the Worm

A primary goal of studying sensation in *C. elegans* is to improve understanding of sensation generally and with particular regard to our own senses. In this section, we review the sensory repertoire of *C. elegans* hermaphrodites, which have the ability to detect chemical, mechanical and thermal stimuli. By morphological criteria (White et al. 1986), *C. elegans* hermaphrodites have at least 70 sensory neurons, some of which are organized into sensory organs. Sixty of these neurons have ciliated endings that are believed to be the loci of transduction. The 60 ciliated sensory

---

[1] The nomenclature for *C. elegans* neurons was developed by White et al. (1986) and consists of a three-letter name followed by a positional designation (L=left, R=right, D=dorsal, V=ventral). The reader is directed to the Wormatlas website, http://www.wormatlas.org (Hall et al. 2006), for details.

neurons are the only ciliated cells in *C. elegans*; their function relies on a group of proteins that includes homologues of several proteins associated with inherited human diseases (Barr 2005; Inglis et al. 2006).

The gross anatomy of *C. elegans* sensory organs and sensory neurons is shown schematically in Fig. 10.1.

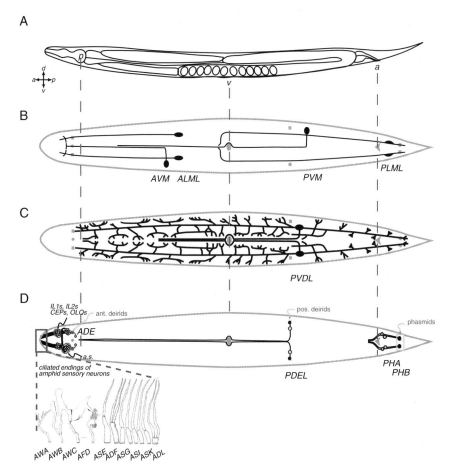

Fig. 10.1 Gross and fine anatomy of sensory neurons in the *Caenorhabditis elegans* hermaphrodite. **a** Lateral view of an adult hermaphrodite. Anterior is to the left and ventral is down. Major anatomical landmarks are labeled: *p* pharynx, *v* vulva, *a* anus. *Ovals* eggs in the uterus. The view in panels **b–d** was derived by slicing a virtual worm along its dorsal midline, removing the pharynx, body wall muscles, intestine and gonad, and laying the cuticle flat. In this view, the ventral midline defines the central axis and the dorsal midline defines the outer rim. **b** The touch receptor neurons. **c** The PVD neurons, showing the extensively branched processes. **d** Ciliated sensory neurons. The ciliated endings of the amphid neurons are shown in detail at the bottom left (adapted from Ward et al. 1975, with permission)

10.1.3   Methods Used to Study Sensory Transduction Genes in C. elegans

Candidate transduction genes have been identified by screening for mutants deficient in the sensory-dependent behaviors outlined above. Over the years, this strategy has generated hundreds of mutant alleles that affect the development or function of sensory neurons in *C. elegans*. Genetic screens for mutants with defects in particular behaviors cannot detect genes that function redundantly or that are essential for normal development, however. Neuron-specific transcriptional profiling has emerged as a complementary approach that has been used to identify genes up-regulated in particular subsets of sensory neurons (Colosimo et al. 2004; Zhang et al. 2002). New methods for in vivo (Goodman et al. 1998; Lockery and Goodman 1998) and in vitro (Christensen et al. 2002; Bianchi and Driscoll 2006) patch clamp recording as well as Ca^{2+} imaging using genetically encoded fluorescent Ca^{2+} indicators (Kerr 2006; Suzuki et al. 2003) now complement these genetic and genomic techniques and make genetic dissection of sensory transduction an obtainable goal.

10.1.4   Ion Channel Families That Sense in the Worm

Members of three classes of ion channels are known to contribute to sensation in *C. elegans*: cyclic-nucleotide gated (CNG) channels (chemosensation, thermosensation, oxygen sensation), TRP channels (chemosensation, mechanosensation), and DEG/ENaC channels (mechanosensation). Here, we briefly review what is known about each of these protein families in *C. elegans*.

The *C. elegans* genome contains at least four genes predicted to encode CNG channels (*tax-4* ZC84.2, *tax-2* F36F2.5, *cng-1* F14H8.6, *cng-3* F38E11.7). Two of these genes, *tax-4* and *tax-2*, can mutate to alter chemo-avoidance, chemotaxis, aerotaxis and thermotaxis (Coburn and Bargmann 1996; Gray et al. 2004; Komatsu et al. 1996). Like CNG channels in vertebrate olfactory receptor neurons and photoreceptors, the channels formed by TAX-4 and TAX-2[2] are likely to act downstream of receptors that (potentially through a signaling cascade) alter the concentration of cyclic nucleotides in the cytoplasm.

Channels belonging to the TRP superfamily are more numerous [there are 22 genes predicted to encode TRP channels in the genome (Goodman and Schwarz 2003)], but no less flexible in their apparent roles in sensation. For example, one TRP channel subunit, OSM-9, is needed for avoidance of noxious chemicals, high osmolarity, and mechanical stimuli delivered to the nose (Colbert et al. 1997). It also plays a role in adaptation to chemical stimuli. Another TRP channel protein, PKD-2, functions in sensory rays in the male tail and is needed for successful mating

---

[2] Standard *C. elegans* nomenclature is to use lowercase italics for genes and uppercase roman to refer to their products. Using this scheme, TAX-2 is the protein encoded by the *tax-2* gene.

(Barr and Sternberg 1999). It is not known whether sensory stimuli activate putative sensory TRP channels directly or indirectly.

The *C. elegans* genome contains 28 genes predicted to encode DEG/ENaC channel proteins (Goodman and Schwarz 2003). This set of genes includes *mec-4* and *mec-10* (see Sect. 10.2.1) that encode pore-forming subunits of a channel complex responsible for detecting mechanical energy (O'Hagan et al. 2005). The large size of this gene family suggests a high degree of molecular redundancy. Consistent with this idea, except for *mec-4*, null or knockout alleles of DEG/ENaC family members appear to result in either a wild-type phenotype or only subtle behavioral defects. For example, loss-of-function mutants in family member *unc-8* have only mild defects in sinusoidal locomotion (Tavernarakis et al. 1997) and *mec-10* knockout mutants are only partially touch-insensitive (R. O'Hagan, M. Chalfie, and M. B. Goodman, unpublished), while no phenotypes have been reported for loss-of-function mutations in *deg-1*, *del-1* and *unc-105*.

10.2 Mechanosensation and Mechanotransduction

Humans and other vertebrates can detect mechanical energy delivered in a sound wave, in a touch, and in the bending, tilting or twisting of heads and limbs. Alterations in blood pressure, renal function and bladder filling also generate mechanical stimuli that are sensed by specialized sensory neurons. The extent to which the same ion channels underlie these mechanical senses remains a significant open question. In *C. elegans*, members of the DEG/ENaC and TRP ion channel superfamilies have emerged from genetic screens as the best candidate mechano-electrical transduction (MeT) channels. Two DEG/ENaCs, MEC-4 and MEC-10, are pore-forming subunits of the ion channel expressed in touch receptor neurons (see Fig. 10.1b) and are opened by low-intensity touch (O'Hagan et al. 2005). The two proteins, which are 53% identical, are co-expressed in touch receptor neurons. In wild type animals, MEC-4 and two channel accessory subunits, MEC-2 and MEC-6, assemble into discrete puncta along the length of sensory dendrites (Chelur et al. 2002; Zhang et al. 2004). In this section, we review the properties of channels formed by MEC-4 and MEC-10 in vivo and in heterologous cells. Additional candidate MeT channels are discussed, but much less is known about their properties. The genetic strategies used to discover candidate MeT channel proteins have been extensively reviewed (Ernstrom and Chalfie 2002; Goodman 2006; Goodman and Schwarz 2003; O'Hagan and Chalfie 2006; Syntichaki and Tavernarakis 2004) and will not be considered here.

10.2.1 Somatosensation

In *C. elegans*, touches delivered to the body wall evoke avoidance behaviors. Responses to low-intensity stimuli ($<100\,\mu N$) are mediated by the touch receptor neurons, while responses to high-intensity stimuli ($>100\,\mu N$) can be sensed by the PVD neurons (Goodman 2006). All of these neurons innervate the outermost tissue layer (the hypodermis) and are believed to be sensitive to mechanical stimuli throughout their dendrites (Fig. 10.1a,b). In electron micrographs, touch receptor neurons are associated with a specialized, osmophilic extracellular matrix and contain a striking, cross-linked bundle of 15-protofilament microtubles that runs the length of the process (Chalfie and Sulston 1981; Chalfie and Thomson 1979).

10.2.1.1 No channel is an Island

In addition to *mec-4* and *mec-10*, there are many genes that, when mutant, disrupt avoidance of low-intensity touch. These genes encode ion channel accessory proteins (*mec-2* and *mec-6*), components of the microtubule cytoskeleton (*mec-7* β-tubulin and *mec-12* α-tubulin), the extracellular matrix (*mec-5* collagen, *mec-9* and *mec-1*) and soluble proteins (*mec-14*, *mec-17* and *mec-18*). Except for *mec-5*, all of these genes are expressed in touch receptor neurons. It is widely believed that they specify a structural framework essential for mechanotransduction in vivo.

At least three of these genes are needed for the proper subcellular localization of MEC-4 in *C. elegans* touch receptor neurons: *mec-6*, *mec-1* and *mec-5* (Chelur et al. 2002; Emtage et al. 2004). *mec-6* encodes a protein related to mammalian paraoxonases (Chelur et al. 2002). Based on protein threading against a 3-D crystal structure of human PON1 (Fokine et al. 2003), the extracellular domain of MEC-6 is likely to fold into a six-bladed β-propeller (A. Brown and M. B. Goodman, unpublished). In *mec-6* mutants, MEC-4 puncta are absent from touch receptor cell sensory dendrites. Regulating the subcellular localization of MEC-4 in touch receptor cells is not the only function of MEC-6, however. It also increases the size of currents carried by MEC-4-dependent channels expressed in *Xenopus* oocytes (Chelur et al. 2002). *mec-6* may have similar effects on additional DEG/ENaC family members, since it can suppress defects caused by gain-of-function mutations in *deg-1* (Chalfie and Wolinsky 1990) and *unc-8* (Shreffler et al. 1995) and is expressed widely in the nervous system (Chelur et al. 2002).

MEC-1 is a large, secreted protein expressed by touch receptor cells and composed of several Kunitz-like and EGF-like repeats (Emtage et al. 2004). Although the exact function of these domains is unknown, *mec-1* mutants show defects in touch receptor neurite localization and in the formation of the adjacent extracellular matrix (Emtage et al. 2004). MEC-5 is an unusual collagen expressed and secreted by the surrounding hypodermal cell (Du et al. 1996). Like MEC-4, MEC-1 and MEC-5 are distributed in puncta along the touch receptor neuron dendrites. MEC-4 puncta are disrupted in both *mec-1* and *mec-5* mutants (Emtage et al. 2004). It seems that MEC-1 and MEC-5 organize the channel rather than the other way around, since the ECM puncta are preserved in *mec-6* mutants in which the channel puncta are absent (Emtage et al. 2004).

10.2.1.2   MEC-4 and MEC-10 Channels in Heterologous Cells

In *Xenopus laevis* oocytes, MEC-4 and MEC-10 form a heteromultimeric ion channel. Current amplitude is dramatically enhanced by co-expression with either MEC-2 or MEC-6 (Chelur et al. 2002; Goodman et al. 2002). MEC-2 is an integral membrane protein related to human stomatin (Huang et al. 1995). MEC-4 generates a current without MEC-10, but MEC-10 without MEC-4 does not (Goodman et al. 2002). Like other DEG/ENaC channels, the resulting channel is blocked by the diuretic amiloride and is sodium-selective. Although channel stoichiometry remains unknown, MEC-4 and MEC-10 interact both physically and functionally (Brown et al. 2007; Goodman et al. 2002) and are likely to form heteromultimers in touch receptor neurons in vivo.

Wild-type MEC-4 channels expressed in Xenopus oocytes have a single-channel conductance of $31\,pS$ and a low (< 0.05) steady-state open probability, P_o (Brown et al. 2007). This low P_o is not surprising since the machinery responsible for delivering force to these channels in vivo is likely to be missing from oocytes. Mutations at the so-called degenerin site, which is occupied by alanine in wild type MEC-4 and MEC-10, cause degeneration in vivo (Chalfie and Wolinsky 1990; Driscoll and Chalfie 1991) and increase open probability in oocytes (Brown et al. 2007). Open probability and mean open times, but not surface expression, increase with the volume of the side-chain at this site (Brown et al. 2007).

10.2.1.3   MEC-4 and MEC-10 Channels in Native Cells

At least 40 recessive, loss-of-function alleles of mec-4 have been recovered in genetic screens (Chalfie and Au 1989; Driscoll and Chalfie 1991; Hong et al. 2000, Royal et al. 2005). Ten affect residues in or near the highly conserved second transmembrane domain, reinforcing the idea that this domain is critical for channel function. By contrast, genetic screens recovered only six alleles of *mec-10* (Chalfie and Au 1989), all of which encode missense mutations (Huang and Chalfie 1994). Deleting the *mec-10* gene has a modest effect on behavioral responses to low-intensity touch (R. O'Hagan, M. B. Goodman, and M. Chalfie, unpublished) and decreases, but does not abolish, native mechanoreceptor currents (O'Hagan 2005). Taken together, these data indicate that MEC-10 is not essential for mechanosensation or for mechanotransduction, and suggest that the six loss-of-function alleles of *mec-10* eliminate behavioral responses to touch by decreasing the activity of MEC-4/MEC-10 heteromeric channels in vivo.

10.2.1.4   Physiology of Touch Cells

To study the activation of native MeT channels and their properties, we must return to the worm itself and analyze cellular responses to mechanical stimuli applied to the cuticle. As revealed by in vivo whole-cell patch clamp recording (O'Hagan et al.

2005), forces as small as 100 nN open native MeT channels. Larger forces evoke larger currents, which reach saturation near 2 μN. These forces are applied to the cuticle; the efficiency and mechanism of force transfer from the cuticle to the touch receptor neuron are unknown. Whether or not the directionality of applied force is a significant factor is an open question.

Channels open following a delay of less than 1 ms, strongly suggesting that force opens channels directly. Unexpectedly, MeT channels open in response to both the application and withdrawal of force steps. These response dynamics resemble those of mammalian Pacinian corpuscles (Lowenstein and Mendelson 1965), which are specialized to detect vibrations, and suggest that *C. elegans* neurons could also be vibration sensors. In support of this idea, bursts of vibratory stimuli ('buzz') evoke larger somatic Ca^{2+} transients than constant stimuli of comparable duration ('poke') (Suzuki et al. 2003).

Mutations that eliminate *mec-4*, *mec-2*, and *mec-6* abolish both touch-evoked Ca^{2+} transients (Suzuki et al. 2003) and mechanoreceptor currents (O'Hagan et al. 2005). These data demonstrate that these genes are required for the formation of functional MeT channels and are consistent with the idea that they are subunits of the native MeT channel. Unambiguous identification of the pore-forming subunits of an ion channel requires the demonstration that mutations in the pore region alter channel permeation properties, however. Such a demonstration was provided by analysis of missense mutations in *mec-4(u2)* and *mec-10(u20)*, both of which alter a conserved glycine residue near the pore and render native MeT channels less selective for Na^+ ions (O'Hagan et al. 2005).

Only one other *C. elegans* DEG/ENaC channel has been studied in vivo and in heterologous cells. UNC-105, which is expressed in body wall muscle, forms amiloride-sensitive Na^+ channels that are blocked by extracellular Ca^{2+} and Mg^{2+} ions in Xenopus oocytes (Garcia-Anoveros et al. 1998). Gain-of-function mutations in *unc-105* cause animals to be hypercontracted (Park and Horvitz 1986), depolarize body wall muscle, and reveal an amiloride-sensitive current in resting muscle (Jospin et al. 2002). All of these phenotypes are suppressed by mutations in the LET-2 collagen (Jospin et al. 2002; Liu et al. 1996), suggesting that, like MEC-4-dependent channels in touch receptor neurons, UNC-105 relies on extracellular collagen for proper assembly.

10.2.1.5 Models for Transduction – Tethered vs Membrane Tension

In order for mechanical stimuli applied to the body wall to open MeT channels in touch receptor neurons, cuticle deformation must be physically coupled to the channel. In touch receptor neurons, the unusual extracellular matrix (ECM) and microtubules are excellent candidates for delivering force. By analogy to the classic model of hair cell transduction, where the tip link serves to deliver tension (Pickles et al. 1984), a tethered channel model was the first widely voiced model of MeT channel activation in touch receptor neurons. In this model, force applied to the cuticle is transmitted via the ECM to MeT channels, which are anchored to the microtubules intracellularly (Gu et al. 1996). In this way, external forces could generate differential displacements of the extracellular and intracellular domains of the channel, tugging it open and leading to signaling.

Integral to the tethered model is the idea that MeT channels are bound to the ECM and to microtubules, either directly or though intermediates. Recent observations cast doubt on the idea that channels are tethered to microtubules. For example, while microtubules are needed for behavioral responses to low-intensity touch, they are dispensable for touch-evoked channel activation in vivo. Loss of *mec-7* or *mec-12*, which eliminates the 15-protofilament microtubules, reduces, but does not abolish mechanoreceptor currents (O'Hagan 2005; O'Hagan et al. 2005).

There is no theoretical requirement for tethering. Instead, changes in membrane tension could be sufficient to open channels. In this alternative model, the microtubule bundle could stiffen the sensory dendrite, exert force on the membrane, or both. In support of these alternatives, the microtubule bundle extends a dense lattice of filaments to the plasma membrane (J. Cueva and M. B. Goodman, unpublished). Similarly, the ECM could play a role in converting cuticle deformation into changes in membrane tension. In this way, the lipids of the membrane could deliver force to MeT channels. This is reminiscent of the mechanism in the well-studied MscL and MscS channels of bacteria (see Chap. 2 by Blount et al. this volume).

10.2.1.6 Transduction Candidates in PVD

Although little is known about the MeT channels that are activated by high-intensity touch in PVD, several genes that are expressed in touch neurons are also expressed by PVD: *mec-3*, *mec-6*, *mec-9*, and *mec-10* (Chelur et al. 2002; Du et al. 1996; Huang and Chalfie 1994; Way and Chalfie 1988). Loss of *mec-3* abolishes behavioral responses to high-intensity touch (Way and Chalfie 1988) and eliminates the higher-order dendritic branches in PVD (Tsalik et al. 2003). Taken together, these data suggest that PVD's branches (Fig. 10.1c) are critical for function, and support the idea that they are the locus of mechanotransduction. PVD also expresses OSM-9 (Colbert et al. 1997), a candidate TRP mechanosensor needed for responses to nose touch (see Sect. 10.2.2).

10.2.2 Nose Touch

Animals recoil upon collision with small objects in their environment. Such responses persist in animals that lack sensitivity to touch along the body wall and rely on eight ciliated neurons that innervate sensilla in the head (in descending order of importance): the pair of ASH neurons, the pair of FLP neurons, and the four OLQ neurons (Kaplan and Horvitz 1993). The six IL1 neurons are also involved, but sense stimuli delivered to the side of the nose rather than the front (Hart et al. 1995). This section focuses on ASH, since almost nothing is known about the physiology of FLP, OLQ and IL1. Nonetheless, some familiar genes are expressed in these neurons: FLP and IL1 express *mec-6* (Chelur et al. 2002); FLP and OLQ express *osm-9* (Colbert et al. 1997); FLP also expresses *mec-10* and two additional DEG/ENaCs, *del-1* and *unc-8* (Tavernarakis et al. 1997). Given the overlap in the expression of putative mechanotransduction channel subunits, it is tempting to speculate that similar mechanisms underlie mechanotransduction in ASH, FLP and IL1.

ASH innervates the bilaterally symmetric amphid sensilla (Fig. 10.1d), where its cilium is exposed to the environment within the amphid channel. Nose touch evokes somatic Ca^{2+} transients in ASH (Hilliard et al. 2005), which expresses two TRPV channel genes (*osm-9* and *ocr-2*) as well as two DEG/ENaC genes (*deg-1* and *unc-8*). Loss of *osm-9* eliminates response to nose touch (Colbert et al. 1997) and touch-evoked somatic Ca^{2+} transients (Hilliard et al. 2005). Because both proteins are needed to deliver these TRPV channels to the cilium (Tobin et al. 2002), OSM-9 is likely to form a heteromeric channel with OCR-2, although this has yet to be determined directly.

Force could open putative OSM-9/OCR-2 channels directly or indirectly, via a second messenger. Polyunsaturated fatty acids (PUFAs) are emerging as possible second messengers that activate TRPVs. In support of this idea, loss of *fat-3*, an enzyme critical for the synthesis of PUFAs, decreases behavioral responses to nose touch, while exogenous PUFAs increase somatic Ca^{2+} in ASH (Kahn-Kirby et al. 2004). Other ion channels, including tandem-pore K channels (Fink et al. 1998) and mammalian TRPV channels (Matta et al. 2007), are also regulated by PUFAs.

10.2.3 Proprioception

Feedback on muscle contraction is critical for coordinated movement in mammals and *C. elegans* alike. This feedback is provided by proprioceptors. In *C. elegans*, candidate proprioceptors include stretch receptors in body wall muscle and long, undifferentiated neurites in the motor neurons that innervate those muscles. An unexpected proprioceptor has come to light recently – the DVA interneurons (Li et al. 2006). The molecule implicated as a potential transducer is the TRP-4 channel – a TRPN channel homologous to NompC, which is needed for mechanotransduction in *Drosophila* bristles (Walker et al. 2000) and zebrafish lateral line hair cells (Sidi et al. 2003).

10.2.4 Male-Specific Mechanotransduction

The tail of *C. elegans* males is richly endowed with ciliated sensory neurons, many of which are likely to be mechanosensors that detect contact with the hermaphrodite body wall and vulva (reviewed in Goodman 2006). Defects in the *lov-1* and *pkd-2* genes result in males who cannot locate the hermaphrodite vulva during mating (Barr et al. 2001; Barr and Sternberg 1999). Mammalian homologues of these genes are mutated in familial polycystic kidney disease and are thought to mediate mechanotransduction by primary cilia that line the lumen of kidney tubules (Nauli et al. 2003). Though it has yet to be tested directly, these data strongly support the idea that LOV-1 and PKD-2 are subunits of a MeT channel in sensory neurons in *C. elegans* males.

10.3   Thermotransduction

C. elegans is exquisitely sensitive to thermal stimuli, responding to temperature changes smaller than 0.1°C (Clark et al. 2006b; Hedgecock and Russell 1975; Luo et al. 2006). Despite significant advances in our understanding of *C. elegans* thermotransduction, how worms achieve this remarkable sensitivity to temperature remains a mystery. In vertebrates and fruit flies, TRP channels directly activated by temperature play a key role in thermotransduction (reviewed by Dhaka et al. 2006). Interestingly, the evidence suggests that in worms, temperature acts on a cGMP signaling pathway that is closely related to vertebrate visual transduction. The study of thermotransduction in *C. elegans* promises to yield insights into alternative molecular mechanisms for detecting temperature, and may even shed light on how visual structures evolved from a simpler sensory system.

10.3.1   C. elegans *Temperature-Guided Behaviors*

C. elegans responds to thermal stimuli with stereotyped patterns of locomotion that are readily observed in the laboratory. Following cultivation at a constant temperature (T_c), worms placed on a thermal gradient perform two distinct behaviors. Near T_c, animals track isotherms, crawling along trajectories of constant temperature for extended periods of time. The worm's thermal sensitivity is evident during this behavior: trajectories deviate from tracked isotherms by less than 0.1°C (Hedgecock and Russell 1975; Luo et al. 2006; Mori and Ohshima 1995; Ryu and Samuel 2002). A second behavior, thermotaxis, is observed at temperatures warmer than T_c, where animals actively crawl down the thermal gradient in the direction of T_c (Clark et al. 2006b; Hedgecock and Russell 1975; Ito et al. 2006; Ryu and Samuel 2002). A biased random walk strategy, the suppression of turns in response to decreases in temperature and the up-regulation of turning in response to increases in tempera-ture, is likely to underlie this behavior (Clark et al. 2006b; Ryu and Samuel 2002; Zariwala et al. 2003). Whether worms also perform thermotaxis up thermal gradients is currently under debate; some groups have reported observing thermotaxis in this direction at temperatures cooler than T_c (Hedgecock and Russell 1975; Ito et al. 2006), while others do not detect it (Clark et al. 2006b; Ryu and Samuel 2002).

A third behavior, proposed to be distinct from thermotaxis, is triggered in response to temperatures warmer than 33°C. Such temperatures likely represent a noxious stimulus [temperatures >25°C are generally harmful to *C. elegans* (Fay 2006)] and elicit an avoidance response (Wittenburg and Baumeister 1999).

10.3.2 A Neural Circuit for Detecting and Processing Temperature

Mori and Ohshima (1995) deduced a neural circuit required for isothermal tracking by analyzing the effects of ablating specific neurons. A pair of sensory neurons, AFD, appears to provide the sensory input. AFD makes >95% of its synapses onto AIY, an interneuron which is critical for the generation of wild type isothermal tracking. A second interneuron, AIZ, which receives input from AIY, is also required. AIY and AIZ synapse onto motor neurons and are likely to be responsible for linking sensory encoding by AFD to behavior.

Consistent with the laser ablation studies, mutations in *ceh-14*, *ttx-3* and *lin-11*, LIM homeobox genes required for proper development of AFD, AIY and AIZ respectively, disrupt *C. elegans'* responses to thermal gradients (Cassata et al. 2000; Hobert et al. 1997, 1998). Similarly, mutations in the *otd/Otx* homolog *ttx-1*, required for specifying the AFD cell-fate, result in a phenotype reminiscent of AFD-ablated animals (Satterlee et al. 2001). Avoidance of noxious temperatures likely engages a distinct neural circuit, since *ttx-1*, *ttx-3* and *lin-11* do not disrupt thermal avoidance. Genetic and behavioral evidence predicts *C. elegans* thermal nociceptors are ciliated neurons located at the animal's head and tail (Wittenburg and Baumeister 1999).

10.3.3 cGMP Signaling is Critical for AFD Thermotransduction

To date, AFD is the only sensory neuron with an identified role in *C. elegans'* behavioral responses to temperature. *tax-4* and *tax-2*, which encode subunits of a CNG channel, are expressed in AFD and are required for thermotaxis and isothermal tracking (Coburn and Bargmann 1996; Hedgecock and Russell 1975; Komatsu et al. 1996; Mori and Ohshima 1995). When co-expressed in heterologous cells, TAX-4 and TAX-2 form a heteromeric cation channel preferentially activated by cGMP (Komatsu et al. 1999). The expression pattern of TAX-4 and TAX-2 in vivo, visualized by tagging the proteins with GFP, suggests that both are localized to ciliated endings (Coburn and Bargmann 1996; Komatsu et al. 1996).

Ca^{2+} dynamics in AFD, monitored by a genetically encoded fluorescent Ca^{2+} indicator, confirm that AFD responds to thermal stimuli. Warming elevates intracellular Ca^{2+} concentration whereas cooling decreases intracellular Ca^{2+} in AFD (Clark et al. 2006a; Kimura et al. 2004). Temperature changes as small as 0.05°C elicit robust Ca^{2+} transients, indicating that AFD is sensitive enough to account for the behavioral precision of isothermal tracking (Clark et al. 2006a). Mutations in *tax-4* abolish calcium transients, suggesting that activation of the TAX-4 channel is required for thermotransduction in AFD (Kimura et al. 2004). Consistent with these data, whole-cell patch clamp recordings in AFD reveal that membrane potential and membrane current vary in response to temperature ramps and that these responses require *tax-4* (D. Ramot and M. B. Goodman, unpublished).

In support of the hypothesis that thermotransduction in AFD involves cGMP signaling, three guanylate cyclases, *gcy-8*, *gcy-18* and *gcy-23*, are expressed exclusively in AFD and are required for thermotaxis. As observed with TAX-4 and TAX-2, GFP-tagged GCY-8, GCY-18 and GCY-23 proteins are localized to the sensory cilia. Although deletion mutants of each of these guanylate cyclases show essentially normal thermotaxis, double and triple mutants are defective in the behavior, suggesting that the cyclases function redundantly (Inada et al. 2006). Whether these guanylate cyclases are thermosensitive or if they operate downstream of a temperature receptor is not known. In addition, the requirement for a cGMP-gated channel and guanylate cyclases in AFD thermotransduction predicts the involvement of an as yet unidentified cGMP-degrading phosphodiesterase(s).

A small number of additional genes acting in AFD and required for normal thermotaxis have been identified. These include *tax-6*, a calcineurin A subunit (Kuhara et al. 2002), *cmk-1*, a Ca^{2+}/Calmodulin-dependent protein kinase I (Satterlee et al. 2004), and *ttx-4*, a novel protein kinase C epsilon/eta ortholog (Okochi et al. 2005). How these proteins contribute to thermosensory processing in AFD is not well understood, but their importance for thermotaxis suggests Ca^{2+}, calmodulin and diacylglycerol play key roles in AFD signal transduction.

10.3.4   Subcellular Localization of the Transduction Apparatus

Three lines of evidence point to the localization of AFD thermotransduction at the neuron's ciliated tips. AFD's sensory endings are quite elaborate, with a large number of microvilli extending around a short cilium (Fig. 10.1d) (Ward et al. 1975). Abnormalities in the morphology of these endings may underlie the behavioral deficits of two mutants, *ceh-14* and *ttx-1*. Notably, a reduced surface area-to-volume ratio of the finger-like endings is the only detectable deficit in *ceh-14* animals, suggesting that the structure of this sensory specialization may be critical for wild-type behavioral responses (Cassata et al. 2000; Satterlee et al. 2001). Second, as reviewed above, the TAX-4/TAX-2 ion channel required for AFD's temperature response is localized to the ciliated tips. Third, in technically elegant experiments, Clark, Chung and colleagues used femtosecond laser pulses to selectively sever the AFD dendrite, separating the ciliated tip from the cell body. This manipulation resulted in behavioral deficits that were very similar to laser ablation of the AFD soma. In addition, temperature-dependent Ca^{2+} transients were retained at the sensory tips following the surgery, but were abolished at the soma (Chung et al.

2006; Clark et al. 2006a). Taken together, these data support a model in which temperature is detected at the AFD sensory tip.

10.3.5 Similarities to Vertebrate Vision

In several respects, AFD offers interesting parallels to vertebrate rod photoreceptors. TAX-4 and TAX-2 display high sequence similarity with the α and β subunits of the human rod photoreceptor CNG channel (Coburn and Bargmann 1996; Komatsu et al. 1996). TTX-1, required for specification of the AFD cell fate, is an OTD/OTX homolog. OTD/OTX proteins are critical for patterning sensory organs, including visual structures (Hirth and Reichert 1999). Svendsen and McGhee (1995) also reported similarity of regulatory mechanisms between AIY, the neuron downstream of AFD, and bipolar cells of the vertebrate retina. Finally, it is tempting to compare the intricate finger-like ciliated endings of AFD and the rod outer segment. It is not known whether the high surface area-to-volume ratio of the AFD fingers contributes to signal amplification, as it does in rods (Pugh and Lamb 1993), but the defect observed in *ceh-14* mutants supports this possibility.

10.4 Chemosensation and Chemotransduction

In *C. elegans*, genetic studies indicate that volatile and soluble chemicals exert their effects on chemosensory neurons by modulating the activity of either CNG or TRP channels. Neither channel type is uniquely associated with a particular chemosensory modality. The CNG channel formed by TAX-4 and TAX-2 is essential for the function of sensory neurons that detect both volatile and soluble chemicals (Coburn and Bargmann 1996). Similarly, a TRP channel subunit encoded by *osm-9* is needed for volatile chemotaxis and avoidance of noxious soluble chemicals (Colbert et al. 1997) and high levels of oxygen (Chang et al. 2006; Rogers et al. 2006). These observations suggest that receptor proteins, rather than transduction channels establish the sensory repertoire of chemosensory neurons. Current evidence indicates that such receptor proteins are either G-protein-coupled receptors like the ODR-10 diacetyl receptor (Sengupta et al. 1996) or receptor guanylate cyclases like DAF-11 (Birnby et al. 2000).

10.4.1 CNG Channels in Chemotransduction

TAX-4 and TAX-2 are co-expressed in a subset of chemosensory neurons: the AWC and AWB olfactory neurons and the ASE, ASG, ASK, ASI, and ASJ gustatory neurons (Coburn and Bargmann 1996; Komatsu et al. 1996). As reviewed in the previous section, TAX-4 and TAX-2 are localized to ciliated endings, and when heterologously expressed form a heteromeric channel preferentially activated by cGMP (Komatsu et al. 1999). The latter finding suggests that receptor proteins modify intracellular cGMP levels either directly or by altering the activity of guanylate cyclases or phosphodiesterases. Although two other CNG channel subunits, *cng-1* and *cng-3*, are expressed in overlapping subsets of chemosensory neurons, mutations that disrupt these genes have no detectable effects on either chemotaxis or chemo-avoidance (Cho et al. 2004, 2005).

Many questions remain about the physiology of CNG-channel mediated chemotransduction in *C. elegans*. For example, do chemical stimuli open or close channels? To what extent are signals amplified by an enzymatic cascade that links receptor activation to channel opening (or closing)? How fast is chemotransduction? How do these parameters vary among chemosensory neurons? New techniques for monitoring neuronal responses in *C. elegans* such as in vivo and in vitro Ca^{2+} imaging and patch-clamp electrophysiology will allow researchers to seek answers to these questions in the very near future.

10.4.2   TRP Channels in Chemotransduction

Defects in the *osm-9* gene affect the function of a subset of chemosensory and mechanosensory neurons as well as gene expression (Colbert et al. 1997). OSM-9 is a member of the TRPV subfamily of ion channel proteins. The homologous subfamily in mammals includes the capsaicin-receptor TRPV1. In *C. elegans*, this subfamily includes four additional genes: *ocr-1*, *ocr-2*, *ocr-3* and *ocr-4*. Each of the *ocr* genes is expressed in a subset of sensory neurons and all *ocr* genes are co-expressed with *osm-9* (Tobin et al. 2002). This suggests that OSM-9 forms heteromeric channels with the products of each of the *ocr* genes, though there is no direct biochemical or functional evidence for this prediction.

The AWA neurons rely on both *osm-9* and *ocr-2* for their olfactory functions (Tobin et al. 2002). Genetic epistasis studies suggest that a putative OSM-9/OCR-2 channel is activated downstream of G-protein-coupled odorant receptors and functions as a sensory transduction channel (reviewed by Kahn-Kirby and Bargmann 2006). In addition to its role as a putative sensory transduction channel in AWA, *osm-9* also contributes to adaptation to prolonged application of volatile and soluble stimuli sensed by the ASE and AWC neurons (Colbert et al. 1997).

10.4.3   Oxygen Sensing and Aerotaxis

Worms avoid both high (>12%) and low (<4%) oxygen concentrations (Dusenbery 1980; Gray et al. 2004). *C. elegans* requires oxygen for survival, and may shun high O_2 concentrations to limit cellular oxidative damage or as a strategy for detecting bacteria that locally deplete O_2. The ion channels required for hyperoxia avoidance include the usual *C. elegans* suspects: the CNG channel TAX-4/TAX-2 (Gray et al. 2004) and the TRP channel subunits OCR-2 and OSM-9 (Rogers et al. 2006). However, how hypoxia avoidance is mediated is not well understood.

A complex, distributed neural network is responsible for *C. elegans* aerotaxis (Chang et al. 2006). TAX-4 and TAX-2 function in URX, AQR and PQR to modulate avoidance of high O_2 concentrations. These sensory neurons extend dendrites to the tip of the nose (URX) and into the body cavity (AQR, PQR), and are thus in a position to monitor both external and internal O_2 concentrations (White et al. 1986). The likely O_2 sensor is GCY-35, a soluble guanylate cyclase (sGC), since GCY-35 activity in URX, AQR and PQR is required for avoidance of hyperoxia, and the GCY-35 heme domain binds molecular oxygen (Gray et al. 2004). Mutations in a second sGC, GCY-36, also disrupt hyperoxia and can be rescued by expression of *gcy-36* in URX, AQR and PQR (Cheung et al. 2005). It has been proposed that GCY-35 and GCY-36 function together as a heterodimer (Cheung et al. 2004).

OCR-2 and OSM-9 act in a separate set of neurons, ASH and ADL, to promote avoidance of high O_2 (Chang et al. 2006; Rogers et al. 2006). ODR-4, a transmembrane protein critical for localization of odorant receptors to cilia, is also required in ASH and ADL for aerotaxis (Rogers et al. 2006). How these neurons sense O_2 is unknown. It is possible that they do not directly sense O_2, but rather modulate the activity of O_2-sensing cells.

Although many questions remain regarding O_2 sensation in *C. elegans*, the identification of a sGC that appears to function as an O_2 receptor promises to yield exciting new insights into how this vital environmental stimulus is detected and processed.

10.5 Polymodal Channels and Nociception

Nociceptors are activated by high-intensity stimuli that have the potential to cause tissue damage. They can detect thermal, mechanical and chemical stimuli (i.e., they are polymodal) and are responsible for our ability to sense pain. The ASH neurons in *C. elegans* are also polymodal (they detect chemical, mechanical, and osmotic stimuli; see Hilliard et al. 2005) and may provide a simplified model of vertebrate nociceptors. An intact *osm-9* gene is required for ASH-mediated behavioral responses to all three kinds of stimuli (Colbert et al. 1997) as well as for stimulus-evoked somatic Ca^{2+} transients (Hilliard et al. 2005). This suggests a single TRPV channel can be activated by diverse physical stimuli. It is not known whether or not such activation is direct or receptor-mediated. It is possible that some stimuli activate such channels directly while others act via receptors.

10.6   Conclusion

Despite its small size and compact nervous system, *C. elegans* is able to detect and respond to a variety of sensory stimuli. Ion channels responsible for converting sensory stimuli into neuronal responses are being identified and characterized by a remarkable convergence of methods in classical genetics, molecular genetics and cellular neurophysiology and biophysics. Indeed, the first answer to the question of what proteins form mechanoelectrical transduction channels in metazoans was obtained in *C. elegans* (O'Hagan et al. 2005). The proteins in question, MEC-4 and MEC-10, belong to the superfamily of DEG/ENaC proteins and the channel that they form is likely to be directly activated by mechanical force. Related proteins may serve similar functions in other mechanosensory neurons. Other channel proteins implicated in *C. elegans* sensory transduction include TAX-4 and TAX-2, CNG channel subunits, and OSM-9, a TRPV channel subunit. All three have key roles in sensing soluble and volatile chemicals as well as oxygen (Sect. 10.4), while OSM-9 and the two CNG channel subunits are required for some forms of mechanosensation (Sect. 10.2) and for temperature sensation (Sect. 10.3), respectively. Though much has been learned, many fundamental questions remain open. Do mechanical stimuli open TRPV channels directly, as seems to be the case for the MEC-4/MEC-10 channel complex? Do chemical stimuli act on TRPV directly or by activation of specific membrane receptors? The answer to the latter question may depend on the cellular context. Additional questions for the near future include: How do worms detect temperature changes <0.1°C? What is the molecular identity of the thermosensor in AFD? Technical advances in *C. elegans* cellular physiology are offering new leads for answering these questions regarding the fundamental role of ion channels in sensation.

Acknowledgments We thank the members of the Goodman laboratory for many fruitful discussions. Work in the laboratory is supported by grants from the National Institute of Neurological Disorders and Stroke, NIH, USA (NS047715) and the American Heart Association and by fellowships from the Alfred P. Sloan, Donald B. and Delia E. Baxter Foundations, the McKnight Endowment for Neuroscience, the Esther A. and Joseph Klingenstein Fund to M.B.G and from the American Heart Association to A.L.B.

References

Bargmann CI (2006) Chemosensation. In: The *C. elegans* Research Community (ed) WormBook, doi/10.1895/wormbook.1.123.1, http://www.wormbook.org
Barr MM (2005) *Caenorhabditis elegans* as a model to study renal development and disease: sexy cilia. J Am Soc Nephrol 16:305–312
Barr MM, Garcia LR (2006) Male mating behavior. In: The *C. elegans* Research Community (ed) WormBook, doi/10.1895/wormbook.1.78.1, http://www.wormbook.org
Barr MM, Sternberg PW (1999) A polycystic kidney-disease gene homologue required for male mating behaviour in *C. elegans*. Nature 401:386–389
Barr MM, DeModena J, Braun D, Nguyen CQ, Hall DH, Sternberg PW (2001) The *Caenorhabditis elegans* autosomal dominant polycystic kidney disease gene homologs *lov-1* and *pkd-2* act in the same pathway. Curr Biol 11:1341–1346

Bianchi L, Driscoll M (2006) Culture of embryonic *C. elegans* cells for electrophysiological and pharmacological analyses. In: The *C. elegans* Research Community (ed) WormBook, doi/10.1895/wormbook.1.122.1, http://www.wormbook.org

Birnby DA, Link EM, Vowels JJ, Tian H, Colacurcio PL, Thomas JH (2000) A transmembrane guanylyl cyclase (DAF-11) and Hsp90 (DAF-21) regulate a common set of chemosensory behaviors in *Caenorhabditis elegans*. Genetics 155:85–104

Brown AL, Fernandez-Illescas SM, Liao Z, Goodman MB (2007) Gain-of-function mutations in the MEC-4 DEG/ENaC sensory mechanotransduction channel alter gating and drug blockade. J Gen Physiol 129:161–173

Cassata G, Kagoshima H, Andachi Y, Kohara Y, Durrenberger MB, Hall DH, Burglin TR (2000) The LIM homeobox gene *ceh-14* confers thermosensory function to the AFD neurons in *Caenorhabditis elegans*. Neuron 25:587–597

Chalfie M, Au M (1989) Genetic control of differentiation of the *Caenorhabditis elegans* touch receptor neurons. Science 243:1027–1033

Chalfie M, Sulston J (1981) Developmental genetics of the mechanosensory neurons of *Caenorhabditis elegans*. Dev Biol 82:358–370

Chalfie M, Thomson JN (1979) Organization of neuronal microtubules in the nematode *Caenorhabditis elegans*. J Cell Biol 82:278–289

Chalfie M, Wolinsky E (1990) The identification and suppression of inherited neurodegeneration in *Caenorhabditis elegans*. Nature 345:410–416

Chang AJ, Chronis N, Karow DS, Marletta MA, Bargmann CI (2006) A distributed chemosensory circuit for oxygen preference in *C. elegans*. PLoS Biol 4:e274

Chelur DS, Ernstrom GG, Goodman MB, Yao CA, Chen L, O'Hagan R, Chalfie M (2002) The mechanosensory protein MEC-6 is a subunit of the *C. elegans* touch-cell degenerin channel. Nature 420:669–673

Cheung BH, Arellano-Carbajal F, Rybicki I, de Bono M (2004) Soluble guanylate cyclases act in neurons exposed to the body fluid to promote *C. elegans* aggregation behavior. Curr Biol 14:1105–1111

Cheung BH, Cohen M, Rogers C, Albayram O, de Bono M (2005) Experience-dependent modulation of *C. elegans* behavior by ambient oxygen. Curr Biol 15:905–917

Cho SW, Choi KY, Park CS (2004) A new putative cyclic nucleotide-gated channel gene, *cng-3*, is critical for thermotolerance in *Caenorhabditis elegans*. Biochem Biophys Res Commun 325:525–531

Cho SW, Cho JH, Song HO, Park CS (2005) Identification and characterization of a putative cyclic nucleotide-gated channel, CNG-1, in *C. elegans*. Mol Cell 19:149–154

Christensen M, Estevez A, Yin X, Fox R, Morrison R, McDonnell M, Gleason C, Miller DM 3rd, Strange K (2002) A primary culture system for functional analysis of *C. elegans* neurons and muscle cells. Neuron 33:503–514

Chung SH, Clark DA, Gabel CV, Mazur E, Samuel AD (2006) The role of the AFD neuron in *C. elegans* thermotaxis analyzed using femtosecond laser ablation. BMC Neurosci 7:30

Clark DA, Biron D, Sengupta P, Samuel AD (2006a) The AFD sensory neurons encode multiple functions underlying thermotactic behavior in *Caenorhabditis elegans*. J Neurosci 26:7444–7451

Clark DA, Gabel C, Lee TM, Samuel A (2006b) Short-term adaptation and temporal processing in the cryophilic response of *Caenorhabditis elegans*. J Neurophysiol 97:1903–1910

Coburn CM, Bargmann CI (1996) A putative cyclic nucleotide-gated channel is required for sensory development and function in *C. elegans*. Neuron 17:695–706

Colbert HA, Smith TL, Bargmann CI (1997) OSM-9, a novel protein with structural similarity to channels, is required for olfaction, mechanosensation, and olfactory adaptation in *Caenorhabditis elegans*. J Neurosci 17:8259–8269

Colosimo ME, Brown A, Mukhopadhyay S, Gabel C, Lanjuin AE, Samuel AD, Sengupta P (2004) Identification of thermosensory and olfactory neuron-specific genes via expression profiling of single neuron types. Curr Biol 14:2245–2251

Dhaka A, Viswanath V, Patapoutian A (2006) TRP ion channels and temperature sensation. Annu Rev Neurosci 29:135–161

Driscoll M, Chalfie M (1991) The *mec-4* gene is a member of a family of *Caenorhabditis elegans* genes that can mutate to induce neuronal degeneration. Nature 349:588–593

Du H, Gu G, William CM, Chalfie M (1996) Extracellular proteins needed for *C. elegans* mechanosensation. Neuron 16:183–194

Dusenbery DB (1980) Appetitive response of the nematode *Caenorhabditis elegans* to oxygen. J Comp Physiol 133:333–336

Emtage L, Gu G, Hartwieg E, Chalfie M (2004) Extracellular proteins organize the mechanosensory channel complex in *C. elegans* touch receptor neurons. Neuron 44:795–807

Ernstrom GG, Chalfie M (2002) Genetics of sensory mechanotransduction. Annu Rev Genet 36:411–453

Fay D (2006) Genetic mapping and manipulation In: The *C. elegans* Research Community (ed) WormBook, doi/10.1895/wormbook.1.90.1, http://www.wormbook.org

Fink M, Lesage F, Duprat F, Heurteaux C, Reyes R, Fosset M, Lazdunski M (1998) A neuronal two P domain K+ channel stimulated by arachidonic acid and polyunsaturated fatty acids. EMBO J 17:3297–3308

Fokine A, Morales R, Contreras-Martel C, Carpentier P, Renault F, Rochu D, Chabriere E (2003) Direct phasing at low resolution of a protein copurified with human paraoxonase (PON1). Acta Crystallogr D Biol Crystallogr 59:2083–2207

Garcia-Anoveros J, Garcia JA, Liu JD, Corey DP (1998) The nematode degenerin UNC-105 forms ion channels that are activated by degeneration- or hypercontraction-causing mutations. Neuron 20:1231–1241

Girard LR, Fiedler TJ, Harris TW, Carvalho F, Antoshechkin I, Han M, Sternberg PW, Stein LD, Chalfie M (2006) WormBook: the online review of *Caenorhabditis elegans* biology. Nucleic Acids Res 35:D472–D475

Goodman MB (2006) Mechanosensation. In: The *C. elegans* Research Community (ed) WormBook, doi/10.1895/wormbook.1.62.1, http://www.wormbook.org

Goodman MB, Schwarz EM (2003) Transducing touch in *Caenorhabditis elegans*. Annu Rev Physiol 65:429–452

Goodman MB, Hall DH, Avery L, Lockery SR (1998) Active currents regulate sensitivity and dynamic range in *C. elegans* neurons. Neuron 20:763–772

Goodman MB, Ernstrom GG, Chelur DS, O'Hagan R, Yao CA, Chalfie M (2002) MEC-2 regulates *C. elegans* DEG/ENaC channels needed for mechanosensation. Nature 415:1039–1042

Gray JM, Karow DS, Lu H, Chang AJ, Chang JS, Ellis RE, Marletta MA, Bargmann CI (2004) Oxygen sensation and social feeding mediated by a *C. elegans* guanylate cyclase homologue. Nature 430:317–322

Gu G, Caldwell GA, Chalfie M (1996) Genetic interactions affecting touch sensitivity in *Caenorhabditis elegans*. Proc Natl Acad Sci USA 93:6577–6582

Hall DH, Lints R, Altun Z (2006) Nematode neurons: anatomy and anatomical methods in *Caenorhabditis elegans*. Int Rev Neurobiol 69:1–35

Hart AC, Sims S, Kaplan JM (1995) Synaptic code for sensory modalities revealed by *C. elegans* GLR-1 glutamate receptor. Nature 378:82–85

Hedgecock EM, Russell RL (1975) Normal and mutant thermotaxis in the nematode *Caenorhabditis elegans*. Proc Natl Acad Sci USA 72:4061–4065

Hilliard MA, Apicella AJ, Kerr R, Suzuki H, Bazzicalupo P, Schafer WR (2005) in vivo imaging of *C. elegans* ASH neurons: cellular response and adaptation to chemical repellents. EMBO J 24:63–72

Hirth F, Reichert H (1999) Conserved genetic programs in insect and mammalian brain development. Bioessays 21:677–684

Hobert O, Mori I, Yamashita Y, Honda H, Ohshima Y, Liu Y, Ruvkun G (1997) Regulation of interneuron function in the *C. elegans* thermoregulatory pathway by the *ttx-3* LIM homeobox gene. Neuron 19:345–357

Hobert O, D'Alberti T, Liu Y, Ruvkun G (1998) Control of neural development and function in a thermoregulatory network by the LIM homeobox gene *lin-11*. J Neurosci 18:2084–2096

Hong K, Mano I, Driscoll M (2000) in vivo structure-function analyses of *Caenorhabditis elegans* MEC-4, a candidate mechanosensory ion channel subunit. J Neurosci 20:2575–2588

Huang M, Chalfie M (1994) Gene interactions affecting mechanosensory transduction in *Caenorhabditis elegans*. Nature 367:467–470

Huang M, Gu G, Ferguson EL, Chalfie M (1995) A stomatin-like protein necessary for mechano-sensation in *C. elegans*. Nature 378:292–295

Inada H, Ito H, Satterlee J, Sengupta P, Matsumoto K, Mori I (2006) Identification of guanylyl cyclases that function in thermosensory neurons of *Caenorhabditis elegans*. Genetics 172:2239–2252

Inglis PN, Ou G, Leroux MR, Scholey JM (2006) The sensory cilia of *Caenorhabditis elegans*. In: The *C. elegans* Research Community (ed) WormBook, doi/10.1895/wormbook.1.126.1, http://www.wormbook.org

Ito H, Inada H, Mori I (2006) Quantitative analysis of thermotaxis in the nematode *Caenorhabditis elegans*. J Neurosci Methods 154:45–52

Jospin M, Mariol MC, Segalat L, Allard B (2002) Characterization of K$^+$ currents using an in situ patch clamp technique in body wall muscle cells from *Caenorhabditis elegans*. J Physiol 544:373–384

Kahn-Kirby AH, Bargmann CI (2006) TRP channels in *C. elegans*. Annu Rev Physiol 68:719–736

Kahn-Kirby AH, Dantzker JL, Apicella AJ, Schafer WR, Browse J, Bargmann CI, Watts JL (2004) Specific polyunsaturated fatty acids drive TRPV-dependent sensory signaling in vivo. Cell 119:889–900

Kaplan JM, Horvitz HR (1993) A dual mechanosensory and chemosensory neuron in *Caenorhabditis elegans*. Proc Natl Acad Sci USA 90:2227–2231

Kerr RA (2006) Imaging the activity of neurons and muscles. In: The *C. elegans* Research Community (ed) WormBook, doi/10.1895/wormbook.1.113.1, http://www.wormbook.org

Kimura KD, Miyawaki A, Matsumoto K, Mori I (2004) The *C. elegans* thermosensory neuron AFD responds to warming. Curr Biol 14:1291–1295

Komatsu H, Mori I, Rhee JS, Akaike N, Ohshima Y (1996) Mutations in a cyclic nucleotide-gated channel lead to abnormal thermosensation and chemosensation in *C. elegans*. Neuron 17:707–718

Komatsu H, Jin YH, L'Etoile N, Mori I, Bargmann CI, Akaike N, Ohshima Y (1999) Functional reconstitution of a heteromeric cyclic nucleotide-gated channel of *Caenorhabditis elegans* in cultured cells. Brain Res 821:160–168

Kuhara A, Inada H, Katsura I, Mori I (2002) Negative regulation and gain control of sensory neurons by the *C. elegans* calcineurin TAX-6. Neuron 33:751–763

Li W, Feng Z, Sternberg PW, Xu XZ (2006) A *C. elegans* stretch receptor neuron revealed by a mechanosensitive TRP channel homologue. Nature 440:684–687

Liu J, Schrank B, Waterston RH (1996) Interaction between a putative mechanosensory membrane channel and a collagen. Science 273:361–364

Lockery SR, Goodman MB (1998) Tight-seal whole-cell patch clamping of *Caenorhabditis elegans* neurons. Methods Enzymol 293:201–217

Lowenstein W, Mendelson M (1965) Components of receptor adaptation in a Pacinian corpuscle. J Physiol 177:377–397

Luo L, Clark DA, Biron D, Mahadevan L, Samuel AD (2006) Sensorimotor control during iso-thermal tracking in *Caenorhabditis elegans*. J Exp Biol 209:4652–4662

Matta JA, Miyares RL, Ahern GP (2007) TRPV1 is a novel target for omega-3 polyunsaturated fatty acids. J Physiol 578:397–411

Mori I, Ohshima Y (1995) Neural regulation of thermotaxis in *Caenorhabditis elegans*. Nature 376:344–348

Nauli SM, Alenghat FJ, Luo Y, Williams E, Vassilev P, Li X, Elia AE, Lu W, Brown EM, Quinn SJ, Ingber DE, Zhou J (2003) Polycystins 1 and 2 mediate mechanosensation in the primary cilium of kidney cells. Nat Genet 33:129–137

O'Hagan R (2005) Components of a mechanotransduction complex in *Caenorhabditis elegans* touch receptor neurons: an in vivo electrophysiological study. PhD Thesis, Columbia University

O'Hagan R, Chalfie M (2006) Mechanosensation in *Caenorhabditis elegans*. Int Rev Neurobiol 69:169–203

O'Hagan R, Chalfie M, Goodman MB (2005) The MEC-4 DEG/ENaC channel of *Caenorhabditis elegans* touch receptor neurons transduces mechanical signals. Nat Neurosci 8:43–50

Okochi Y, Kimura KD, Ohta A, Mori I (2005) Diverse regulation of sensory signaling by *C. elegans* nPKC-epsilon/eta TTX-4. EMBO J 24:2127–2137

Park EC, Horvitz HR (1986) *C. elegans unc-105* mutations affect muscle and are suppressed by other mutations that affect muscle. Genetics 113:853–867

Pickles JO, Comis SD, Osborne MP (1984) Cross-links between stereocilia in the guinea pig organ of Corti, and their possible relation to sensory transduction. Hear Res 15:103–112

Pierce-Shimomura JT, Morse TM, Lockery SR (1999) The fundamental role of pirouettes in *Caenorhabditis elegans* chemotaxis. J Neurosci 19:9557–9569

Pugh EN Jr, Lamb TD (1993) Amplification and kinetics of the activation steps in phototransduction. Biochim Biophys Acta 1141:111–149

Rogers C, Persson A, Cheung B, de Bono M (2006) Behavioral motifs and neural pathways coordinating O_2 responses and aggregation in *C. elegans*. Curr Biol 16:649–659

Royal DC, Bianchi L, Royal MA, Lizzio M Jr, Mukherjee G, Nunez YO, Driscoll M (2005) Temperature-sensitive mutant of the *Caenorhabditis elegans* neurotoxic MEC-4(d) DEG/ENaC channel identifies a site required for trafficking or surface maintenance. J Biol Chem 280:41976–41986

Ryu WS, Samuel AD (2002) Thermotaxis in *Caenorhabditis elegans* analyzed by measuring responses to defined thermal stimuli. J Neurosci 22:5727–5733

Satterlee JS, Sasakura H, Kuhara A, Berkeley M, Mori I, Sengupta P (2001) Specification of thermosensory neuron fate in *C. elegans* requires *ttx-1*, a homolog of *otd/Otx*. Neuron 31:943–956

Satterlee JS, Ryu WS, Sengupta P (2004) The CMK-1 CaMKI and the TAX-4 cyclic nucleotide-gated channel regulate thermosensory neuron gene expression and function in *C. elegans*. Curr Biol 14:62–68

Sengupta P, Chou JH, Bargmann CI (1996) *odr-10* encodes a seven transmembrane domain olfactory receptor required for responses to the odorant diacetyl. Cell 84:899–909

Shreffler W, Magardino T, Shekdar K, Wolinsky E (1995) The *unc-8* and *sup-40* genes regulate ion channel function in *Caenorhabditis elegans* motorneurons. Genetics 139:1261–1272

Sidi S, Friedrich RW, Nicolson T (2003) NompC TRP channel required for vertebrate sensory hair cell mechanotransduction. Science 301:96–99

Suzuki H, Kerr R, Bianchi L, Frokjaer-Jensen C, Slone D, Xue J, Gerstbrein B, Driscoll M, Schafer WR (2003) in vivo imaging of *C. elegans* mechanosensory neurons demonstrates a specific role for the MEC-4 channel in the process of gentle touch sensation. Neuron 39:1005–1017

Svendsen PC, McGhee JD (1995) The *C. elegans* neuronally expressed homeobox gene *ceh-10* is closely related to genes expressed in the vertebrate eye. Development 121:1253–1262

Syntichaki P, Tavernarakis N (2004) Genetic models of mechanotransduction: the nematode *Caenorhabditis elegans*. Physiol Rev 84:1097–1153

Tavernarakis N, Shreffler W, Wang S, Driscoll M (1997) *unc-8*, a DEG/ENaC family member, encodes a subunit of a candidate mechanically gated channel that modulates *C. elegans* locomotion. Neuron 18:107–119

Tobin D, Madsen D, Kahn-Kirby A, Peckol E, Moulder G, Barstead R, Maricq A, Bargmann C (2002) Combinatorial expression of TRPV channel proteins defines their sensory functions, subcellular localization in *C. elegans* neurons. Neuron 35:307–318

Tsalik EL, Niacaris T, Wenick AS, Pau K, Avery L, Hobert O (2003) LIM homeobox gene-dependent expression of biogenic amine receptors in restricted regions of the *C. elegans* nervous system. Dev Biol 263:81–102

Walker RG, Willingham AT, Zuker CS (2000) A Drosophila mechanosensory transduction channel. Science 287:2229–2234

Ward S, Thomson N, White JG, Brenner S (1975) Electron microscopical reconstruction of the anterior sensory anatomy of the nematode *Caenorhabditis elegans*. J Comp Neurol 160:313–337

Way JC, Chalfie M (1988) *mec-3*, a homeobox-containing gene that specifies differentiation of the touch receptor neurons in *C. elegans*. Cell 54:5–16

White JG, Southgate E, Thomson JN, Brenner S (1986) The structure of the nervous system of the nematode *C. elegans*. Philos Trans R Soc B 314:1–340

Wittenburg N, Baumeister R (1999) Thermal avoidance in *Caenorhabditis elegans*: an approach to the study of nociception. Proc Natl Acad Sci USA 96:10477–10482

Zariwala HA, Miller AC, Faumont S, Lockery SR (2003) Step response analysis of thermotaxis in *Caenorhabditis elegans*. J Neurosci 23:4369–4377

Zhang S, Arnadottir J, Keller C, Caldwell GA, Yao CA, Chalfie M (2004) MEC-2 is recruited to the putative mechanosensory complex in *C. elegans* touch receptor neurons through its stomatin-like domain. Curr Biol 14:1888–1896

Zhang Y, Chou JH, Bradley J, Bargmann CI, Zinn K (1997) The *Caenorhabditis elegans* seven-transmembrane protein ODR-10 functions as an odorant receptor in mammalian cells. Proc Natl Acad Sci USA 94:12162–12197

Zhang Y, Ma C, Delohery T, Nasipak B, Foat BC, Bounoutas A, Bussemaker HJ, Kim SK, Chalfie M (2002) Identification of genes expressed in *C. elegans* touch receptor neurons. Nature 418:331–335

Chapter 11
Epithelial Sodium and Acid-Sensing Ion Channels

Stephan Kellenberger

Abstract The epithelial Na$^+$ channel (ENaC) and acid-sensing ion channels (ASICs) are non-voltage-gated Na$^+$ channels that form their own subfamilies within the ENaC/degenerin ion channel family. ASICs are sensors of extracellular pH, and ENaC, whose main function is trans-epithelial Na$^+$ transport, can sense extra- and intra-cellular Na$^+$. In aldosterone-responsive epithelial cells of the kidney, ENaC plays a critical role in the control of sodium balance, blood volume and blood pressure. In airway epithelia, ENaC has a distinct role in controlling fluid reabsorption at the air–liquid interface, thereby determining the rate of mucociliary transport. In taste receptor cells of the tongue, ENaC is involved in salt taste sensation. ASICs have emerged as key sensors for

Pharmacology and Toxicology Department, University of Lausanne, Rue du Bugnon 27, CH-1005 Lausanne, Switzerland, Stephan.Kellenberger@unil.ch extracellular protons in central and peripheral neurons. Although not all of their physiological and pathological functions are firmly established yet, there is good evidence for a role of ASICs in the brain in learning, expression of fear, and in neurodegeneration after ischaemic stroke. In sensory neurons, ASICs are involved in nociception and mechanosensation. ENaC and ASIC subunits share substantial sequence homology and the conservation of several functional domains. This chapter summarises our current understanding of the physiological functions and of the mechanisms of ion permeation, gating and regulation of ENaC and ASICs.

11.1   Introduction

The epithelial Na^+ channel (ENaC) and acid-sensing ion channels (ASICs) constitute the two main mammalian sub-families of the ENaC/degenerin (DEG) family of ion channels. ENaC mediates Na^+ transport in epithelia of mammals and is essential for Na^+ homeostasis; ASICs are H^+-gated cation channels in the nervous system, and the degenerins in *Caenorrhabditis elegans* participate in mechanotransduction in neurons. All members of the ENaC/DEG ion channel family share substantial sequence homology and a common subunit organisation, and form Na^+-selective or -preferring ion channels. In this chapter, I will summarise the present knowledge of the fundamental roles of ENaC and ASICs, and the relationship between their biophysical properties and structural characteristics. The *C. elegans* degenerins are covered in a separate chapter (see Chap. 10 by Brown et al., this volume). The ASICs are sensors of pH that transduce a decrease in extracellular pH into a change of neuronal activity. In contrast, the principal function of ENaC is ion transport. There is, however, evidence that ENaC does sense the intra- and extra-cellular Na^+ concentration, which regulates its activity. ENaC-mediated Na^+ entry can depolarise the membrane thereby inducing, at least in some cell types, downstream cellular events.

11.2   Overview of the ENaC/DEG Channel Family

Figure 11.1 shows a phylogenetic tree of the most relevant ENaC/DEG sequences available. Five different branches can be distinguished. The four main subfamilies comprise the mammalian ASICs and ENaC, the *C. elegans* degenerins and the Drosophila ripped pocket and pickpocket proteins. Of the mammalian channels, human and rat subunits are represented to allow an estimate of cross-species conservation. The ASICs have different splice variants, distinguished by letters (a, b, c). In some of the older literature, ASICs appear under different names: BNaC2 (for ASIC1a, gene ACCN2), ASICβ (ASIC1b); MDEG, MDEG1, BNC1

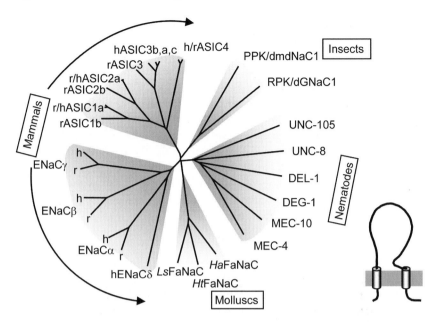

Fig. 11.1 Phylogenetic tree of the epithelial Na⁺ channel/degenerin (ENaC/DEG) family, showing the organisation into subfamilies of related sequences. ENaC/DEG proteins of invertebrates can be divided into three groups: (1) degenerins from *Caenorrhabditis elegans*; (2) Drosophila channels RPK/dGNaC1 and PPK/dmdNaC1 (14 *ppk* genes have been identified); and (3) FaNaC, the FMRFamide-gated ion channel expressed in molluscs. The channels from vertebrates are divided into two groups: ENaC and acid-sensing ion channels (ASICs). *r* Rat, *h* human, *PPK* pickpocket, *RPK* ripped pocket, *HaFaNaC Helix aspersa* FaNaC, *HtFaNaC Helisoma trivolvis* FaNaC, *LsFaNaC Lymnaea stagnalis* FaNaC. The topology of an ENaC/DEG subunit is illustrated on the right

(ASIC2a, ACCN1); MDEG2 (ASIC2b); DRASIC, hTNaC1 (ASIC3, ACCN3); and SPASIC (ASIC4, ACCN4). ASICs have also been cloned from fishes (Coric et al. 2005 – not represented in Fig. 11.1). Degenerins form the channel part of mechanotransduction complexes. They comprise six members of the ~23 predicted ENaC/DEG proteins identified in the *C. elegans* genome. Twenty-four ENaC/DEG family members are predicted in Drosophila. The function of RPK and of several PPKs has been analyzed. While RPK likely has a mechanosensor function, PPK11 and PPK19 are critical for salt taste sensation (Liu et al. 2003). The peptide-gated Na⁺ channel FaNaC of molluscs comprises the fifth subfamily. FaNaC is a receptor channel for an excitatory peptide neurotransmitter in the mollusc nervous system (Lingueglia et al. 2006). The amino acid sequence identity between the different ENaC/DEG subfamilies is ~15–30%. For a more exhaustive representation of the ENaC/DEG family, see e.g. Kellenberger and Schild (2002).

11.3   Localisation and Physiological Role

11.3.1   The Epithelial Na⁺ Channel

ENaC is a constitutively active channel that is located in the apical membrane of polarised cells of tight epithelia. ENaC mediates trans-epithelial Na$^+$ transport in a two-step process that involves passive entry of apical Na$^+$ through ENaC, and active outward transport across the basolateral membrane by the Na$^+$-K$^+$-ATPase (Fig. 11.2a). ENaC is functionally and structurally very different from voltage-gated Na$^+$ channels. In addition to the ENaC functions discussed below, ENaC expression occurs in other organs such as the ear and the skin, as discussed in Kellenberger and Schild (2002) or in more specialised recent reviews.

11.3.1.1   Kidney

The kidneys filter approximately 25,000 mEq Na$^+$/day from the plasma; 95% of the filtered Na$^+$ is reabsorbed before reaching the distal part of the nephron where ENaC is expressed and where its expression and function are under tight control of hormones such as aldosterone and vasopressin. ENaC is the major pathway for Na$^+$ absorption in the aldosterone-sensitive distal nephron (ASDN), to which the late portion of the distal convoluted tubule, the connecting tubule and the collecting duct contribute. In the ASDN, ENaC is present at the apical membrane of principal, but not intercalated, cells. The sub-cellular localisation of ENaC subunits changes drastically with plasma aldosterone levels in response to changes in dietary Na$^+$ (reviewed in Loffing and Schild 2005). Low Na$^+$ intake leads to an increase in plasma aldosterone concentration, which promotes insertion of intracellular ENaC subunits into the apical membrane and leads to an increase in ENaC currents. Fine tuning of Na$^+$ reabsorption in the ASDN is essential for maintaining a balance between daily sodium intake and renal sodium excretion. The electrogenic entry of Na$^+$ ions through ENaC depolarises the apical membrane and increases the electrochemical driving force for K$^+$ secretion into the tubule (Fig. 11.2a). Thus, in the ASDN, Na$^+$ absorption, which is mediated and "sensed" by ENaC, determines the extent of K$^+$ secretion. The importance of ENaC for Na$^+$, K$^+$ and fluid homeostasis is emphasised by the observation that gain-of-function mutations or loss-of-function mutations of ENaC lead, respectively, to extracellular volume expansion and hypertension (Liddle syndrome) or to a renal salt wasting syndrome (pseudohypoaldosteronism type 1) associated with alterations in K$^+$ homeostasis (Lifton et al. 2001). These observations proved the essential role of ENaC in electrolyte homeostasis and blood pressure control.

11.3.1.2   Airway Epithelia

In adult rats and humans, the α, β and γ ENaC subunits are highly expressed in small and medium-sized airways, where they are important for Na$^+$ absorption

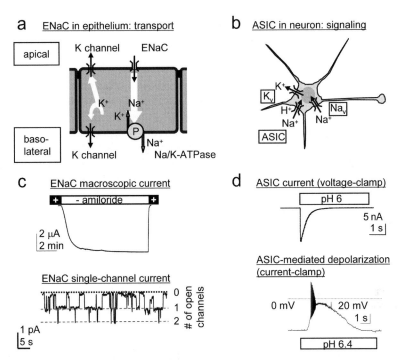

Fig. 11.2 Function of ENaC and ASICs. **a** Scheme illustrating ion transport across an epithelial cell (principal cell) lining the tubule of the distal nephron. ENaC at the apical membrane (urinary pole) allows entry of Na^+ ions into the cell that are then extruded across the basolateral membrane (blood side) by the Na^+/K^+-ATPase. The apical K^+ channel (ROMK) is responsible for K^+ secretion. **b** Scheme illustrating the mechanism by which ASIC activation can affect neuronal signaling. Activation of ASICs by H^+ allows Na^+ entry, leading to a depolarisation that, by activating voltage-gated Na^+ channels (Na_v), induces action potentials. Potassium channels (K_v) are important for the re-polarisation of the membrane potential. **c** *Upper panel* Trace of a macroscopic ENaC current (baseline obtained in the presence of 5 μM amiloride). *Lower panel* Experimental trace from a patch containing two active ENaC channels. Traces are from *Xenopus* oocytes expressing αβγ ENaC, measured under voltage-clamp. **d** Voltage-clamp measurement from a Chinese hamster ovary cell stably expressing ASIC1a, holding potential −60 mV (*upper panel*), and current-clamp measurement of action potential induction by extracellular acidification in a cultured hippocampal neuron (*lower panel*). Extracellular pH was changed from pH 7.4 to the pH indicated

(Farman et al. 1997). Apical Na^+ transport by ENaC is important for the maintenance of the composition and the volume of the airway surface liquid, which forms a thin layer of liquid – the periciliary liquid layer (PCL) – that coats the airway epithelium (Hummler et al. 1996; Randell and Boucher 2006). The thickness of the PCL is controlled by two opposing processes: ENaC-mediated Na^+ absorption, and CFTR-dependent Cl^- secretion. The PCL allows the cilia at the surface of the epithelium to move in an outward direction particles and microbes that have entered the airways.

11.3.1.3   Gastrointestinal Tract and Other Tissues

Salt taste guides the ingestion of NaCl and other required minerals, thus serving an essential function in ion and water homeostasis. There is evidence in mammals for two types of salt taste receptors: one that responds specifically to Na^+ and one or more that respond to various cations. There is strong evidence that the Na^+-selective type is mediated by ENaC and is regulated by aldosterone. The α, β and γ ENaC subunits are present in rat fungiform papilla taste cells. Injection of rats with aldosterone increases apical taste cell membrane immunoreactivity to β and γ ENaC and the number of fungiform taste cells with amiloride-sensitive currents, and enhances the magnitude of the Na^+ current (DeSimone and Lyall 2006). In various species, taste nerve responses to NaCl are significantly inhibited by amiloride or its analogue benzamil. In rodents the amiloride-sensitive part of the chorda tympani response to NaCl is typically 70% or more of the total response. In humans however, only ~20% of the Na^+ taste intensity or the Na^+-evoked lingual surface potential was inhibited by amiloride in two independent studies (Smith and Ossebaard 1995; Feldman et al. 2003). As mentioned, ENaC cell surface expression in circumvallate taste cells is aldosterone-dependent. It is likely that, under the high salt intake of the modern western diet, ENaC cell-surface expression in circumvallate taste cells is reduced due to low circulating aldosterone levels.

The colon is a tight epithelium absorbing Na^+ by electrogenic transport that is sensitive to amiloride and stimulated by aldosterone (Loffing and Schild 2005). The contribution of the distal colon to the maintenance of Na^+ homeostasis is not clear, and differences may exist between species.

11.3.2   Acid-Sensing Ion Channels

ASICs are expressed in neurons and are activated by extracellular acidification (Waldmann et al. 1997; Waldmann and Lazdunski 1998; Krishtal 2003; Wemmie et al. 2006). ASIC activation is transient; all ASICs respond with a rapidly activating and subsequently desensitising (inactivating) current to extracellular acidification (Fig. 11.2d, upper panel). Of the known ASIC subunits, ASIC1a, -1b, -2a and -3 can form functional channels by themselves, while ASIC2b is functional only when associated with other ASIC subunits; no channel function has been detected so far for ASIC4. Activation of ASICs is expected to shift the membrane potential of the neuron towards more positive potentials, thereby generating action potentials by activation of voltage-gated Na^+ channels (Fig. 11.2b). This activating role of ASICs in neuronal signaling has been experimentally confirmed in central and peripheral neurons. An experimental trace of such an experiment is shown in Fig. 11.2d, lower panel. The probability of inducing action potentials by extracellular acidification is positively correlated to the ASIC expression level and the acidity of the stimulus (Baron et al. 2002; Vukicevic and Kellenberger 2004; Poirot et al. 2006). More extensive analysis, however, also showed that the number of action potentials during an acidification was negatively correlated with the ASIC-mediated depolarisation, illustrating a potentially modulatory role of ASICs (Deval et al. 2003; Vukicevic and Kellenberger 2004; Poirot et al. 2006). In actively signalling hippocampal neurons, ASIC activation was shown to inhibit action potential bursts (Vukicevic and Kellenberger 2004). These experiments thus show that the electrical response of a neuron to acidification depends on the extent of the pH change, the expression level of ASICs present in the neuron, as well as on the momentary signaling activity of the neuron. In the following, I describe the main physiological and pathological roles of ASICs. For a more detailed description, see Wemmie et al. (2006).

11.3.2.1 Central Nervous System

Expression of ASIC1a, -2a, -2b and -4 in the central nervous system (CNS) has been demonstrated by Northern blot, in situ hybridisation and immunohistochemical analysis. ASIC3 is expressed in spinal cord, but not in the rodent brain. ASIC1a, -2a and -2b have a similar widespread distribution pattern in the brain, with the highest expression levels in the hippocampus, the cerebellum, the cortex, the main olfactory bulb, the habenula and the amygdala complex. ASIC currents have been measured from neurons of the hippocampus, the cortex, the cerebellum and the amygdala. The sub-cellular localisation of ASICs in CNS neurons is currently unclear. Some studies suggested highest ASIC densities in synapses, likely in the post-synaptic membrane, while other studies have found similar ASIC expression throughout the neuron (Wemmie et al. 2002; de la Rosa et al. 2003). ASICs in the synapse would likely be activated by the acidification of the synaptic cleft induced by neurotransmitter release.

Based on ASIC1a knockout mice and the use of the ASIC1a-specific inhibitor Psalmotoxin 1 (present in the venom of the spider *Psalmopoeus cambridgei*), several roles of ASIC1a in the brain have been identified. Loss of ASIC1a decreased long-term potentiation in hippocampus and spatial memory in the Morris water maze, as well as cerebellum-dependent learning as assessed by eye-blink conditioning (Wemmie et al. 2002). Although the effects of ASIC loss in behavioural experiments were modest, these results indicate that ASIC1a modulates synaptic plasticity and contributes to learning and memory. Consistent with the high ASIC1a expression in the amygdala complex, which is important for fear-related behaviour, ASIC1a knockout mice showed reduced fear behaviour, while overexpression of ASIC1a enhanced context fear conditioning (Wemmie et al. 2003, 2004). Recent studies have generated interest in ASICs as possible therapeutic targets for stroke. Severe cerebral ischaemia can induce a pH decrease to 6.3 and below (Siesjo et al. 1996), and it is known that the increase in the intracellular calcium concentration substantially contributes to neuronal death following ischaemia. Because ASIC1a homomultimers conduct Ca^{2+} (see Sect. 11.5.1). ASIC1a activation might contribute to the cell damage and death associated with cerebral ischaemia. The hypothesis that ASIC1a contributes to ischemic toxicity was recently tested in cells heterologously expressing ASIC1a and in hippocampal neurons (Xiong et al. 2004;

Gao et al. 2005). Neurons lacking ASIC1a and cells treated with amiloride or *P. cambridgei* venom resisted acidosis-induced injury. In a mouse model in which middle cerebral artery occlusion causes ischemic stroke, disrupting ASIC1a reduced infarct volume by 60% (Xiong et al. 2004). Because pH remains low for several hours after a stroke, ASIC1a inhibitors might be beneficial for some time after an ischemic event.

11.3.2.2 Peripheral Nervous System

All known ASIC subunits are found in the peripheral nervous system (PNS). ASIC expression has been demonstrated in neurons innervating the skin, heart, gut and muscle. They have also been detected in the eye, ear, taste buds and bone (reviewed by Wemmie et al. 2006). ASIC subunits have been detected in dorsal root ganglion (DRG) neurons of all diameter ranges, with the subunit composition varying among DRG neurons, producing heterogeneity in H^+-evoked currents (Benson et al. 2002; Poirot et al. 2006). Immunocytochemistry studies demonstrated the presence of ASIC2a and ASIC3 in specialised cutaneous nerve endings and in small free nerve endings in the epidermis (Price et al. 2000, 2001). Extracellular pH falls to <7 with inflammation, ischaemia, exercise, haematoma and in tumours (Cobbe and Poole-Wilson 1980; Issberner et al. 1996). Inflammation has been shown to increase ASIC subunit transcript levels in DRG neurons (Voilley et al. 2001). In contrast, ASIC subunit transcript levels and ASIC current densities in DRG neurons are decreased in an animal model of neuropathic pain (Poirot et al. 2006). The expression in sensory neurons together with their functional properties make ASICs good candidates to serve as H^+-gated nociceptors. In human skin, ASICs appear to be the major pH sensor for pH ≥ 6 (Ugawa et al. 2002; Jones et al. 2004). Studies with ASIC knockout mice, however, gave a mixed picture with regard to the potential role of ASICs in acid-induced nociception. Different studies indicated a role of mainly ASIC3 in inflammatory and acid-induced pain. However, two of these studies showed a pro-nociceptive function of ASIC3 (Price et al. 2001; Sluka et al. 2003), while two other studies indicated an inhibitory role of ASICs in pain sensation (Chen et al. 2002; Mogil et al. 2005). The similarity between ASICs and the degenerins of *C. elegans* that are part of mechanoreceptor complexes suggested that ASICs might contribute to mechanosensation. While experiments with ASIC2 null animals showed conflicting results in two studies (Price et al. 2000; Roza et al. 2004), there is evidence from recordings on nerve-skin preparations and behavioural studies for a role of ASIC3 as a mediator of cutaneous low-threshold mechanotransduction (Price et al. 2001). However, mice expressing a dominant-negative ASIC3 subunit (leading to loss of all ASIC currents) were more sensitive to mechanical pain and showed increased mechanical hypersensitivity as compared to wild-type controls (Mogil et al. 2005). Detailed analysis of the involvement of ASICs in visceral mechanosensation suggested an inhibitory role for ASIC1, a location-dependent function of ASIC2 and an activatory function of ASIC3 (Wemmie et al. 2006).

In summary, it is clear that functional ASICs are abundant in the PNS and that their expression and function is regulated. There is evidence for potential roles of

ASICs in pain and mechanosensation. However, the situation appears to be quite complex; some of the results obtained in different studies under similar conditions are contradictory and it is currently not clear whether ASICs act predominantly as positive sensors or as modulators in pain and mechanosensation. This may be due to two reasons: (1) many of the effects shown on pain and mechanosensation due to inactivation of the gene were relatively modest. (2) ASICs in the PNS are present in neurons that likely have different functions, and most of the functional ASICs in the PNS are composed of several different ASIC subunits. Deleting one ASIC subunit is therefore expected to affect ASIC currents with different properties and different, possibly even opposing, functions.

11.4   Structural Aspects

11.4.1   Primary Structure and Membrane Topology of Channel Subunits

The initial cloning of ENaC by functional expression showed that the functional channel is a heteromeric complex of homologous subunits (Canessa et al. 1994). Each ENaC subunit contains ~650 amino acids. ENaC subunits share 27–37% identity with each other. The homology between human and rat ENaC orthologues is ~85%. ASIC subunits are ~540 amino acids long and show 45–60% identity among each other. The homology between human and rat ASIC orthologues is close to 100% for ASIC1, -2 and -4, and ~83% for ASIC3. ENaC and ASIC subunits were predicted to have intracellular N- and C-termini, two transmembrane segments and a large extracellular loop that makes up more than 50% of the protein mass (Fig. 11.1). This predicted subunit topology has been experimentally verified for α ENaC (reviewed in Kellenberger and Schild 2002) and for ASIC2a (Saugstad et al. 2004). In Fig. 11.3a, a linear representation of an ENaC/DEG subunit is shown, and the homology of the amino acid sequence along the subunit is indicated, allowing comparison of all ENaC/DEG family members as well as comparison within the ENaC and ASIC subfamilies.

11.4.2   Multimeric Channels and Subunit Stoichiometry

The three homologous ENaC subunits, α, β and γ, are required for maximal expression of ENaC activity in *Xenopus* oocytes (Canessa et al. 1994). The functional ENaC resulting from co-expression of $\alpha\beta\gamma$ ENaC has the same functional properties as the highly selective epithelial Na$^+$ channels found in the distal nephron, indicating that the highly selective epithelial Na$^+$ channel of the distal nephron is formed by ENaC α, β and γ subunits (Garty and Palmer 1997). In the absence of a known channel

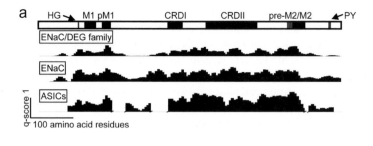

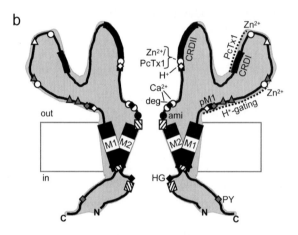

Fig. 11.3 Primary subunit structure and model of an ENaC/ASIC channel. **a** *Top* Linear representation of an ENaC/ASIC subunit with the two transmembrane segments *M1* and *M2*, the Cys-rich domains *CRDI* and *CRDII*, the *HG*, post-M1 (*pM1*) and *PY* motifs. The lower part of panel **a** illustrates conservation of the amino acid sequence across the ENaC/DEG family and within the ENaC and the ASIC sub-family. The height of the black column at a given position is proportional to the degree of conservation (obtained from column score of alignments in Clustal X, using a sliding window of ten amino acids; gaps in the ASIC figure correspond to gaps of ASIC in the alignment). **b** Model of an ENaC or ASIC channel, showing two of the four subunits of a functional channel around the central ion pore in a side view. On the subunits showing the structural domains M1 and M2, CRDI and II, and pM1, diverse features are indicated as *open symbols* if they are unique to ASICs, *gray-filled* symbols if they are unique to ENaC and *black symbols* if they are shared by ENaC and ASICs. A *rhombus* indicates a site that is important for channel gating and/or for its activity, *circles* mark binding sites and *triangles* mark protease cleavage sites, as discussed in the text. Extended regions shown to be important for H[+]-gating and binding of Psalmotoxin 1 (*PcTx1*) are indicated in addition by *dotted lines*. Domains that participate in the formation of the ion pore are indicated by *hatched bars*. The two Ca[2+] binding sites shown are involved in ASIC1a pore block by Ca[2+] (Paukert et al. 2004), but not in the regulation of ASIC1 gating by Ca[2+]. *deg* Degenerin site, *ami* amiloride binding site

structure at atomic resolution, biochemical analysis of the channel complex or functional measurement of the contribution of mutations in the different subunits to channel inhibition by various reagents was used to determine ENaC subunit stoichiometry. This suggested that ENaC is a tetramer formed by two α and one each of β and γ subunits (Firsov et al. 1998), consistent with the tetrameric stoichiometry determined for FaNaC, another ENaC/DEG family member (Coscoy et al. 1998). In contrast to these results, several studies suggested an 8–9 subunit stoichiometry for ENaC (Snyder et al. 1998). As discussed below (Sect. 11.5.2), there is good evidence that the homologous ENaC subunits are arranged around the central channel pore, lining the permeation pathway. From our knowledge of the three-dimensional structure of other cation channels that have 4–5 subunits, I consider it very unlikely for a highly Na^+-selective channel to contain nine subunits that all contribute to the narrow channel pore.

ASIC subunits can form, with similar efficiency, functional channels composed of one type of subunits (homomultimers) or of different types of ASIC subunits (heteromultimers). It is assumed that functional ASICs are tetramers, in analogy to ENaC and FaNaC. However, the subunit stoichiometry of ASICs has not yet been addressed experimentally.

11.5 Ion Conduction and Channel Pore

11.5.1 Biophysical and Pharmacological Properties

The ENaC current is highly selective for the small inorganic ions Na^+ and Li^+, with a Na^+/K^+ selectivity ratio ≥ 100 (Palmer 1982; Kellenberger et al. 1999a). The unitary conductance of ENaC measured in the presence of 140 mM Na^+ is 5 pS, and 9–10 pS in the presence of 100–150 mM Li^+. The pyrazinoylguanidine derivatives amiloride and benzamil, and the pteridine compound triamterene decrease Na^+ absorption and K^+ secretion in the ASDN by inhibiting ENaC (IC_{50} amiloride = 0.1 μM, see Table 11.1) (Canessa et al. 1994). Amiloride and triamterene are used clinically as K^+-sparing diuretics. Early work on the effect of amiloride on electrogenic Na^+ absorption in epithelia showed that amiloride interacts competitively with permeant Na^+ or Li^+ ions, suggesting that it acts as an ENaC pore blocker to prevent the flow of permeant Na^+ ions through the channel pore (Palmer 1984). Site-directed mutagenesis studies confirmed that the amiloride binding site is located in the ion pathway close to the selectivity filter (see Sect. 11.5.2).

The transient current of neuronal and recombinantly expressed ASICs displays a Na^+/K^+ selectivity ratio of ~10. ASIC single-channel Na^+ conductance has been determined for several homomeric and heteromeric ASIC channels, and ranges from 10 to 15 pS. ASIC1a homomultimeric channels are permeable to Ca^{2+}, in contrast to all other ASICs and ENaC. The P_{Na}/P_{Ca} permeability ratio is ~17 (Bassler et al. 2001; Sutherland et al. 2001). The extracellular Ca^{2+} concentration is also a modulator of the pH dependence of ASIC gating (see Sect. 11.6.2) and an inhibitor of unitary ASIC1a currents (IC_{50} ~4 mM), likely due to a partial pore block (Paukert et al. 2004). The Ca^{2+} permeability is likely important for several of the physiological and pathological functions of ASIC1a.

Table 11.1 Properties of $\alpha\beta\gamma$ epithelial Na$^+$ channel (ENaC) and of homomultimeric acid-sensing ion channels (ASICs). In the case of ASICs, the parameters shown refer to peak currents from recombinantly expressed channels. P_{Na}/P_K Na$^+$/K$^+$ permeability ratio, *IC50* concentration for half-maximal inhibition, *pH0.5* pH for half-maximal ASIC activation, *pHIn0.5* pH for half-maximal ASIC inactivation

Name	P_{Na}/P_K	IC50 amiloride (µM)	pH0.5	pHIn0.5
$\alpha\beta\gamma$ ENaC	≥100	0.1	–	–
ASIC1a	~10	10	6.2–6.8	7.2–7.4
ASIC1b	~10	21–23	5.9	~7.0
ASIC2a	~10	28	4.4	~6.0
ASIC3	~10	16–63	6.2–6.7	~7.0

Transient ASIC currents are inhibited by amiloride, although with IC$_{50}$ values in the range of 10 µM–~60 µM, thus at ≥100-fold higher concentrations than ENaC (Table 11.1).

11.5.2 Structure-Function Relationship of the Ion Permeation Pathway

The analysis of permeability properties and the voltage-dependence of block by non-permeant ions of the epithelial Na$^+$ channel of tight epithelia had suggested that the outer pore entrance forms a funnel-like structure that narrows down to the selectivity filter, which discriminates among cations based on their size (Palmer 1982). Therefore, the selectivity filter of ENaC likely constitutes the narrowest part of the pore, allowing only the small monovalent cations Na$^+$, Li$^+$ and H$^+$ to pass through the channel. Mutagenesis studies indicate that the ENaC selectivity filter is formed by a stretch of three amino acid residues, G/S-X-S (X being any amino acid residue) located on each subunit at the extracellular start of the second transmembrane segment (M2) (Kellenberger et al. 1999a, 1999b; Snyder et al. 1999) (Fig. 11.3b). Mutations of selectivity filter residues affect the unitary conductance and/or ion selectivity. Of the residues constituting the selectivity filter, the effects of amino acid substitutions are most dramatic for the third selectivity filter position of the α subunit, the conserved Ser589. Mutation of this residue allows larger ions such as K$^+$, Rb$^+$, Cs$^+$, NH$_4^+$ and divalent cations to pass through the channel, consistent with an enlargement of the pore at the selectivity filter (Kellenberger et al. 1999a, 2001). By analogy to the structure of the bacterial K$^+$ channel KcsA, it was expected that the distal part of the M2 segment may line the transmembrane part of the ENaC pore. However, currently there is no clear evidence for such a role of the M2 segment. The amiloride binding site is four amino acid residues upstream of the first selectivity filter residue (Fig. 11.3b). Mutation of a single amiloride binding site residue on the α, β or γ subunit decreases ENaC affinity for amiloride inhibition by up to 1,000-fold (Schild et al. 1997).

In the ASIC primary structure the ENaC selectivity filter sequence G-A-S is conserved, and the pre-M2 region shows a very high homology between ENaC and

ASICs (Fig. 11.3). Despite this sequence homology, however, ASIC Na^+/K^+ selectivity is lower than that of ENaC. The importance of the G-A-S sequence for ASIC ion selectivity has not been investigated; however, other regions of the channel protein have been shown to likely contribute to the ion pathway. The analysis of ASIC2b/2a chimera identified residues in the cytoplasmic N-terminus, whose mutation changed the Na^+/K^+ permeability ratio by up to 7-fold (Coscoy et al. 1999). The same region was shown to play a role in the selectivity of ASIC1 towards divalent cations (Bassler et al. 2001). A screening of the N-terminal cytoplasmic region and of M1 of ASIC1a showed inhibition by sulfhydryl reagents of engineered Cys residues in the cytoplasmic N-terminal part close to and at the intracellular start of M1. An engineered Cys residue at the intracellular start of M1 was accessible to sulfhydryl reagents from the cytoplasmic as well as from the extracellular side, indicating that it participates in the formation of the channel pore (Pfister et al. 2006).

In conclusion, functional studies have revealed differences in ion selectivity between ENaC and ASICs. The region important for ion selectivity in ENaC is conserved in ASICs and, so far, mutagenesis studies have identified additional regions in the ASIC protein that contribute to the ion pore.

11.6 Channel Gating and Regulation

11.6.1 Channel Gating

ENaC is a constitutively active channel that switches between the open and closed state with dwell times of the order of seconds (Fig. 11.2c). The ENaC open probability (Po) varies from <0.1 to >0.9 (Palmer and Frindt 1996).

ASICs are normally closed and when activated by protons display in general a transient peak current that lasts from hundreds of milliseconds to seconds, followed by channel inactivation (termed "desensitisation" or "open-channel inactivation") despite the continued presence of a low extracellular pH solution. ASICs of different composition differ in their pH dependence of activation, as illustrated by the pH values of half-maximal activation (pH 0.5) in Table 11.1, and in the time course of open channel inactivation. Under certain pH conditions, inactivation of the current carried by some ASIC multimers (ASIC3 homomultimers and the heteromultimers ASIC2a/2b, ASIC2a/3 and ASIC2b/3) is incomplete, leaving a sustained residual inward current that follows the fast transient current (reviewed in Kellenberger and Schild 2002). Surprisingly, the sustained currents in ASIC-expressing cells exhibit biophysical and pharmacological characteristics different from the transient currents – mainly a lower Na^+/K^+ permeability ratio and a lower apparent affinity for amiloride block. So far, only one study has investigated the sustained current over a time period longer than about 1 min (Yagi et al. 2006). Homomeric ASIC3 or heteromeric ASIC2a/3 produced a sustained current lasting over tens of minutes when activated by acidification to pH 7–6.7. A similar sustained current was also found in sensory neurons innervating the heart. For ASICs that do not mediate a sustained current, a prolonged acidification to pH values that are not acidic enough to generate a detectable ASIC current can inactivate the ASICs and reduce the number of ASICs that can be activated in a given situation. This form of inactivation, termed "steady-state inactivation" has an extremely steep pH dependence and occurs at physiological pH ranges. The pH of half-maximal inactivation, pHIn 0.5 for different ASICs is indicated in Table 11.1.

11.6.2   Regulation of Channel Expression and Function

11.6.2.1   ENaC

ENaC stability at the apical cell surface is controlled by the ubiquitin-protein ligase Nedd4-2, which binds to conserved proline-rich motifs in the C-terminus of β and γ ENaC (Staub et al. 1996). Binding of Nedd4-2 reduces the expression and the activity of ENaC at the cell surface by promoting ubiquitylation of the channel, endocytosis, and subsequent degradation by the proteasome and the lysosome. In addition, ENaC cell surface expression is regulated by the aldosterone-induced protein serum and glucocorticoid-induced kinase 1 (Staub and Verrey 2005). The control of ENaC Po is still poorly understood. A number of extracellular factors have been shown to modulate ENaC activity, including extracellular Na^+ (through a phenomenon called "self-inhibition"), several other organic or inorganic cations that appear to interfere with self-inhibition, and serine proteases. Soluble proteases, such as trypsin, elastase or furin and the membrane-bound "channel-activating proteases" 1–3, increase ENaC currents by a mechanism that likely involves protein cleavage (Vallet et al. 1997; Hughey et al. 2004; Rossier 2004). From the cytoplasmic side, ENaC activity is regulated by both pH, in that intracellular acidification is inhibitory, and by Na^+ concentration, by a mechanism termed "feedback inhibition". In the context of ASICs it may be interesting to note that extracellular pH changes do not affect $\alpha\beta\gamma$ ENaC function, but that extracellular acidification increases $\delta\beta\gamma$ ENaC currents (Ji and Benos 2004).

The fact that ENaC Po depends on both extra- and intra-cellular Na^+ concentration indicates that ENaC has binding sites for Na^+ and can sense Na^+ concentration. Increasing the extracellular or intracellular Na^+ concentration decreases ENaC Po and may be a protecting mechanism to prevent cell overload with Na^+ ions and to distribute Na^+ reabsorption along the distal nephron. Self-inhibition has been demonstrated in amphibian epithelia and *Xenopus* oocytes expressing ENaC (Garty and Palmer 1997; Horisberger and Chraibi 2004). The time course for ENaC self-inhibition is too slow to represent Na^+ ions entering the channel pore and too fast to be related to changes in the intracellular Na^+ concentration. In ENaC expressed in *Xenopus* oocytes, self-inhibition is observable as a fast decline in the inward current after an initial peak value observed either following rapid removal of amiloride or after a sudden increase in extracellular Na. It is an intrinsic property of the channel and strongly suggests the presence of an allosteric binding site for Na^+ ions in the extracellular loop. The apparent Na^+ affinity of this site was estimated as $\geq 100\,mM$ (Chraibi and Horisberger 2002). A number of agents that increase the ENaC-mediated Na^+ transport rate seem to act, at least in part, by relieving the channel self-inhibition: these include sulfhydryl reagents, detergents, trypsin and other proteases such as furin (Chraibi and Horisberger 2002; Sheng et al. 2006).

Feedback inhibition was first proposed by MacRobbie and Ussing (1961), who noted that cells of Na^+ absorbing epithelium, such as frog skin expressing an apical amiloride-sensitive Na^+ channel, did not swell when Na^+ extrusion by the Na^+/K^+-ATPase at the basolateral cell was blocked. They proposed that the rise in $[Na^+]_i$ upon inhibition of the Na/K-ATPase downregulates Na^+ permeability of the apical membrane to prevent the cell becoming overloaded with Na^+ ions. The presence of this feedback regulation, which inhibits ENaC when Na^+ enters the cell, has since been confirmed in cortical collecting duct cells and in *Xenopus* oocytes expressing ENaC (Silver et al. 1993; Anantharam et al. 2006). Recent studies indicate that feedback inhibition relies on reducing both channel open probability and channel density at the cell surface to limit Na^+ entry when $[Na]_i$ is high (Kellenberger et al. 1998; Anantharam et al. 2006). These studies also provide evidence that feedback inhibition is reduced in ENaC containing mutations associated with a severe form of hereditary hypertension, the Liddle syndrome.

11.6.2.2 ASICs

Extracellular protons are the only ASIC activators known to date; however, many different ways of ASIC regulation exist. The peptide FMRFamide and related mammalian peptides have been shown to increase ASIC peak current amplitudes and to slow the inactivation time course (reviewed by Lingueglia et al. 2006). Recently, it has been shown that intracellular acidification reduces ASIC currents in cortical neurons by shifting the pH dependence of activation to more acidic pH (Wang et al. 2006). The action of serine proteases on recombinantly expressed ASICs or ASICs present in cultured neurons leads to a shift in pH dependence of activation and inactivation to more acidic pH, thereby adapting the pH dependence of ASIC gating to the more acidic environment that occurs e.g. during ischemia (Poirot et al. 2004).

The extracellular Ca^{2+} concentration influences ASIC function, and there is evidence from studies carried out mainly with ASIC1 and ASIC3 for competition between Ca^{2+} and H^+ for binding sites on ASICs. For ASIC1a, due to this competition, Ca^{2+} shifts the pH dependence of steady-state inactivation and of activation to more acidic pH (Babini et al. 2002). Reducing the extracellular Ca^{2+} concentration increases ASIC3 currents, probably due to a shift in the activation curve. In ischemic conditions, the extracellular concentration of lactate increases. Lactate is a weak Ca^{2+} chelator, an increase of the concentration of which leads to a reduction in free Ca^{2+} concentration and thereby an increase of ASIC3 current amplitudes (Immke and McCleskey 2001). Zinc is present in the nervous system and is a known modulator of almost any ion channel. Zinc shifts the pH dependence of activation of ASIC2-containing channels to slightly less acidic pH, thereby increasing ASIC current amplitude. In addition, zinc slows the time course of open-channel inactivation of ASIC2-containing channels. The EC_{50} for these effects is $100\,\mu M$ for ASIC1a/2a channels (Baron et al. 2001). Zinc acts in an inhibitory manner on homomeric ASIC1a and ASIC3 with, in the case of ASIC1a, an IC_{50} of 7 nM (Chu et al. 2004, Poirot et al. 2006).

Two different toxins are known to inhibit ASICs. Psalmotoxin 1 (present in the venom of the spider *P. cambridgei*) inhibits exclusively ASIC1a homomeric channels, with an IC_{50} of ~1 nM (Escoubas et al. 2000). Channel inhibition occurs by a shift of the pH dependence of steady-state inactivation of ASIC1a to higher pH in the presence of the toxin, leading to inactive channels at pH 7.4 (Chen et al. 2005). The sea anemone toxin APETx2 inhibits ASIC3-containing channels by an unknown mechanism, with an IC_{50} of ~60 nM (Diochot et al. 2004). The identification of many different proteins that interact with ASICs and that can direct their sub-cellular localisation, phosphorylation status or have other functions, indicate that ASICs are part of macromolecular complexes (Wemmie et al. 2006).

11.6.3   Gating and Regulatory Domains

11.6.3.1   Extracellular Domains

The mutation of a conserved Ala residue in degenerins of *C. elegans* causes degeneration of the touch receptor cells that express them, with morphological features that are consistent with an abnormal cation leak into the cell (Driscoll and Chalfie 1991). The corresponding residues in other family members are small residues, Ser in ENaC and Gly in ASICs, located seven amino acid residues upstream of the amiloride binding site (labeled "deg" for "degenerin site" in Fig. 11.3b). It has been shown for most of the ENaC/DEG family members that this conserved residue plays an important role in channel gating. Mutation of this residue to a larger residue increases the ENaC current by several-fold and induces a sustained current in the absence of extracellular acidification in ASICs (Kellenberger and Schild 2002).

The post-M1 domain is highly conserved among all ENaC/DEG family members (Fig. 11.3, "pM1"). A mutation that causes muscle hypercontraction consistent with a channel gain-of-function mutation has been identified in this region of the degenerin UNC-105 (Liu et al. 1996). A loss-of-function mutation causing pseudo-hypoaldosteronism type 1 was found in the post-M1 region of α ENaC (Firsov et al. 1999). In ASICs, a short sequence in the proximity of the post-M1 domain co-determines the time course of open channel inactivation (Coric et al. 2003).

Sequence comparison between ASICs of different species, some of them not being pH-gated, and the construction and functional analysis of chimera, identified a region encompassing residues D78–E136 and the short stretch of D351–E359 in ASIC1a as necessary for pH-dependent gating (Coric et al. 2005) (Fig. 11.3b). The binding site of Psalmotoxin 1 also maps to two domains, F157–V186 and to D349 and surrounding residues (Salinas et al. 2006), suggesting an involvement of these domains in pH-dependent gating of ASICs. Our laboratory has recently shown that exposure to trypsin, which leads to a shift in the pH dependence of activation and inactivation, cleaves the channel protein in the N-terminal half of the extracellular loop (Δ in Fig. 11.3b). This site cannot be cleaved in the inactivated channel, indicating that this region of the channel protein undergoes conformational changes during channel gating (Vukicevic et al. 2006). Amino acid residues contributing to the binding sites for the modulatory or inhibitory action of cations such as Zn^{2+} and Ca^{2+} have been identified on ASIC subunits (circles in Fig. 11.3) and identify domains exposed to the extracellular solution that are likely involved in either channel gating or mediating pore block.

11.6.3.2 Intracellular Domains

A His-Gly motif in the intracellular N-terminus is conserved in all ENaC/degenerin family members (Fig. 11.3b, "HG"). A loss-of-function mutation in the HG motif of β ENaC is one of the mutations causing pseudohypoaldosteronism type 1. This and mutations in corresponding residues of α ENaC result in an important reduction in ENaC Po, without effects on ENaC cell surface expression (Grunder et al. 1997). Mutation of some residues in close proximity to the HG motif result in a similar inhibition. Furthermore, this region of the N-terminus of $\alpha\beta\gamma$ ENaC subunits is rich in Cys residues. These residues are responsible for the high ENaC sensitivity to inhibition by a variety of intracellular sulfhydryl reagents, including methanethiosulfonates, metal divalent cations, and oxidising agents (Kellenberger et al. 2005). These reagents inhibit ENaC activity from the cytosolic side by inducing long and only slowly reversible channel closures.

The C-termini of β and γ ENaC subunits contain a proline-rich motif that binds the ubiquitin-protein ligase Nedd4-2. Binding of Nedd4-2 reduces the expression and the activity of ENaC at the cell surface by promoting ubiquitylation of the channel, endocytosis, and subsequent degradation by the proteasome and the lysosome (Staub and Verrey 2005). Some interactions of proteins that bind to ASICs occur via the C-terminal PDZ binding domains of ASIC1 and -2 (Wemmie et al. 2006).

11.7 Conclusions

The important role of ENaC in the kidney in the control of Na^+ homeostasis and blood pressure, and of ENaC in the lung in the regulation of airway surface liquid composition and volume, is well documented. Molecular and functional studies have allowed the identification of functional units in ENaC and ASIC channels, and a better understanding of their basic function on the molecular and cellular level. ASICs are pH sensors in the nervous system and there are clear indications for an involvement of ASICs of the CNS in the expression of fear, in memory formation and in neurodegeneration after ischemia. ASICs of the PNS are likely involved in pain and mechanosensation. In the last few years much has been learnt about the localisation of ASICs and about their biochemical, physiological and pathological effects. ENaC is the target of K^+-sparing diuretics, and is a target for the development of drugs aimed at treating cystic fibrosis. ASICs may be interesting drug targets for treating pain, ischemic stroke and psychiatric disease. Many unanswered questions still remain regarding the physiological functions of ASICs and the regulation, function and structure of ENaC and ASICs.

Note added in proof After submission of this book chapter, Gouaux and colleagues published the structure of ASIC1 at a resolution of 1.9Å. They confirm the predicted topology of ASIC subunits and show that functional ASIC channels are trimers.

Jasti J, Furukawa H, Gonzales EB, Gouaux E. (2007) Structure of acid-sensing ion channel 1 at 1.9Å resolution and low pH. Nature 449:316–323

Acknowledgements I thank the Swiss National Science Foundation for funding.

References

Anantharam A, Tian Y, Palmer LG (2006) Open probability of the epithelial sodium channel is regulated by intracellular sodium. J Physiol 574:333–347

Babini E, Paukert M, Geisler HS, Grunder S (2002) Alternative splicing and interaction with di- and polyvalent cations control the dynamic range of acid-sensing ion channel 1 (ASIC1). J Biol Chem 277:41597–41603

Baron A, Schaefer L, Lingueglia E, Champigny G, Lazdunski M (2001) Zn^{2+} and H^+ are coactivators of acid-sensing ion channels. J Biol Chem 276:35361–35367

Baron A, Waldmann R, Lazdunski M (2002) ASIC-like, proton-activated currents in rat hippocampal neurons. J Physiol 539:485–494

Bassler EL, Ngo-Anh TJ, Geisler HS, Ruppersberg JP, Grunder S (2001) Molecular and functional characterization of acid-sensing ion channel (ASIC) 1b. J Biol Chem 276:33782–33787

Benson CJ, Xie JH, Wemmie JA, Price MP, Henss JM, Welsh MJ, Snyder PM (2002) Heteromultimers of DEG/ENaC subunits form H^+-gated channels in mouse sensory neurons. Proc Natl Acad Sci USA 99:2338–2343

Canessa CM, Schild L, Buell G, Thorens B, Gautschi I, Horisberger J-D, Rossier BC (1994) Amiloride-sensitive epithelial Na^+ channel is made of three homologous subunits. Nature 367:463–467

Chen C-C, Zimmer A, Sun W-H, Hall J, Brownstein MJ, Zimmer A (2002) A role for ASIC3 in the modulation of high-intensity pain stimuli. Proc Natl Acad Sci USA 99:8992–8997

Chen X, Kalbacher H, Grunder S (2005) The tarantula toxin psalmotoxin 1 inhibits acid-sensing ion channel (ASIC) 1a by increasing its apparent H^+ affinity. J Gen Physiol 126:71–79

Chraibi A, Horisberger AD (2002) Na self inhibition of human epithelial Na channel: temperature dependence and effect of extracellular proteases. J Gen Physiol 120:133–145

Chu XP, Wemmie JA, Wang WZ, Zhu XM, Saugstad JA, Price MP, Simon RP, Xiong ZG (2004) Subunit-dependent high-affinity zinc inhibition of acid-sensing ion channels. J Neurosci 24:8678–8689

Cobbe SM, Poole-Wilson PA (1980) Tissue acidosis in myocardial hypoxia. J Mol Cell Cardiol 12:761–770

Coric T, Zhang P, Todorovic N, Canessa CM (2003) The extracellular domain determines the kinetics of desensitization in acid-sensitive ion channel 1. J Biol Chem 278:45240–45247

Coric T, Zheng D, Gerstein M, Canessa CM (2005) Proton-sensitivity of ASIC1 appeared with the rise of fishes by changes of residues in the region that follows TM1 in the ectodomain of the channel. J Physiol 568.3:725–735

Coscoy S, Lingueglia E, Lazdunski M, Barbry P (1998) The phe-met-arg-phe-amide-activated sodium channel is a tetramer. J Biol Chem 273:8317–8322

Coscoy S, de Weille JR, Lingueglia E, Lazdunski M (1999) The pre-transmembrane 1 domain of acid-sensing ion channels participates in the ion pore. J Biol Chem 274:10129–10132

De la Rosa DA, Krueger SR, Kolar A, Shao D, Fitzsimonds RM, Canessa CM (2003) Distribution, subcellular localization and ontogeny of ASIC1 in the mammalian central nervous system. J Physiol 546:77–87

DeSimone JA, Lyall V (2006) Taste receptors in the gastrointestinal tract III. Salty and sour taste: sensing of sodium and protons by the tongue. Am J Physiol Gastrointest Liver Physiol 291: G1005–G1010

Deval E, Baron A, Lingueglia E, Mazarguil H, Zajac JM, Lazdunski M (2003) Effects of neuropeptide SF and related peptides on acid sensing ion channel 3 and sensory neuron excitability. Neuropharmacology 44:662–671

Diochot S, Baron A, Rash LD, Deval E, Escoubas P, Scarzello S, Salinas M, Lazdunski M (2004) A new sea anemone peptide, APETx2, inhibits ASIC3, a major acid-sensitive channel in sensory neurons. EMBO J 23:1516–1525

Driscoll M, Chalfie M (1991) The *mec-4* gene is a member of a family of *Caenorhabditis elegans* genes that can mutate to induce neuronal degeneration. Nature 349:588–593

Escoubas P, DeWeille JR, Lecoq A, Diochot S, Waldmann R, Champigny G, Moinier D, Ménez A, Lazdunski M (2000) Isolation of a tarantula toxin specific for a class of proton-gated Na⁺ channels. J Biol Chem 275:25116–25121

Farman N, Talbot CR, Boucher R, Fay M, Canessa C, Rossier B, Bonvalet JP (1997) Noncoordinated expression of α, β, and γ subunit mRNAs of epithelial Na⁺ channel along rat respiratory tract. Am J Physiol 41:C 131–C 141

Feldman GM, Mogyorosi A, Heck GL, DeSimone JA, Santos CR, Clary RA, Lyall V (2003) Salt-evoked lingual surface potential in humans. J Neurophysiol 90:2060–2064

Firsov D, Gautschi I, Merillat AM, Rossier BC, Schild L (1998) The heterotetrameric architecture of the epithelial sodium channel (ENaC). EMBO J 17:344–352

Firsov D, Robert-Nicoud M, Gruender S, Schild L, Rossier BC (1999) Mutational analysis of cysteine-rich domains of the epithelium sodium channel (ENaC) – identification of cysteines essential for channel expression at the cell surface. J Biol Chem 274:2743–2749

Gao J, Duan B, Wang DG, Deng XH, Zhang GY, Xu L, Xu TL (2005) Coupling between NMDA receptor and acid-sensing ion channel contributes to ischemic neuronal death. Neuron 48:635–646

Garty H, Palmer LG (1997) Epithelial sodium channels – function, structure, and regulation. Physiol Rev 77:359–396

Grunder S, Firsov D, Chang SS, Jaeger NF, Gautschi I, Schild L, Lifton RP, Rossier BC (1997) A mutation causing pseudohypoaldosteronism type 1 identifies a conserved glycine that is involved in the gating of the epithelial sodium channel. EMBO J 16:899–907

Horisberger JD, Chraibi A (2004) Epithelial sodium channel: a ligand-gated channel? Nephron Physiol 96:p37–41

Hughey RP, Bruns JB, Kinlough CL, Harkleroad KL, Tong Q, Carattino MD, Johnson JP, Stockand JD, Kleyman TR (2004) Epithelial sodium channels are activated by furin-dependent proteolysis. J Biol Chem 279:18111–18114

Hummler E, Barker P, Gatzy J, Beermann F, Verdumo C, Schmidt A, Boucher RC, Rossier BC (1996) Early death due to defective neonatal lung liquid clearance in αENaC-deficient mice. Nat Genet 12:325–328

Immke DC, McCleskey EW (2001) Lactate enhances the acid-sensing Na⁺ channel on ischemia-sensing neurons. Nat Neurosci 4:869–870

Issberner U, Reeh PW, Steen KH (1996) Pain due to tissue acidosis: a mechanism for inflammatory and ischemic myalgia? Neurosci Lett 208:191–194

Ji HL, Benos DJ (2004) Degenerin sites mediate proton activation of δβγ-epithelial sodium channel. J Biol Chem 279:26939–26947

Jones NG, Slater R, Cadiou H, McNaughton P, McMahon SB (2004) Acid-induced pain and its
modulation in humans. J Neurosci 24:10974–10979

Kellenberger S, Schild L (2002) Epithelial sodium channel/degenerin family of ion channels: a
variety of functions for a shared structure. Physiol Rev 82:735–767

Kellenberger S, Gautschi I, Rossier BC, Schild L (1998) Mutations causing Liddle syndrome
reduce sodium-dependent downregulation of the epithelial sodium channel in the *Xenopus*
oocyte expression system. J Clin Invest 101:2741–2750

Kellenberger S, Gautschi I, Schild L (1999a) A single point mutation in the pore region of the
epithelial Na^+ channel changes ion selectivity by modifying molecular sieving. Proc Natl Acad
Sci USA 96:4170–4175

Kellenberger S, Hoffmann-Pochon N, Gautschi I, Schneeberger E, Schild L (1999b) On the
molecular basis of ion permeation in the epithelial Na^+ channel. J Gen Physiol 114:13–30

Kellenberger S, Auberson M, Gautschi I, Schneeberger E, Schild L (2001) Permeability properties
of ENaC selectivity filter mutants. J Gen Physiol 118:679–692

Kellenberger S, Gautschi I, Pfister Y, Schild L (2005) Intracellular thiol-mediated modulation of
epithelial sodium channel activity. J Biol Chem 280:7739–7747

Krishtal O (2003) The ASICs: signaling molecules? Modulators? Trends Neurosci 26:477–483

Lifton RP, Gharavi AG, Geller DS (2001) Molecular mechanisms of human hypertension. Cell
104:545–556

Lingueglia E, Deval E, Lazdunski M (2006) FMRFamide-gated sodium channel and ASIC channels:
a new class of ionotropic receptors for FMRFamide and related peptides. Peptides 27:1138–1152

Liu JD, Schrank B, Waterston RH (1996) Interaction between a putative mechanosensory membrane
channel and a collagen. Science 273:361–364

Liu L, Leonard AS, Motto DG, Feller MA, Price MP, Johnson WA, Welsh MJ (2003) Contribution
of Drosophila DEG/ENaC genes to salt taste. Neuron 39:133–146

Loffing J, Schild L (2005) Functional domains of the epithelial sodium channel. J Am Soc
Nephrol 16:3175–3181

MacRobbie EAC, Ussing HH (1961) Osmotic behaviour of the epithelial cells of frog skin. Acta
Physiol Scand 53:348–365

Mogil JS, Breese NM, Witty M-F, Ritchie J, Rainville M-L, Ase A, Abbadi N, Stucky CL, Seguela
P (2005) Transgenic expression of a dominant-negative asic3 subunit leads to increased sensitivity to mechanical and inflammatory stimuli. J Neurosci 25:9893–9901

Palmer LG (1982) Ion selectivity of the apical membrane Na channel in the toad urinary bladder.
J Membr Biol 67:91–98

Palmer LG (1984) Voltage-dependent block by amiloride and other monovalent cations of apical
Na channels in the toad urinary bladder. J Membr Biol 80:153–165

Palmer LG, Frindt G (1996) Gating of Na channels in the rat cortical collecting tubule: effects of
voltage and membrane stretch. J Gen Physiol 107:35–45

Paukert M, Babini E, Pusch M, Grunder S (2004) Identification of the Ca^{2+} blocking site of acid-
sensing ion channel (ASIC) 1: implications for channel gating. J Gen Physiol 124:383–394

Pfister Y, Gautschi I, Takeda AN, van Bemmelen M, Kellenberger S, Schild L (2006) A gating
mutation in the internal pore of ASIC1a. J Biol Chem 281:11787–11791

Poirot O, Vukicevic M, Boesch A, Kellenberger S (2004) Selective regulation of acid-sensing ion
channel 1 by serine proteases. J Biol Chem 279:38448–38457

Poirot O, Berta T, Decosterd I, Kellenberger S (2006) Distinct ASIC currents are expressed in rat
putative nociceptors and are modulated by nerve injury. J Physiol 576:215–234

Price MP, Lewin GR, McIlwrath SL, Cheng C, Xie J, Heppenstall PA, Stucky CL, Mannsfeldt AG,
Brennan TJ, Drummond HA, Qiao J, Benson CJ, Tarr DE, Hrstka RF, Yang B, Williamson RA,
Welsh MJ (2000) The mammalian sodium channel BNC1 is required for normal touch sensation.
Nature 407:1007–1011

Price MP, McIlwrath SL, Xie JH, Cheng C, Qiao J, Tarr DE, Sluka KA, Brennan TJ, Lewin GR, Welsh MJ (2001) The DRASIC cation channel contributes to the detection of cutaneous touch and acid stimuli in mice. Neuron 32:1071–1083

Randell SH, Boucher RC (2006) Effective mucus clearance is essential for respiratory health. Am J Respir Cell Mol Biol 35:20–28

Rossier BC (2004) The epithelial sodium channel: activation by membrane-bound serine proteases. Proc Am Thorac Soc 1:4–9

Roza C, Puel J-L, Kress M, Baron A, Diochot S, Lazdunski M, Waldmann R (2004) Knockout of the ASIC2 channel in mice does not impair cutaneous mechanosensation, visceral mechanonociception and hearing. J Physiol 558:659–669

Salinas M, Rash LD, Baron A, Lambeau G, Escoubas P, Lazdunski M (2006) The receptor site of the spider toxin PcTx1 on the proton-gated cation channel ASIC1a. J Physiol 570:339–354

Saugstad JA, Roberts JA, Dong J, Zeitouni S, Evans RJ (2004) Analysis of the membrane topology of the acid-sensing ion channel 2a. J Biol Chem 279:55514–55519

Schild L, Schneeberger E, Gautschi I, Firsov D (1997) Identification of amino acid residues in the α, β, γ subunits of the epithelial sodium channel (ENaC) involved in amiloride block and ion permeation. J Gen Physiol 109:15–26

Sheng S, Carattino MD, Bruns JB, Hughey RP, Kleyman TR (2006) Furin cleavage activates the epithelial Na^+ channel by relieving Na^+ self-inhibition. Am J Physiol Renal Physiol 290: F1488–F1496

Siesjo BK, Katsura K, Kristian T (1996) Acidosis-related damage. Adv Neurol 71:209–233; discussion 234–206

Silver RB, Frindt G, Windhager EE, Palmer LG (1993) Feedback regulation of Na channels in rat CCT. I. Effects of inhibition of Na pump. Am J Physiol 264:F557–F564

Sluka KA, Price MP, Breese NA, Stucky CL, Wemmie JA, Welsh MJ (2003) Chronic hyperalgesia induced by repeated acid injections in muscle is abolished by the loss of ASIC3, but not ASIC1. Pain 106:229–239

Smith DV, Ossebaard CA (1995) Amiloride suppression of the taste intensity of sodium chloride: evidence from direct magnitude scaling. Physiol Behav 57:773–777

Snyder PM, Cheng C, Prince LS, Rogers JC, Welsh MJ (1998) Electrophysiological and biochemical evidence that DEG/ENaC cation channels are composed of nine subunits. J Biol Chem 273:681–684

Snyder PM, Olson DR, Bucher DB (1999) A pore segment in DEG/ENaC Na^+ channels. J Biol Chem 274:28484–28490

Staub O, Verrey F (2005) Impact of Nedd4 proteins and serum and glucocorticoid-induced kinases on epithelial Na^+ transport in the distal nephron. J Am Soc Nephrol 16:3167–3174

Staub O, Dho S, Henry P, Correa J, Ishikawa T, McGlade J, Rotin D (1996) WW domains of Nedd4 bind to the proline-rich PY motifs in the epithelial Na^+ channel deleted in Liddle's syndrome. EMBO J 15:2371–2380

Sutherland SP, Benson CJ, Adelman JP, McCleskey EW (2001) Acid-sensing ion channel 3 matches the acid-gated current in cardiac ischemia-sensing neurons. Proc Natl Acad Sci USA 98:711–716

Ugawa S, Ueda T, Ishida Y, Nishigaki M, Shibata Y, Shimada S (2002) Amiloride-blockable acid-sensing ion channels are leading acid sensors expressed in human nociceptors. J Clin Invest 110:1185–1190

Vallet V, Chraibi A, Gaeggeler HP, Horisberger JD, Rossier BC (1997) An epithelial serine protease activates the amiloride-sensitive sodium channel. Nature 389:607–610

Voilley N, de Weille J, Mamet J, Lazdunski M (2001) Nonsteroid anti-inflammatory drugs inhibit both the activity and the inflammation-induced expression of acid-sensing ion channels in nociceptors. J Neurosci 21:8026–8033

Vukicevic M, Kellenberger S (2004) Modulatory effects of acid-sensing ion channels (ASICs) on action potential generation in hippocampal neurons. Am J Physiol Cell Physiol 287:C682–C690

Vukicevic M, Weder G, Boillat A, Boesch A, Kellenberger S (2006) Trypsin cleaves acid-sensing ion channel 1a in a domain that is critical for channel gating. J Biol Chem 281:714–722

Waldmann R, Lazdunski M (1998) H$^+$-gated cation channels – neuronal acid sensors in the NaC/DEG family of ion channels. Curr Opin Neurobiol 8:418–424

Waldmann R, Champigny G, Bassilana F, Heurteaux C, Lazdunski M (1997) A proton-gated cation channel involved in acid-sensing. Nature 386:173–177

Wang WZ, Chu XP, Li MH, Seeds J, Simon RP, Xiong ZG (2006) Modulation of acid-sensing ion channel currents, acid-induced increase of intracellular Ca^{2+}, and acidosis-mediated neuronal injury by intracellular pH. J Biol Chem 281:29369–29378

Wemmie JA, Chen JG, Askwith CC, Hruska-Hageman AM, Price MP, Nolan BC, Yoder PG, Lamani E, Hoshi T, Freeman JH, Welsh MJ (2002) The acid-activated ion channel ASIC contributes to synaptic plasticity, learning, and memory. Neuron 34:463–477

Wemmie JA, Askwith CC, Lamani E, Cassell MD, Freeman JH, Welsh MJ (2003) Acid-sensing ion channel 1 is localized in brain regions with high synaptic density and contributes to fear conditioning. J Neurosci 23:5496–5502

Wemmie JA, Coryell MW, Askwith CC, Lamani E, Leonard AS, Sigmund CD, Welsh MJ (2004) Overexpression of acid-sensing ion channel 1a in transgenic mice increases acquired fear-related behavior. Proc Natl Acad Sci USA 101:3621–3626

Wemmie JA, Price MP, Welsh MJ (2006) Acid-sensing ion channels: advances, questions and therapeutic opportunities. Trends Neurosci 29:578–586

Xiong ZG, Zhu XM, Chu XP, Minami M, Hey J, Wei WL, MacDonald JF, Wemmie JA, Price MP, Welsh MJ, Simon RP (2004) Neuroprotection in ischemia: blocking calcium-permeable acid-sensing ion channels. Cell 118:687–698

Yagi J, Wenk HN, Naves LA, McCleskey EW (2006) Sustained currents through ASIC3 ion channels at the modest pH changes that occur during myocardial ischemia. Circ Res 99:501–509

Chapter 12
P2X$_3$ Receptors and Sensory Transduction

Charles Kennedy

Abstract It has been known for many years that exogenously administered adenosine 5′-triphosphate (ATP) evokes acute pain, but the physiological and pathophysiological roles of endogenous ATP in nociceptive signalling are only now becoming clear. ATP produces its effects through P2X and P2Y receptors, and the P2X3 receptor is of notable importance. It shows a selective expression, at high levels in nociceptive sensory neurons, where it forms functional receptors on its own and in combination with the P2X2 receptor. Recent studies have used gene knockout methods, antisense oligonucleotides, small interfering RNA technologies, and a novel selective P2X$_3$ antagonist, A-317491, to show that P2X$_3$ receptors play a prominent role in both chronic inflammatory and neuropathic pain. Several other P2X subunits also appear to be expressed in sensory neurons and there is evidence for functional P2X$_{1/5}$ or P2X$_{2/6}$ heteromers in some of these. These data indicate that P2X receptors, particularly the P2X$_3$ subtype, could be targetted in the search for new, effective analgesics.

Strathclyde Institute of Pharmacy and Biomedical Sciences, University of Strathclyde, Glasgow G4 0NR, UK, c.kennedy@strath.ac.uk

12.1   Introduction

Specialised sensory nerve cells in mammals are activated by a wide range of mechanical, chemical and temperature-related stimuli. Subsets of these cells, nociceptors, are activated when such stimuli are potentially noxious and transmit the signal via the spinal cord to cortico-limbic structures in the brain. We then consciously experience the signal as pain. Over many years there has been a steady progress in our understanding of the transduction mechanisms by which non-noxious and acute noxious stimuli act and the pathways along which the sensory nerves carry the signal to the brain (see Millan 1999 for review). The roles of endogenous agents such as prostaglandins, substance P and bradykinin in inducing acute pain or transmitting the signal are now quite well characterised. In addition, a variety of drugs, such as morphine and the non-steroidal anti-inflammatory drugs, are available for the treatment of acute pain. In contrast, our understanding of the mechanisms that underlie chronic pain have lagged significantly behind and the ability to treat chronic compared with acute pain is much more limited.

In this chapter, I will review recent evidence that endogenous adenosine 5′-triphosphate (ATP) plays an important role in mediating certain types of nociceptive and non-nociceptive sensory signaling. The ability of ATP to evoke acute pain has been known for nearly 40 years (see Chizh and Illes 2000; Burnstock 2001; Kennedy et al. 2003 for reviews), but the receptors and mechanisms by which ATP acts are only now becoming clear. ATP acts via P2X and P2Y receptors; of especial importance here is the $P2X_3$ receptor, which is expressed selectively at high levels in sensory neurons, where it forms functional receptors on its own and in combination with the $P2X_2$ receptor (Chen et al. 1995; Lewis et al. 1995). Recent reports using a variety of technologies indicate that $P2X_3$ receptors are likely to be involved in chronic inflammatory and neuropathic pain. Additionally, there is evidence for functional $P2X_{1/5}$ or $P2X_{2/6}$ heteromers in some sensory neurons. Although these studies are at only a relatively early stage, the data published thus far are encouraging and support further research into the possibility that P2X receptors, particularly the $P2X_3$ subtype, could be targeted in the search for new, effective analgesics.

12.2   P2X Receptor Subtypes

12.2.1   Discovery of Subtypes

ATP acts via P2 receptors (originally called P_2 purinoceptors), and Burnstock and Kennedy (1985) proposed subdivision into P_{2X} and P_{2Y} subtypes on the basis of two broad patterns of agonist activity. At the P_{2X} purinoceptor in tissues such as the guinea pig vas deferens and urinary bladder, α,β-methyleneATP was more potent than ATP, which was equipotent with 2-methylthioATP, at evoking contraction.

In contrast, at the P$_{2Y}$ purinoceptor in the guinea pig taenia-coli and rabbit portal vein, 2-methylthioATP was more potent than ATP, which in turn was more potent than α,β-methyleneATP at causing relaxation. In 1985, potent and selective P$_2$ antagonists were not available, but repeated administration of α,β-methyleneATP could selectively desensitise the P$_{2X}$, but not the P$_{2Y}$ purinoceptor. Over the next few years it became clear that P$_{2X}$ purinoceptors are ligand-gated cation channels, whereas P$_{2Y}$ purinoceptors are G protein-coupled receptors (Kennedy 1990). Thus, a division in molecular structure backed up the original pharmacological division. This was perhaps not surprising. It had been clear for some time that P$_{2X}$ purinoceptors were involved in the fast neurotransmitter actions of ATP, thus implying that they were ion channels. In contrast, cellular responses mediated via P$_{2Y}$ purinoceptors occurred over a slower time-course, implying the involvement of second messengers. This subdivision was confirmed in the 1990s with the cloning of multiple subtypes of both ligand-gated cation channels and G protein-coupled receptors that responded to adenine and/or uracil nucleotides. In recognition of the fact that pyrimidines as well as purines can act at some subtypes, the term P2 purinoceptor was replaced by P2 receptor. Furthermore, it was agreed that all ligand-gated cation channels would now be known as P2X receptors and that all G protein-coupled receptors were to be named P2Y receptors.

12.2.2   Homomeric P2X Receptors

A number of reviews (Khakh et al. 2001; North 2002; Vial et al. 2004) describe the properties of the recombinant P2X receptors in detail, therefore I will give only a brief description of their properties here. To date, seven P2X receptor subunits have been cloned (P2X$_{1-7}$) and they form a new structural family of ligand-gated cation channels. The P2X$_1$–P2X$_6$ receptors have 379–472 amino acids, with a predicted tertiary structure of two transmembrane segments, a large extracellular loop and intracellular C and N termini. The P2X$_7$ receptor (595 amino acids) has a similar structure, but with a much larger intracellular C terminus. This is very different from that of the other known ligand-gated ionotropic receptors, such as the nicotinic, glutamate, glycine, GABA$_A$ or 5HT$_3$ receptors. All seven subunits have ten conserved cysteine residues in the extracellular loop, which are thought to form disulphide bonds and so influence the tertiary structure of the receptors. Each subunit also has several potential N-linked glycosylation sites in the extracellular loop. Trafficking of the subunits to the cell surface requires that at least some of these sites are glycosylated. Mutagenesis experiments have also identified a number of amino acid residues in the extracellular loop that appear to be involved in the binding of the agonist ATP. Positively charged arginine and lysine residues close to the external opening of the ionic channel are thought to interact with the negative charges of the phosphate chain of ATP, whilst phenylalanine residues may interact with the adenine ring (Vial et al. 2004).

Each P2X subunit can form functional homomultimeric ion channels when expressed in mammalian cell lines or *Xenopus* oocytes, although the P2X$_6$ subunit needs to be N-linked glycosylated before significant functional expression is achieved (Jones et al. 2004). At present we do not know the crystal structure of any P2X receptor and so the number of subunits that combine to form a receptor is not clear, but a variety of biochemical approaches suggest that functional receptors are trimers. The subunit composition of most native receptors has also still to be determined, but initial observations suggest that heteromultimers are needed to mimic the functional characteristics of many native P2X receptors. Regardless, all P2X receptors are permeable to Na$^+$, K$^+$ and Ca^{2+} and so, when activated, cause depolarisation and excite cells. The Ca^{2+} permeability of the homomeric channels varies quite widely, with P2X$_1$ and P2X$_4$ receptors being the most permeable at 4–5 times more permeable than the P2X$_3$ receptor (Egan and Khakh 2004).

The homomeric receptors display a range of functional phenotypes (Table 12.1). The P2X$_1$ and P2X$_3$ receptors show much faster desensitisation than the other subtypes. Until recently, they were also thought to be the only subunits to be activated by α,β-methyleneATP; however, some species homologues of the P2X$_4$ receptor and the glycosylated P2X$_6$ receptor are also sensitive (see below). One point to note is that the P2X receptor was defined originally by an agonist potency profile of α,β-methyleneATP >> 2-methylthioATP ≥ ATP, but it is now clear that the potency of ATP and 2-methylthioATP in intact tissues is decreased greatly (100- to 1,000-fold) by breakdown by ectonucleotidases. When agonist action is studied in the absence of breakdown, ATP and 2-methylthioATP are in fact slightly more potent than α,β-methyleneATP at the P2X$_1$ and P2X$_3$ receptors (Kennedy and Leff 1995b). In general, there is a lack of selective and potent antagonists for P2X receptors, but this is not now the case for the P2X$_1$ and P2X$_3$ subtypes. Both are inhibited by the non-selective antagonists suramin and PPADS, but TNP-ATP is selective for the P2X$_1$ and P2X$_3$ over other subtypes, whilst IP$_5$I is highly selective for the P2X$_1$ receptor and A-317491 is likewise for the P2X$_3$ receptor.

Table 12.1 Properties of homomeric P2X receptors. See Jones et al. (2000, 2004), Khakh et al. (2001), North (2002), and Egan and Khakh (2004) for details

	P2X$_1$	P2X$_2$	P2X$_3$	P2X$_4$	P2X$_5$	P2X$_6$	P2X$_7$
Desensitisation	Fast	Slow	Fast	Slow	Slow	Slow	Slow
Ca^{2+} permeability (%)	12.4	5.7	2.7	11.0	4.5	–	4.6
Sensitive to α,β-methyleneATP?	Yes	No	Yes	Yes[a]	No	Yes	No
Sensitivity to:							
Suramin	1 μM	10 μM	3 μM	>300 μM	4 μM	–	500 μM
PPADS	1 μM	1 μM	1 μM	>300 μM	3 μM	22 μM	50 μM
TNP-ATP	6 nM	1 μM	1 nM	15 μM	0.83 μM	15 μM	>30 μM
IP5I	3 nM	3 μM	–	–	–	3 nM	–
A-317491	11 μM	47 μM	22 nM	>100 μM	–	–	>100 μM

[a]Species-dependent

The P2X$_2$ and P2X$_5$ receptors are insensitive to α,β-methyleneATP and currents activated by ATP desensitise slowly. Both receptors, like the P2X$_1$ and P2X$_3$ receptors, are inhibited by suramin and PPADS, but as yet there are no selective antagonists for either subtype. The P2X$_4$ and P2X$_6$ receptors also show slow desensitisation when activated by ATP. Both were also initially reported to be insensitive to α,β-methyleneATP, but more recent studies show that this is not the case. The pharmacological activity of α,β-methyleneATP at the P2X$_4$ is species-dependent, being an antagonist at the rat receptor, but a partial agonist at the mouse and human variants, where it produced a maximum response that was 29% and 24% respectively of that to ATP (Jones et al. 2000). Early studies on the recombinant P2X$_6$ receptor were hampered by very low functional expression levels, but this was overcome in a recent study by developing a subclonal cell line in which the receptor was N-linked glycosylated (Jones et al. 2004). Under these conditions, α,β-methyleneATP was a full and potent agonist at the P2X$_6$ receptor. The antagonist profile of the P2X$_4$ receptor is also species-dependent. PPADS is a moderately potent antagonist at mouse and human receptors, but has little or no effect at the rat variant (Jones et al. 2000). All three were insensitive to suramin. The P2X$_6$ receptor is also suramin-insensitive, but moderately sensitive to PPADS (Jones et al. 2004).

Finally, the P2X$_7$ receptor is insensitive to α,β-methyleneATP and has a low sensitivity to ATP. The potency of ATP is increased greatly by reducing the extracellular concentration of Ca^{2+} or Mg^{2+}. This receptor also has a low sensitivity to suramin and PPADS, but can be inhibited by $2',3'$-dialdehydeATP and KN-62. The most remarkable feature of the P2X$_7$ receptor is that the initial opening of a non-selective cationic channel is followed by the opening of pores that allow much larger molecules, such as the dye YO-PRO-1, to pass through. A similar property has also been described for P2X$_2$ and P2X$_4$ receptors. Pore formation has been widely studied and it was long thought that the P2X$_7$ channel itself dilated in a time-dependent manner to allow passage of the dye molecules, but more recently it became clear that a separate protein mediates this effect, following activation by the P2X$_7$ receptor; Pelegrin and Surprenant (2006) have identified pannexin-1 as the pore protein. Pannexin-1 is a recently identified hemi-channel protein that co-immunoprecipitates with the P2X$_7$ receptor in macrophages. Selective inhibition of pannexin-1 had no effect on P2X$_7$ ion channel formation, but inhibited subsequent dye influx. Similar studies are required to determine if pannexin-1 also mediates the pore dilation described for P2X$_2$ and P2X$_4$ receptors.

12.2.3 Heteromeric P2X Receptors

To date, functional P2X$_{2/3}$, P2X$_{1/5}$, P2X$_{4/6}$, P2X$_{2/6}$, P2X$_{1/2}$ and P2X$_{1/4}$ heteromers with distinct biophysical and pharmacological properties have been demonstrated (Table 12.2). Other combinations are also possible, as all subunits, apart from the P2X$_7$, can coimmunoprecipitate with at least two others when coexpressed in HEK293 cells (Torres et al. 1999). The P2X$_{2/3}$ is the best characterised of the

Table 12.2 Properties of heteromeric P2X receptors. See Khakh et al. (2001), Brown et al. (2002), North (2002), Egan and Khakh (2004), and Nicke et al. (2005) for details

	$P2X_{2/3}$	$P2X_{1/5}$	$P2X_{4/6}$	$P2X_{2/6}$	$P2X_{1/2}$	$P2X_{1/4}$
Desensitisation	Slow	Slow	Slow	Slow	Fast	Slow
Ca^{2+} permeability (%)	3.5	3.3	11.3	7.7	–	–
Sensitive to α,β-methyleneATP	Yes	Yes	Yes	No	Yes	Yes
Sensitivity to:						
Suramin	1 μM	1.6 μM	10 μM	6 μM	–	–[a]
PPADS	1 μM	0.6 μM	–	–	–	–
TNP-ATP	7 nM	0.4 μM	–	–	–	–[a]
IP5I	3 nM	–	–	–	–	–
A-317491	9 nM	–	–	–	–	–

[a]Single concentrations of antagonist tested and so antagonist potency cannot be accurately stated

heteromers. It has the pharmacological properties of the $P2X_3$ subunit (activated by α,β-methyleneATP, antagonised by TNP-ATP and A-317491) and the biophysical properties (slow desensitisation) of the $P2X_2$ subunit. The $P2X_{1/5}$ heteromer combines the properties of the individual subunits in a similar manner. It is activated by α,β-methyleneATP ($P2X_1$) and desensitises slowly ($P2X_5$). However, it is much less sensitive than the $P2X_1$ homomer to the antagonist TNP-ATP. The $P2X_{4/6}$ heteromer is similar to the $P2X_{1/5}$ in that it is activated by α,β-methyleneATP and desensitises slowly. The remaining known heteromers have not been studied in detail. The $P2X_{2/6}$ receptor is activated by ATP, but not α,β-methyleneATP, and antagonised by suramin. The $P2X_{1/2}$ combination appears only in a small subset of cells when the recombinant receptors are coexpressed in *Xenopus* oocytes and is activated by ATP and α,β-methyleneATP (Brown et al. 2002). The effects of antagonists have not been reported. Finally, the $P2X_{1/4}$ receptor is also activated by ATP and α,β-methylene-ATP and antagonised by suramin and TNP-ATP (Nicke et al. 2005). As will be discussed in more detail in the following sections, ATP-activated currents through the $P2X_{2/3}$ heteromer resemble closely currents seen in the cell bodies of many sensory neurons. Similarly, responses that resemble the $P2X_{1/5}$ and $P2X_{4/6}$ heteromers are also seen in the central terminals of some sensory cells. The possible expression of the other heteromers in sensory neurons remains to be studied.

12.3   P2X Receptors in Sensory Nerves

When the $P2X_3$ receptor was cloned and found to be expressed selectively at high levels in nociceptive sensory neurons, it was proposed that it could play a role in acute pain (Kennedy and Leff 1995a; Burnstock and Wood 1996), but studies on $P2X_3$-knockout mice showed no change in responsiveness to acute, noxious heat and mechanical stimuli (Cockayne et al. 2000; Souslova et al. 2000). Burnstock (1999) proposed an alternative acute role for ATP and P2X$_3$ receptors in mechanosensory transduction in visceral tubes and sacs, such as ureters, the urinary bladder and the gut, in which distension of the epithelial cells that line these tissues induces ATP release, which in turn acts at P2X$_3$ and P2X$_{2/3}$ receptors on adjacent sensory nerves, leading to activation of pain centres in the brain. In addition, a number of non-nociceptive, acute sensory transduction roles for ATP and P2X receptors have been proposed in recent years (see Burnstock 2006). Finally, recent studies suggest that P2X$_3$ receptors play an important role in mediating chronic pain (see Kennedy et al. 2003). Therefore, we will look at the distribution of P2X$_3$ receptors in sensory nerves and consider the evidence for each of these proposed roles.

12.3.1   Sensory Nerves

Sensory nerve cell bodies lie in the trigeminal, nodose and dorsal root (DRG) ganglia and their axons project peripherally to tissues and organs throughout the body. Approximately 70% of the nerve cells in the DRG are C cells with small diameter cell bodies and unmyelinated, slow conducting axons, and Aδ cells, which have medium-sized cell bodies and thinly myelinated axons that conduct action potentials more rapidly. The majority of these nerve cells are nociceptors that respond to chemical, thermal and mechanical stimuli; those that respond to all three are classified as polymodal nociceptors (Perl 1992). On the basis of biochemical, anatomical and physiological properties, polymodal C cells present in the adult DRG can be divided into two classes (see Lawson 1992; Snider and McMahon 1998): 40–45% constitutively synthesise the neuropeptides substance P and calcitonin gene-related peptide (CGRP) and express the TrkA receptor for nerve growth factor. They project centrally to the spinal dorsal horn lamina I and outer lamina II. The remaining 55–60% of C cells do not express substance P, CGRP and TrkA, but do express the enzymes thiamine monophosphatase and fluoride-resistant acid phosphatase (FRAP). They can be identified with cellular markers such as LA4 and the isolectin B4 (IB4) and are sensitive to glial cell-derived neurotrophic factor. These cells project to inner lamina II in the spinal dorsal horn. Many, but not all, of both classes of cell also express the TRPV1 receptor (previously known as the VR1 receptor), the site of action of capsaicin.

12.3.2   P2X Receptor Expression in Sensory Nerves

All of the P2X subunits, except for P2X$_7$, are found in sensory nerves, but the P2X$_3$ subunit is by far and away the most prominent of these. Initial reports on the P2X$_3$ receptor showed that its mRNA was selectively expressed at high levels in neurons in the rat DRG and nodose ganglia (Chen et al. 1995; Lewis et al. 1995). This was confirmed in subsequent studies with selective antibodies (Bradbury et al.

1998; Vulchanova et al. 1998; Llewellyn-Smith and Burnstock 1998). These data are consistent with the ability of $P2X_3$ subunits to form homomultimers on its own and heteromultimers with the $P2X_2$ subunit (Lewis et al. 1995). Interestingly, Vacca et al. (2004) showed recently that the $P2X_3$ receptor localises into lipid rafts in rat DRG neurons. These are cholesterol/sphingolipid-rich membrane domains thought to be involved in targeting receptors to specific areas of the plasma membrane. The mRNA (Collo et al. 1996) and protein (Xiang et al. 1998; Barden and Bennett 2000) of other P2X receptor subtypes have also been detected in sensory neurons, but at lower levels, except for the nodose ganglion, where the $P2X_2$ and $P2X_3$ receptors are expressed at similar levels (Xiang et al. 1998). The function of P2X subunits other than the $P2X_2$ and $P2X_3$ in sensory neurons largely remains to be determined.

$P2X_3$-like immunoreactivity is seen in about 40% of rat DRG neurons, which tend to have small or medium-sized cell bodies and to coexpress IB4, FRAP and LA4: only a small minority (14%) coexpressed CGRP (Bradbury et al. 1998; Vulchanova et al. 1998). A similar pattern of expression has also been reported in rat trigeminal ganglia (Eriksson et al. 1998). The $P2X_3$ receptor also coexpresses to a large degree with the TRPV1 receptor and capsaicin pretreatment decreases $P2X_3$ mRNA (Chen et al. 1995) and $P2X_3$-like immunoreactivity (Vulchanova et al. 1998) in the rat DRG by about 70%. Interestingly, Yiangou et al. (2000) demonstrated $P2X_3$-like immunoreactivity in human DRG, mainly in cells that do not express TrkA. The central projections of the $P2X_3$-positive neurons in the rat DRG (Bradbury et al. 1998; Vulchanova et al. 1998) and trigeminal ganglia (Llewellyn-Smith and Burnstock 1998) terminate in inner lamina II of the spinal cord, and $P2X_3$-like immunoreactivity is also seen in this region. It is located in the central terminals of the sensory nerves, rather than in spinal neurons, as it is abolished by section of the dorsal roots (rhizotomy) (Bradbury et al. 1998) or by selective destruction of IB4-positive sensory nerves (Nakatsuka et al. 2003).

12.3.3   *Functional Expression in Sensory Nerves*

P2X agonists evoke inward currents in acutely dissociated rat DRG neurons. In our studies we recorded from small diameter cells under voltage clamp conditions and found that the ATP-induced currents desensitised rapidly (Robertson et al. 1996; Fig. 12.1a). The rank order of agonist potency was 2-methylthioATP = ATP > α,β-methyleneATP (Fig. 12.1b). We then showed that another structural analogue of ATP, β,γ-methyleneATP, evoked similar currents in a stereo-selective manner. The D isomer was active, although about 18 times less potent than ATP, but the L isomer was virtually inactive (Rae et al. 1998; Fig. 12.2). This pharmacological profile is consistent with activation of $P2X_3$ receptors, which was confirmed using neurons from $P2X_3$-knockout mice (Cockayne et al. 2000; Souslova et al. 2000). Further studies have shown that these fast currents are predominant in small diameter neurons that are capsaicin-sensitive and display IB4 staining, indicating that homomeric $P2X_3$ receptors are expressed mainly in small, capsaicin-sensitive

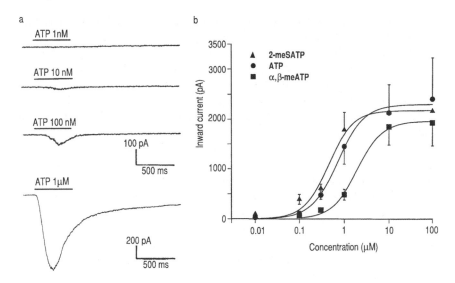

Fig. 12.1 Currents evoked by P2X$_3$ agonists in neurons of the rat dorsal root ganglia (DRG). **a** Traces show fast inward currents evoked by ATP (1 nM–1 µM) in the same cell when applied rapidly for 500 ms, as indicated by the *solid bars*. Note the difference in scale for the current evoked by 1 µM ATP. ATP was applied at 10 min intervals to minimise current rundown. **b** Mean peak inward current amplitude [± standard error of the mean (SEM)] is plotted against log concentration of ATP (●), 2-methylthioATP (▲) and α,β-methyleneATP (■), n = 4–23. Error bars for several points have been omitted for clarity where necessary. Reprinted from Robertson et al. (1996)

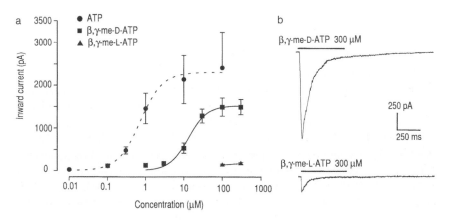

Fig. 12.2 Currents evoked by β,γ-me-D-ATP and β,γ-me-L-ATP in neurons of the rat DRG. **a** Mean peak inward current amplitude (± SEM) is plotted against log concentration ATP (●), β,γ-me-D-ATP (■) and β,γ-me-L-ATP (▲), n = 7–28. **b** Traces show fast inward currents evoked by β,γ-me-D-ATP (*upper*) and β,γ-me-L-ATP (*lower*) (300 µM) when applied rapidly for 500 ms, as shown by the *solid bars*. Reprinted from Rae et al. (1998)

C cells in the rat DRG (Li et al. 1999; Ueno et al. 1999; Petruska et al. 2000a, 2000b; Tsuzuki et al. 2003).

ATP also induces slowly desensitising currents in sensory neurons (Lewis et al. 1995; Burgard et al. 1999; Grubb and Evans 1999). Indeed, in the rat nodose ganglion, the slow current predominates (Lewis et al. 1995). Pharmacological studies indicate that the $P2X_{2/3}$ heteromultimer mediates these currents, which is supported by studies on neurons from $P2X_3$-knockout mice. In these cells, α,β-methyleneATP, an agonist at the $P2X_3$ homomer and the $P2X_{2/3}$ heteromultimer, but not the $P2X_2$ homomer (Lewis et al. 1995), was inactive. In contrast, ATP, an agonist at all three receptors, evoked a slowly desensitising current, consistent with an action at the $P2X_2$ homomer (Cockayne et al. 2000; Souslova et al. 2000). The slowly desensitising currents tend to be seen in medium-sized cells that are capsaicin-insensitive (Li et al. 1999; Ueno et al. 1999; Petruska et al. 2000a, 2000b; Tsuzuki et al. 2003), suggesting that the $P2X_{2/3}$ heteromultimer may be predominant in medium-sized Aδ cells in the rat DRG.

Recent studies show that α,β-methyleneATP can evoke slowly decaying currents independently of $P2X_{2/3}$ receptors in a subset of DRG neurons (Tsuzuki et al. 2003). TNP-ATP, a potent antagonist at $P2X_1$, $P2X_3$ and $P2X_{2/3}$ receptors, abolished rapidly desensitising responses to α,β-methyleneATP in all DRG cells tested. In contrast, TNP-ATP had a variable effect against the slowly decaying currents and was ineffective in 25% of cells. The TNP-ATP-resistant currents were, however, abolished by the non-selective P2X antagonist PPADS. The authors proposed that these currents might be elicited via $P2X_{1/5}$ or $P2X_{4/6}$ heteromers, as both are α,β-methyleneATP-sensitive and TNP-ATP-insensitive. Further analysis showed that the cells expressing these currents were medium sized and capsaicin-insensitive, suggesting that they are Aδ cells. Further studies are needed to identify the P2X subunits that mediate these actions and to explain why these currents were not seen in neurons from $P2X_3$ knockout mice.

As well as sensory nerve cell bodies, functional P2X receptors are also present on their central terminals and may have a neuromodulatory role. Intrathecal ATP and α,β-methyleneATP decreased the pinch pressure threshold and induced mechanical allodynia to Von Frey hairs in the rat paw (Okada et al. 2002). Although the mechanism of action was not determined, ATP is known to increase glutamate release from DRG nerve terminals in lamina II of the spinal cord (Nakatsuka et al. 2003). The increase was transient and inhibited by TNP-ATP, suggesting that it was mediated via $P2X_3$ receptors. P2X receptors are also present on the central terminals of Aδ fibres that project to lamina V. ATP and α,β-methyleneATP caused a long lasting increase in glutamate release from these neurons, which was unaffected by TNP-ATP and may be mediated by $P2X_{1/5}$ or $P2X_{4/6}$ heteromers (Nakatsuka et al. 2001, 2003). P2X-like immunoreactivity has not been studied in this region and further experiments are required to identify which P2X subunits are expressed. Also, as both nociceptive and non-nociceptive Aδ fibres project to lamina V, the modality of the P2X agonist-sensitive Aδ nerves needs to be determined. Interestingly, ARL 67156, which inhibits the breakdown of ATP by ecto-ATPases, potentiated the release of glutamate elicited by stimulation of the dorsal roots (Nakatsuka et al. 2001). These data imply that the physiological role of ATP released from Aδ nerves in lamina V is more likely to be as a presynaptic neuromodulator, than as a fast neurotransmitter.

12.4   Physiological Roles for Sensory P2X Receptors

Exogenously applied ATP induces acute pain (inhibited by suramin, PPADS and TNP-ATP) in humans and animals, but endogenous ATP does not appear to be involved in acute noxious thermal and mechanical pain, as the responses to these stimuli were unchanged in P2X$_3$-knockout mice (Cockayne et al. 2000; Souslova et al. 2000). As noted above, Burnstock (1999) proposed an alternative acute role for ATP and P2X$_3$ receptors in mechanosensory transduction in visceral tubes and sacs, such as ureters, the urinary bladder and the gut. In addition, a number of non-nociceptive, acute sensory transduction roles for ATP and P2X receptors have been proposed in recent years. Here, we will briefly consider an example of each of these. A more extensive description can be found in Burnstock (2006).

12.4.1   Filling of the Urinary Bladder

ATP plays an important role in the expulsion of urine from the urinary bladder; ATP is released as an excitatory co-transmitter with acetylcholine from postsynaptic parasympathetic nerves and acts at postjunctional P2X receptors to evoke contraction of the detrusor smooth muscle (Kennedy 2001). Recent evidence shows that ATP and P2X$_3$ receptors are also likely to play a crucial role in the sensing of the volume of urine present in the resting bladder. Sensory nerves are found throughout the bladder, but are particularly dense in the suburothelial space between the detrusor smooth muscle and the innermost layer of the bladder, which comprises a sheet of epithelial cells (the urothelium). Many of these sensory fibres show P2X$_3$-like immunoreactivity (Lee et al. 2000; Elneil et al. 2001; Birder et al. 2004) and are stimulated by P2X receptor agonists (Rong et al. 2002) Stretching of the urothelium (as occurs as the bladder fills with urine) causes the epithelial cells to release ATP (Fergusson et al, 1997; Wang et al. 2005) and induces action potential discharge in the adjacent mechanosensitive afferent nerves (Vlaskovska et al. 2001). This signal is thought to be interpreted by the brain as increased filling of the bladder. Consistent with this mechanos-transductory role for P2X$_3$ receptors, P2X$_3$ knockout mice have a much lower micturition frequency and much higher bladder capacity (Cockayne et al. 2000) and show much less activation of the bladder's afferent nerves on stretch (Vlaskovska et al. 2001) than wild-type animals.

12.4.2   Sensing of Blood O_2 and CO_2 Levels

In mammals, the carotid body helps maintain blood O_2 levels by inducing changes in ventilation rate when the blood partial pressures of O_2 and CO_2 change. Recent studies indicate that ATP, acting via P2X receptors, plays a crucial role in this process. Stimulation of O_2- or CO_2-sensitive chemoreceptors (glomus cells) in the carotid body causes them to release ATP and acetylcholine onto the adjacent nerve endings of the carotid sensory nerve, whose cell bodies are located in the petrosal ganglion (Zhang et al. 2000; Prasad et al. 2001), which in turn induces depolarisation and action potential firing in the carotid nerve. Consistent with this, both $P2X_2$- and $P2X_3$-like immunoreactivity are present in these sensory nerve cell bodies and endings (Prasad et al. 2001; Rong et al. 2003). The $P2X_2$ receptor appears to be the more important of the two, at least in hypoxia, as an in vitro carotid body preparation prepared from $P2X_2$ knockout mice showed a much smaller response to hypoxia compared with wild-type and $P2X_3$ knockout mice (Rong et al. 2003).

12.5   ATP and $P2X_3$ Receptors in Chronic Neuropathic and Inflammatory Pain

Chronic pain arises in response to prolonged inflammation and tissue injury and is associated with a wide variety of pathological conditions. In such conditions, there is a heightened sensitivity to acute painful stimuli (hyperalgesia) and stimuli that are normally not noxious become so (allodynia). Chronic pain is debilitating and can greatly decrease quality of life, not just due to the pain per se, but also because of the depression that can often ensue. However, great advances have recently been made in our understanding of the mechanisms that underlie chronic pain. A variety of novel, sensory neuron-specific receptors and ion channels, such as TTX-resistant Na^+ channels, H^+-sensitive ion channels and TRPV receptors, have been identified that appear to play a role in initiating and maintaining chronic pain (see for example Snider and McMahon 1998; Caterina and Julius 1999; Millan 1999). The mechanisms that initiate plasticity in sensory nerves in response to chronic stimuli are also beginning to be more clearly understood (Snider and McMahon 1998; Woolf and Salter 2000). These studies indicate that multiple factors play a role in chronic pain and so indicate new targets for development of novel analgesics that are effective in the treatment of chronic pain.

12.5.1   Down Regulation of $P2X_3$ Receptors

Recent studies are consistent with a role of $P2X_3$ receptors in the pain associated with chronic inflammation and neuronal injury. Two research groups downregulated $P2X_3$ receptor expression in rats by delivering antisense oligonucleotides (ASO) specific for P2X$_3$ receptors to lumbar DRG neurons, via an indwelling intrathecal catheter attached to an osmotic mini-pump (Barclay et al. 2002; Honore et al. 2002). Delivery of P2X$_3$ receptor ASO for 7 days significantly reduced the levels of P2X$_3$ mRNA in the DRG and P2X$_3$ protein levels in the DRG and the inner lamina II of the dorsal horn of the spinal cord. The animals showed no changes in overt behaviour and delivery of a missense oligonucleotide had no effect on P2X$_3$ expression. The ASO had no effect on acute inflammatory hyperalgesia induced by carageenan, but the treated animals did show a reduced mechanical hyperalgesia to intraplantar administration of formalin or α,β-methyleneATP into the hind paw. Thus, intrathecal P2X$_3$ ASO appears to be taken up by the central terminals of DRG neurons and transported back to the cell soma, where it inhibits translation of P2X$_3$ mRNA, leading to a decreased expression of P2X$_3$ receptor protein in the central and peripheral terminals of DRG neurons. These studies also showed a pathophysiological role for P2X$_3$ receptors in mediating chronic inflammatory and neuropathic pain. Chronic inflammatory pain was induced by intradermal administration of complete Freund's adjuvant (CFA) into the rat hind paw. Intrathecal application of P2X$_3$ ASO, starting 24 h beforehand, reduced the development of mechanical hyperalgesia over the next 6 days by about 25% (Barclay et al. 2002), whilst the thermal hyperalgesia seen after 2 days was almost abolished (Honore et al. 2002). In either case there was no effect of ASO treatment on contralateral paw responsiveness during the treatment periods.

The sciatic nerve contains the axons of lumbar sensory neurons; the mechanical hyperalgesia induced by its partial ligation (Seltzer model) was significantly reduced when intrathecal P2X$_3$ ASO treatment was started 24 h beforehand. Notably, when ASO application was delayed until 13 days after ligation, it was still effective in reducing the mechanical hyperalgesia within 2 days (Barclay et al. 2002), although mechanical allodynia was unaffected. In contrast, the tactile allodynia induced by L5–L6 spinal nerve ligation (Chung model) was substantially reduced within 2 days of initiation of ASO administration. This effect reverted back to control levels over 7 days after ASO administration was stopped (Honore et al. 2002). Again, in each of these cases, there was no effect of ASO treatment on contralateral paw thresholds during the treatment periods.

A subsequent study by Dorn et al. (2004) used small interfering RNA (siRNA) to downregulate the P2X$_3$ receptor in vivo. A 21-nucleotide-long probe that was specific for the P2X$_3$ receptor was delivered to the lumbar DRG terminals via an indwelling intrathecal catheter attached to an osmotic mini-pump. The expression of P2X$_3$ mRNA was decreased by 40% and P2X$_3$-lir in the spinal cord substantially reduced. In the Setzer model of chronic neuronal injury tactile allodynia and mechanical hyperalgesia were significantly reduced. These results are very similar to those described above from the same group using an ASO (Barclay et al. 2002). Two notable differences were that the siRNA, but not the ASO, inhibited tactile allodynia and that the hyperalgesic response to α,β-methyleneATP was abolished by the siRNA, but decreased only by 50% by the ASO. Thus, siRNA appears to be more potent than ASO in reducing pain. Taken together, the above data are consistent with the proposal that P2X$_3$ receptors play a crucial role in the development and maintenance of chronic inflammatory and neuropathic pain.

12.5.2   A-317491 – a P2X$_3$ Antagonist

The development of a competitive P2X$_3$ antagonist, A-317491 (Jarvis et al. 2002), has enabled the role of P2X$_3$ receptors in pain to be studied more directly. This non-nucleotide is highly selective for the P2X$_3$ homomer and the P2X$_{2/3}$ heteromer over other P2X subtypes. It shows substantial stereo-selectivity (S >> R) and the S-enantiomer has a pA$_2$ = 6.63 at the recombinant rat P2X$_{2/3}$ receptor. After subcutaneous (s.c.) dosing, A-317491 had high systemic availability and a plasma half-life in rats of 11 h. A-317491 had no effect on motor activity or coordination, and general cardiovascular and central nervous system (CNS) activity. Thus, A-317491 is the first selective, stable, competitive P2X$_3$ antagonist to be introduced. At doses of up to 100 µmol/kg s.c., A-317491 had little effect against a range of acute noxious thermal, mechanical and chemical stimuli in rats in vivo. However, inflammatory thermal hyperalgesia induced by intraplantar CFA was rapidly and fully blocked, lasting for 8 h after s.c. administration. Importantly, A-317491 did not display tolerance after twice-daily administration for 4 days. Phase II of formalin-induced inflammatory pain was also effectively inhibited. A-317491 also inhibited chronic neuropathic pain, being most potent against the mechanical allodynia and thermal hyperalgesia induced by chronic constriction of the sciatic nerve (Bennett model), both of which it abolished. Again the effect had a rapid onset and lasted for 5 h. A-317491 was also effective against tactile allodynia induced by L5–L6 spinal nerve ligation, but was less potent than against sciatic nerve ligation. In each of these paradigms, A-317491 had no effect on the contralateral paw responsiveness to pain. A further study by the same group showed that A317491 also reversed the inflammatory mechanical hyperalgesia induced by CFA (Wu et al. 2004). Another group have recently introduced a further P2X$_3$ antagonist, compound A, which, unlike A-317491, has good penetration of the blood–brain barrier, and which is an effective antagonist in vivo when administered i.v. (Sharp et al. 2006). Thus, the data obtained from pharmacological blockade of P2X$_3$ receptors are consistent with those seen following downregulation of P2X$_3$ receptor expression.

12.6   Mechanisms Underlying Chronic Pain

These studies indicate that P2X receptors may be involved in mediating certain types of pain, particularly the P2X$_3$ subtype, either as a homomer or as a heteromultimer with P2X$_2$ subunits. The demonstration of P2X$_3$-like immunoreactivity and functional expression in sensory nerves and the recent reports using gene knockout, ASO and siRNA technologies and the selective P2X$_3$ antagonist, A-317491, all point to a crucial role of ATP and P2X$_3$ receptors in chronic inflammatory and neuropathic pain. P2X$_2$ receptors may also mediate some noxious effects of ATP independently of the P2X$_3$ subunit in capsaicin-sensitive C fibres (Wismer et al. 2003), whilst P2X$_{1/5}$ or P2X$_{4/6}$ heteromers are proposed to be expressed in some Aδ fibres. This improved understanding of the mechanisms that underlie the noxious effects of ATP is encouraging and should help to identify novel analgesics. Several issues still have to be addressed, however.

What are the cellular mechanisms that underlie the increased responsiveness to P2X$_3$ agonists? The simplest explanation is that inflammatory mediators increase the potency and/or efficacy of ATP at P2X$_3$ receptors that are already expressed in C and Aδ cells. Indeed, substance P and bradykinin increase the current carried by recombinant P2X$_3$ and P2X$_{2/3}$ receptors (Paukert et al. 2001), but further studies are needed to determine if this also occurs with native receptors. Alternatively, inflammatory mediators may change the phenotype of sensory neurons, such that cells that are normally insensitive to ATP become responsive, i.e. silent afferents or sleeping nociceptors are activated. It has become increasingly clear in recent years that neuronal plasticity is an important factor in the initiation and maintenance of chronic pain (Snider and McMahon 1998; Woolf and Salter 2000). Consistent with this, the acute inflammation induced by carageenan almost doubled the proportion of C fibres in rat skin that responded to α,β-methyleneATP (Hamilton et al. 2001). This occurred in only 5–6h, which is too fast for retrograde signals to have travelled to the cell body and initiated de novo receptor synthesis, followed by anterograde transport of the receptors to the nerve terminals. Thus, such short-term changes must be due to a change in the properties of preformed receptors. Either inactive receptors already expressed in the sensory nerve terminals become sensitive to ATP, or preformed receptors are rapidly inserted into the terminal membrane.

The long-term changes in responsiveness to P2X$_3$ agonists in chronic pain conditions are likely to involve changes in P2X$_3$ mRNA and protein expression levels. P2X$_3$-like immunoreactivity decreased greatly in small DRG neurons after spinal nerve ligation (Kage et al. 2002) and sciatic nerve section (Bradbury et al. 1998), but increased after chronic constriction of the sciatic nerve (Novakovic et al. 1999) and partial injury or section of the inferior alveolar nerve, which contains axons projecting from the trigeminal ganglia (Eriksson et al. 1998). This might reflect a downregulation of P2X$_3$ receptors in injured cells and an upregulation in uninjured cells. Consistent with this, Tsuzuki et al. (2001) found, using in situ hybridisation, that P2X$_3$ mRNA was decreased in the cell bodies of injured DRG neurons and increased in nearby uninjured cells following section of the tibial and common peroneal nerves, terminal branches of the sciatic nerve, or of the infraorbital nerve, another peripheral projection of the trigeminal ganglia. However, further studies are needed to correlate in detail the relationship between degree of cell injury and P2X$_3$ expression.

Finally, what is the cellular source of ATP that stimulates P2X$_3$ receptors in these chronic conditions? Initially, various cell types, including damaged or stressed cells, were suggested as potential sources of extracellular ATP (Kennedy

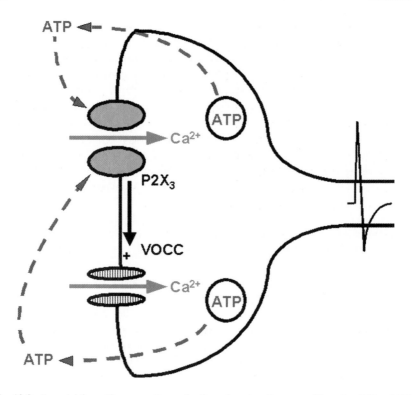

Fig. 12.3 A model for self-regenerating activation of nociceptive nerve fibres by ATP and P2X$_3$ receptors. A schematic representation of a mechanism by which ATP and P2X$_3$ receptors may tonically activate nociceptive nerves in chronic pain conditions is shown. Activation of P2X$_3$ receptors (and perhaps P2X$_{2/3}$ receptors) by extracellular ATP depolarises peripheral and central nociceptive nerve terminals, which opens voltage-operated Ca^{2+} channels (*VOCC*) also present in the terminal. Ca^{2+} influx via the P2X$_3$ receptors and the VOCC induces local, Ca^{2+}-dependent release of ATP into the extracellular space, which then acts on the P2X$_3$ receptors to elicit further depolarisation and Ca^{2+} influx. Reprinted from Kennedy et al. (2003)

and Leff 1995a; Burnstock and Wood 1996). Since P2X$_3$ receptors appear to be functionally expressed along the length of sensory neurons, these cells would have a strong potential to excite pain-sensing nerves. Consistent with this, lysis of keratinocytes was recently shown to excite nociceptors through the release of cytosolic ATP (Cook and McCleskey 2002). Additionally, an intriguing alternative source of ATP is the sensory nerves themselves, as Holton (1959) showed that ATP was released following antidromic stimulation. In this model (Fig. 12.3), stimulation of P2X$_3$ receptors in peripheral terminals initiates local depolarisation and Ca^{2+} influx through the P2X$_3$ receptors themselves and/or via voltage-dependent Ca^{2+} channels that are opened by the depolarisation. The Ca^{2+} influx would then induce local release of ATP, which would in turn feedback in a positive manner to further stimulate the P2X$_3$ receptors and so create a self-regenerating signal. If large enough, the depolarising signal would initiate action potentials that travel along the sensory axon to the spinal cord. A similar mechanism at the central terminals would be consistent with the ability of ATP to potentiate glutamate release from the central sensory terminals in the spinal cord (Nakatsuka et al. 2001, 2003). ATP release is also likely to occur along the length of sensory nerve axons, though it is not clear at present if this release is via Ca^{2+}-dependent exocytosis release or through other mechanisms. Indeed, Grafe et al. (2006) have recently shown that compression of a peripheral nerve trunk induces local release of ATP at a concentration that is likely to stimulate local P2X$_3$ receptors. Together, these terminal and axonal mechanisms would induce a constant, self-regenerating sensory input to the brain, which would be experienced consciously as chronic pain.

Acknowledgements This work was supported by grants from the Medical Research Council, Wellcome Trust and Caledonian Research Foundation.

References

Barclay J, Patel S, Dorn G, Wotherspoon G, Moffatt S, Eunson L, Abdel'al S, Natt F, Hall J, Winter J, Bevan S, Wishart W, Fox A, Ganju P (2002) Functional downregulation of P2X$_3$ receptor subunit in rat sensory neurons reveals a significant role in chronic neuropathic and inflammatory pain. J Neurosci 22:8139–8147

Barden JA, Bennett MR (2000) Distribution of P2X purinoceptor clusters on individual rat dorsal root ganglion cells. Neurosci Lett 287:183–186

Birder LA, Ruan HZ, Chopra B, Xiang Z, Barrick S, Buffington CA, Roppolo JR, Ford AP, de Groat WC, Burnstock G (2004) Alterations in P2X and P2Y purinergic receptor expression in urinary bladder from normal cats and cats with interstitial cystitis. Am J Physiol Renal Physiol 28:F1084–F1091

Bradbury EJ, Burnstock G, McMahon SB (1998) The expression of P2X$_3$ purinoceptor in sensory neurons: effects of axotomy and glial-derived neurotrophic factor. Mol Cell Neurosci 12:256–268

Brown SG, Townsend-Nicholson A, Jacobson KA, Burnstock G, King BF (2002) Heteromultimeric P2X$_{1/2}$ receptors show a novel sensitivity to extracellular pH. J Pharmacol Exp Ther 300:673–680

Burgard EC, Niforatos W, Van Biesen T, Lynch KJ, Touma E, Metzger RE, Kowaluk EA, Jarvis MF (1999) P2X receptor-mediated ionic currents in dorsal root ganglion neurons. J Neurophysiol 82:1590–1598

Burnstock G (1999) Release of vasoactive substances from endothelial cells by shear stress and purinergic mechanosensory transduction. J Anat 194:335–342

Burnstock G (2001) Purine-mediated signalling in pain and visceral perception. Trends Pharmacol Sci 22:182–187

Burnstock G (2006) Purinergic P2 receptors as targets for novel analgesics. Pharmacol Ther 110:433–454

Burnstock G, Kennedy C (1985) Is there a basis for distinguishing two types of P$_2$-purinoceptor? Gen Pharmacol 16:433–440

Burnstock G, Wood JN (1996) Purinergic receptors: their role in nociception and primary afferent neurotransmission. Curr Opin Neurobiol 6:526–532

Caterina MJ, Julius D (1999) Sense and specificity: a molecular identity for nociceptors. Curr Opin Neurobiol 9:525–530

Chen CC, Akopian AN, Sivilotti L, Colquhoun D, Burnstock G, Wood JN (1995) A P2X purinoceptor expressed by a subset of sensory neurons. Nature 377:428–431

Chizh BA, Illes P (2000) P2X receptors and nociception. Pharm Rev 53:553–568

Cockayne DA, Hamilton SG, Zhu QM, Dunn PM, Zhong Y, Novakovic S, Malmberg AB, Cain G, Berson A, Kassotakis L, Hedley L, Lachnit WG, Burnstock G, McMahon SB, Ford AP (2000) Urinary bladder hyporeflexia and reduced pain-related behaviour in P2X$_3$-deficient mice. Nature 407:1011–1015

Collo G, North RA, Kawashima E, Merlo-Pich E, Neidhart S, Surprenant A, Buell G (1996) Cloning of P2X$_5$ and P2X$_6$ receptors and the distribution and properties of an extended family of ATP-gated ion channels. J Neurosci 16:2495–2507

Cook SP, McCleskey EW (2002) Cell damage excites nociceptors through release of cytosolic ATP. Pain 95:41–47

Dorn G, Patel S, Wotherspoon G, Hemmings-Mieszczak M, Barclay J, Natt FJC, Martin P, Bevan S, Fox A, Ganju P, Wishart W, Hall J (2004) siRNA relieves chronic neuropathic pain. Nucleic Acid Res 32:e49

Egan TM, Khakh B (2004) Contribution of calcium ions to P2X channel responses. J Neurosci 24:3413–3420

Elneil S, Skepper JN, Kidd E, Williamson JG, Ferguson DR (2001) Distribution of P2X$_1$ and P2X$_3$ receptors in the rat and human urinary bladder. Pharmacology 63:120–128

Eriksson J, Bongenhielm U, Kidd E, Matthews B, Fried K (1998) Distribution of P2X$_3$ receptors in the rat trigeminal ganglion after inferior alveolar nerve injury. Neurosci Lett 254:37–40

Ferguson DR, Kennedy I, Burton TJ (1997) ATP is released from rabbit urinary bladder epithelial cells by hydrostatic pressure changes – a possible sensory mechanism? J Physiol 505:503–511

Grafe P, Schaffer V, Rucker F (2006) Kinetics of ATP release following compression injury of a peripheral nerve trunk. Purinergic Signalling 2:527–536

Grubb BD, Evans RJ (1999) Characterization of cultured dorsal root ganglion neuron P2X receptors. Eur J Neurosci 11:149–154

Hamilton SG, McMahon SB, Lewin GR (2001) Selective activation of nociceptors by P2X receptor agonists in normal and inflamed rat skin. J Physiol 534:437–445

Holton P (1959) The liberation of adenosine triphosphate on antidromic stimulation of sensory nerves. J Physiol 145:494–504

Honore P, Kage K, Mikusa J, Watt AT, Johnston JF, Wyatt JR, Faltynek CR, Jarvis MF, Lynch K (2002) Analgesic profile of intrathecal P2X$_3$ antisense oligonucleotide treatment in chronic inflammatory and neuropathic pain states in rats. Pain 99:11–19

Jarvis MF, Burgard EC, McGaraughty S, Honore P, Lynch K, Brennan TJ, Subieta A, van Biesen T, Cartmell J, Bianchi B, Niforatos W, Kage K, Yu H, Mikusa J, Wismer CT, Zhu CZ, Chu K, Lee CH, Stewart AO, Polakowski J, Cox BF, Kowaluk E, Williams M, Sullivan J, Faltynek C (2002) A-317491, a novel potent and selective non-nucleotide antagonist of P2X$_3$ and P2X$_{2/3}$ receptors, reduces chronic inflammatory and neuropathic pain in the rat. Proc Natl Acad Sci USA 99:17179–17184

Jones CA, Chessell IP, Simon J, Barnard EA, Miller KJ, Michel AD, Humphrey PPA (2000) Functional characterization of the P2X$_4$ receptor orthologues. Br J Pharmacol 129:388–394

Jones CA, Vial C, Sellers LA, Humphrey PPA, Evans RJ, Chessell IP (2004) Functional regulation of P2X$_6$ receptors by N-linked glycosylation: identification of a novel α,β-methyleneATP-sensitive phenotype. Mol Pharmacol 65:979–985

Kage K, Niforatos W, Zhu CZ, Lynch KJ, Burgard EC, Honore P, Jarvis MF (2002) Alteration of dorsal root ganglion P2X$_3$ receptor expression and function following spinal nerve ligation in the rat. Exp Brain Res 147:511–519

Kennedy C (1990) P$_1$- and P$_2$-purinoceptor subtypes – an update. Arch Int Pharmacodyn 303:30–50

Kennedy C (2001) The role of purines in the peripheral nervous system. In: Abbracchio MP, Williams M (eds) Handbook of experimental pharmacology 151/1; Purinergic and pyrimidinergic signalling I: Molecular, nervous and urogenitary system function. Springer, Berlin, pp 289–304

Kennedy C, Leff P (1995a) Painful connection for ATP. Nature 377:285–386

Kennedy C, Leff P (1995b) How should P$_{2X}$-purinoceptors be characterised pharmacologically? Trends Pharmacol Sci 16:168–174

Kennedy C, Assis TS, Currie A, Rowan EG (2003) Crossing the pain barrier: P2 receptors as targets for novel analgesics. J Physiol 553:683–694

Khakh BS, Burnstock G, Kennedy C, King BF, North RA, Seguela P, Voigt M, Humphrey PPA (2001) International Union of Pharmacology XXIV. Current status of the nomenclature and properties of P2X receptors and their subunits. Pharmacol Rev 53:107–118

Lawson SN (1992) Morphological and biochemical cell types of sensory neurons. In: Scott SA (ed) Sensory neurons: diversity, development and plasticity. Oxford University Press, New York, pp 27–59

Lee HY, Bardini M, Burnstock G (2000) Distribution of P2X receptors in the urinary bladder and the ureter of the rat. J Urol 163:2002–2007

Lewis C, Neidhart S, Holy C, North RA, Buell G, Surprenant A (1995) Coexpression of P2X₂ and P2X₃ receptor subunits can account for ATP currents in sensory neurons. Nature 377:432–435

Li C, Peoples RW, Lanthorn TH, Li ZW, Weight FF (1999) Distinct ATP-activated currents in different types of neurons dissociated from rat dorsal root ganglion. Neurosci Lett 263:57–60

Llewellyn-Smith IJ, Burnstock G (1998) Ultrastructural localization of P2X₃ receptors in rat sensory neurons. Neuroreport 9:2545–2550

Millan MJ (1999) The induction of pain: an integrative review. Prog Neurobiol 57:1–164

Nakatsuka T, Gu JG (2001) ATP P2X receptor-mediated enhancement of glutamate release and evoked EPSCs in dorsal horn neurons of the rat spinal cord. J Neurosci 21:6522–6531

Nakatsuka T, Tsuzuki K, Ling JX, Sonobe H, Gu JG (2003) Distinct roles of P2X receptors in modulating glutamate release at different primary sensory synapses in rat spinal cord. J Neurophysiol 89:3243–3252

Nicke A, Kershensteiner D, Soto F (2005) Biochemical and functional evidence for heteromeric assembly of P2X₁ and P2X₄ subunits. J Neurochem 92:925–33

North RA (2002) Molecular physiology of P2X receptors. Physiol Rev 82:1013–1067

Novakovic SD, Kassotakis LC, Oglesby IB, Smith JA, Eglen RM, Ford AP, Hunter JC (1999) Immunocytochemical localisation of P2X₃ purinoceptors in sensory neurons in naïve rats and following neuropathic injury. Pain 80:273–282

Okada M, Nakagawa T, Minami M, Satoh M (2002) Analgesic effects of intrathecal administration of P2Y nucleotide receptor agonists UTP and UDP in normal and neuropathic pain model rats. J Pharmacol Exp Ther 303:66–73

Paukert M, Osteroth R, Geisler HS, Brändle U, Glowatzki E, Ruppersberg JR, Gründer S (2001) Inflammatory mediators potentiate ATP-gated channels through the P2X₃ subunit. J Biol Chem 276:21077–21082

Pelegrin P, Surprenant AM (2006) Pannexin-1 mediates large pore formation and interleukin-1β release by the ATP-gated P2X₇ receptor. EMBO J 25:5071–5082

Perl ER (1992) Function of dorsal root ganglion neurons: an overview. In: Scott SA (ed) Sensory neurons: diversity, development and plasticity. Oxford University Press, New York, pp 3–23

Petruska JC, Napaporn J, Johnson RD, Gu JG, Cooper BY (2000a) Distribution of P2X₁, P2X₂ and P2X₃ receptor subunits in rat primary afferents: relation to population markers and specific cell types. J Chem Neuroanat 20:141–162

Petruska JC, Napaporn J, Johnson RD, Gu JG, Cooper BY (2000b) Subclassified acutely dissociated cells of rat DRG: histochemistry and patterns of capsaicin-, proton-, and ATP-activated currents. J Neurophysiol 84:2365–2379

Prasad M, Fearon IM, Zhang M, Laing M, Vollmer C, Nurse CA (2001) Expression of P2X₂ and P2X₃ receptor subunits in rat carotid body afferent neurons: role in chemosensory signalling. J Physiol 537:667–677

Rae MG, Rowan EG, Kennedy C (1998) Pharmacological properties of P2X₃ receptors present in neurons of the rat dorsal root ganglia. Br J Pharmacol 124:176–180

Robertson SJ, Rae MG, Rowan EG, Kennedy C (1996) Characterization of a P2X-purinoceptor in cultured neurons of the rat dorsal root ganglia. Br J Pharmacol 118:951–956

Rong W, Spyer KM, Burnstock G (2002) Activation and sensitisation of low and high threshold afferent fibres mediated by P2X receptors in the mouse urinary bladder. J Physiol 541:591–600

Rong W, Gourine AV, Cockayne DA, Xiang Z, Ford APDW, Spyer KM, Burnstock G (2003) Pivotal role of nucleotide P2X$_2$ receptor subunit of the ATP-gated ion channel mediating ventilatory responses to hypoxia. J Neurosci 23:11315–11321

Sharp CJ, Reeve AJ, Collins SD, Martindale JC, Summerfield SG, Sargent BS, Bate ST, Chessell IP (2006) Investigation into the role of P2X$_3$/P2X$_{2/3}$ receptors in neuropathic pain following chronic constriction injury in the rat: an electrophysiological study. Br J Pharmacol 148:845–852

Snider, WD, McMahon SB (1998) Tackling pain at the source: new ideas about nociceptors. Neuron 20:629–632

Souslova V, Cesare P, Ding Y, Akopian AN, Stanfa L, Suzuki R, Carpenter K, Dickenson A, Boyce S, Hill R, Nebenuis-Oosthuizen D, Smith AJ, Kidd EJ, Wood JN (2000) Warm-coding deficits and aberrant inflammatory pain in mice lacking P2X$_3$ receptors. Nature 407:1015–1017

Torres GE, Egan TM, Voigt MM (1999) Hetero-oligomeric assembly of P2X receptor subunits. Specificities exist with regard to possible partners. J Biol Chem 274:6653–6659

Tsuzuki K, Kondo E, Fukuoka T, Yi D, Tsujino H, Sakagami M, Noguchi K (2001) Differential regulation of P2X$_3$ mRNA regulation by peripheral nerve injury in intact and injured neurons in the rat sensory ganglia. Pain 91:351–360

Tsuzuki K, Ase A, Séguéla P, Nakatsuka T, Wang CY, She JX, Gu JG (2003) TNP-ATP-resistant P2X ionic currents on the central terminals and somata of rat primary sensory neurons. J Neurophysiol 89:3235–3242

Ueno S, Tsuda M, Iwanaga T, Inoue K (1999) Cell type-specific ATP-activated responses in rat dorsal root. Br J Pharmacol 126:429–436

Vacca F, Amadio S, Sancesario G, Bernardi G, Volonte C (2004) P2X$_3$ receptor localizes into lipid rafts in neuronal cells. J Neurosci Res 76:653–661

Vial C, Roberts JA, Evans RJ (2004) Molecular properties of ATP-gated P2X receptor ion channels. Trends Pharmacol Sci 25:487–493

Vlaskovska M, Kasakov L, Rong W, Bodin P, Bardini M, Cockayne DA, Ford APDW, Burnstock G (2001) P2X$_3$ knock-out mice reveal a major sensory role for urothelially released ATP. J Neurosci 21:5670–5677

Vulchanova L, Riedl MS, Shuster SJ, Stone LS, Hargreaves KM, Buell G, Surprenant A, North RA, Elde R (1998) P2X$_3$ is expressed by DRG neurons that terminate in inner lamina II. Eur J Neurosci 10:3470–3478

Wang ECY, Lee JM, Ruiz WJ, Balestreire EM, von Bodungen M, Barrick S, Cockayne DA, Birder LA, Apodaca G (2005) ATP and purinergic receptor-dependent membrane traffic in bladder umbrella cells. J Clin Invest 115:2412–2422

Wismer CT, Faltynek CR, Jarvis MF, McGaraughty S (2003) Distinct neurochemical mechanisms are activated following administration of different P2X receptor agonists in the hind paw of a rat. Brain Res 965:187–193

Woolf CJ, Salter MW (2000) Neuronal plasticity: increasing the gain in pain. Science 288:1765–1768

Wu G, Whiteside GT, Lee G, Nolan S, Niosi M, Pearson MS, Ilyin VI (2004) A-317491, a selective P2X$_3$/P2X$_{2/3}$ receptor antagonist, reverses inflammatory mechanical hyperalgesia through action at peripheral receptors in rats. Eur J Pharmacol 504:45–53

Xiang Z, Bo X, Burnstock G (1998) Localisation of ATP-gated P2X receptor immunoreactivity in rat sensory and sympathetic ganglia. Neurosci Lett 256:105–108

Yiangou Y, Facer P, Birch R, Sangameswaran L, Eglen R, Anand P (2000) P2X$_3$ receptor in injured human sensory neurons. Neuroreport 11:993–996

Zhang M, Zhong H, Vollmer C, Nurse CA (2000) Co-release of ATP and ACh mediates hypoxic signalling at rat carotid body chemoreceptors. J Physiol 525:143–158

Chapter 13
Voltage-Gated Calcium Channels in Nociception

Takahiro Yasuda, and David J. Adams(✉)

Abstract Voltage-gated calcium channels (VGCCs) are a large and functionally diverse group of membrane ion channels ubiquitously expressed throughout the central and peripheral nervous systems. VGCCs contribute to various physiological processes and transduce electrical activity into other cellular functions. This chapter provides an overview of biophysical properties of VGCCs, including regulation by auxiliary subunits, and their physiological role in neuronal functions. Subsequently, then we focus on N-type calcium ($Ca_V2.2$) channels, in particular their diversity and specific antagonists. We also discuss the role of N-type calcium channels in nociception and pain transmission through primary sensory dorsal root ganglion neurons (nociceptors). It has been shown that these channels are expressed predominantly in nerve terminals of the nociceptors and that they control neurotransmitter release.

President, Australian Physiological Society (AuPS), Professor and Chair of Physiology, Head of School of Biomedical Sciences, University of Queensland, Brisbane, QLD 4072, Australia, dadams@uq.edu.au

To date, important roles of N-type calcium channels in pain sensation have been elucidated genetically and pharmacologically, indicating that specific N-type calcium channel antagonists or modulators are particularly useful as therapeutic drugs targeting chronic and neuropathic pain.

13.1 Introduction

Voltage-gated calcium channels (VGCCs) contribute to various physiological processes and transduce electrical activity into other cellular functions, such as muscle contraction, neurotransmitter release, endocrine secretion, gene expression, or modulation of membrane excitability. The structure and function of various types of VGCCs have been extensively investigated and comprehensive reviews have been published previously (Bean 1989; Catterall 2000; Hille 2004). Briefly, since the unexpected discovery of a "Ca^{2+} action potential" in crustacean muscle fibres, the VGCC current has been recognised as a ubiquitous component of excitable cells such as muscles and neurons as well as some non-excitable cells. Two types of channels were initially discovered and named high-voltage activated (HVA) and low-voltage activated (LVA) calcium channels. Based on distinct single channel conductance and current inactivation kinetics, they were also designated L-type (Large single channel conductance and Long-lasting current), and T-type (Tiny single channel conductance and Transient currents) channels, respectively. Subsequently, a third type of calcium channels was found to coexist with L- and T-type channels in chick dorsal root ganglion (DRG) neurons. The third "neuronal"-type class was named N-type calcium channels. This type belonged to the HVA calcium channels, requiring strong depolarisation (usually more positive than $-10\,mV$) for their activation. However, N-type calcium channels differ from L-type channels by being susceptible to voltage-dependent inactivation and being insensitive to dihydropyridine L-type calcium channel agonists/antagonists. Thereafter, various isoforms of calcium channels have been molecularly cloned (Ca_v1–Ca_v3 families) and biophysically and/or pharmacologically characterised (L-, P/Q-, N-, R- and T-type) in a wide variety of tissues using isoform-specific venom antagonists. A summary of the diversity and specific antagonists of the different classes of VGCC is given in Table 13.1.

13.2 Calcium Channel Structure, Gene Family and Subunit Composition

VGCC structure and subunit composition has been intensively studied in HVA calcium channels. A calcium channel α_1 subunit protein with auxiliary β and γ subunits was first purified as a dihydropyridine receptor from rabbit skeletal muscle (Curtis and Catterall 1984). Subsequently, a third auxiliary subunit, $\alpha_2\delta$, was also found as

Table 13.1 Voltage-gated calcium channel (VGCC) pharmacology

Class of calcium channel	Isoforms of α_1 subunit	Selective antagonists – small molecules/toxins (*biological source*)	Clinical drugs for pain treatment (marketing/developing company)
L-type	$Ca_v1.1$ (α_{1S}), $Ca_v1.2$ (α_{1C}), $Ca_v1.3$ (α_{1D}), $Ca_v1.4$ (α_{1F})	Dihydropyridines Phenylalkylamines Benzothiazepines Calcicludine (*Dendroaspis angusticeps*)	
P/Q-type	$Ca_v2.1$ (α_{1A})	ω-Agatoxin IVA (*Agelenopsis aperta*) ω-Conotoxin MVIIC (*Conus magus*) (also inhibits N-type) ω-Conotoxin CVIB (*Conus catus*) (also inhibits N-type)	
N-type	$Ca_v2.2(\alpha_{1B})$	ω-Conotoxin GVIA (*Conus geographus*) ω-Conotoxin MVIIA (*Conus magus*) ω-Conotoxin CVID (*Conus catus*) Small molecule (undisclosed structure)	Prialt (Elan) AM336 (CNSBio) – under development MK-6721/NMED-160 (Merck/Neuromed) – under development
R-type	$Ca_v2.3(\alpha_{1E})$	SNX-482 (*Hysterocrates gigas*)	
T-type	$Ca_v3.1$ (α_{1G}), $Ca_v3.2$ (α_{1H}), $Ca_v3.3(\alpha_{1I})$	Mibefradil Ethosuximide Zonisamide Kurtoxin (*Parabuthus transvaalicus*)	Zarontin (Pfizer) Zonegran (Eisai)
Auxiliary subunit	$\alpha_2\delta$ subunit	Gabapentin Pregabalin	Neurotin (Pfizer) Lyrica (Pfizer)

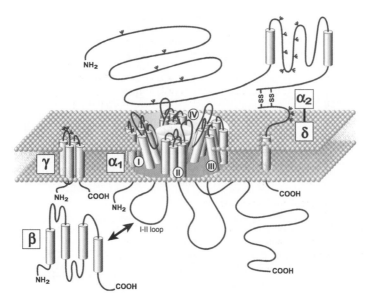

Fig. 13.1 Predicted subunit composition of high-voltage activated (HVA) calcium channels. The α_1 subunit comprises four internally similar domains, each containing six transmembrane (α helices) segments. The β subunit is a cytoplasmic protein that can interact with the I–II loop of the α_1 subunit. The $\alpha_2\delta$ subunit is cleaved post-translationally into two disulfide-linked parts, α_2 and δ, with a single transmembrane segment arising from the δ subunit and the large glycosylated α_2 anchored to the membrane by the δ subunit. The γ subunit is a glycoprotein with four trans-membrane segments

an associated molecule with α_1 subunit (Leung et al. 1987; Takahashi et al. 1987). As shown in Fig. 13.1, it has been proposed that all HVA calcium channels are comprised of a pore-forming α_1 subunit (190–270 kDa), auxiliary β subunit (50–75 kDa) and $\alpha_2\delta$ subunit (~170 kDa) and, in some cases, γ subunit (~25 kDa).

Native LVA calcium channels have yet to be purified, so their subunit assembly is currently unknown. However, electrophysiological studies using recombinant LVA calcium channels expressed with auxiliary subunits have suggested a larger contribution of the $\alpha_2\delta$ subunit to channel function than that of the β subunit (see review by Perez-Reyes 2003, 2006).

13.2.1 Gene Family of α_1 Subunits

To date, ten distinguishable genes encoding VGCC α_1 subunits have been identified (Fig. 13.2). Based on gene similarity, they are divided into three families of $Ca_v1.1$–1.4 (L-type), $Ca_v2.1$–2.3 (non-L-type: P/Q-, N- and R-types) and $Ca_v3.1$–3.3 (T-type). Comparison of the sequences of conserved transmembrane and pore

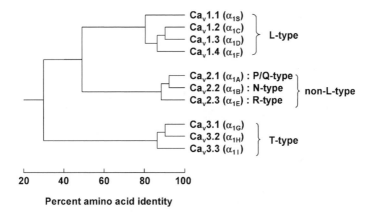

Fig. 13.2 Phylogeny of voltage-gated calcium channel (VGCC) α_1 subunits. A major division exists between the L- and non-L-type (HVA) calcium channels and the T-type (low-voltage activated – LVA) calcium channels. Only the membrane-spanning segments and pore regions (~350 amino acids) are compared (adapted from Ertel et al. 2000)

segments of the α_1 subunit revealed more than 80% intra-family identity among the Ca_v1, Ca_v2 or Ca_v3 families. About 50% inter-family identity is observed between Ca_v1 and Ca_v2 families within the HVA calcium channels, whereas the LVA calcium channels are only distantly related to the HVA channels, with less than 30% identity between Ca_v3 and Ca_v1 or Ca_v2 families (see Ertel et al. 2000). Evidently, these two lineages of calcium channels diverged very early in the evolution of multi-cellular organisms.

13.2.2 Membrane Topology and Functional Motifs of α_1 Subunits

The VGCC α_1 subunit is a protein of about 2,000 amino acid residues and has a similar membrane topology to that of voltage-dependent sodium channels. The α_1 subunit comprises four internally similar domains, each containing six transmembrane (α helices) segments of S1–S6 and therefore a total of 24 transmembrane segments (Fig. 13.3a). The S4 segments of voltage-gated ion channels are well known as voltage sensors that are directly involved in the depolarisation-induced gating charge movement that precedes channel opening (Yang and Horn 1995; Mannuzzu et al. 1996).

The S5 and S6 segments and their extracellular linkers form an hourglass-shaped pore lining the calcium channel. The most important feature of VGCCs is to allow the efficient and selective permeation of extracellular Ca^{2+} ions upon membrane depolarisation. Given that the free Ca^{2+} ion concentration present extracellularly is

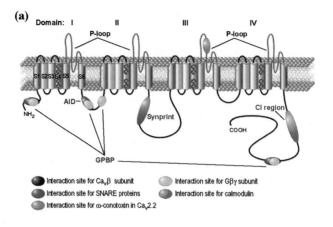

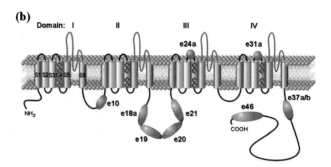

Fig. 13.3 a Schematic membrane topology of the non-L-type channel α₁ subunits (Ca$_v$2) with presumable interaction sites for other protein molecules and toxins. Each domain (*I, II, III, IV*) of the α₁ subunit comprises six transmembrane (α helices) segments (S1–S6). The S4 segment is the voltage sensor and the S5–S6 linker, the P-loop, is involved in Ca^{2+} permeation. *AID* α₁ subunit interaction domain, *GPBP* Gβγ protein-binding pocket, *CI region* Ca^{2+} inactivation region. **b** Alternative splicing sites in the N-type calcium channel α₁ subunit (Ca$_v$2.2). Approximate locations of alternative splicing are indicated in *red*

<1% of the Na$^+$ and Cl$^-$ ion concentration, VGCCs require a highly specialised sieving function. In fact, VGCCs have ~1,000-fold higher permeability to Ca^{2+} ions than to Na$^+$ ions, although the atomic radii of Ca^{2+} and Na$^+$ are similar (~2 Å). The S5–S6 linker, known as the P-region or P-loop, has been demonstrated to be a key determinant of ion selectivity and permeation rate of VGCCs (see reviews by Sather and McCleskey 2003; Hille 2004).

In the cytoplasmic linkers, as well as the N and C termini of the VGCC α₁ subunit, there are many specific motifs that can be subject to phosphorylation or interaction with other protein molecules (Fig. 13.3a for Ca$_v$2 channels) (see reviews by Hofmann et al. 1999; Catterall, 2000). An important motif is the

"α_1 subunit interaction domain (AID)" for β subunits in the cytoplasmic I–II linker of the α_1 subunit (Pragnell et al. 1994; Witcher et al. 1995). The interaction between α_1 and β subunits through this AID is critical for β subunit-induced enhancement of channel expression (Gerster et al. 1999). It is intriguing that there are additional binding sites for β subunits in the N-terminus of $Ca_v1.1$ and $Ca_v2.1$ subunits (Walker et al. 1999) and C-terminus of $Ca_v2.1$ and $Ca_v2.3$ subunits (Qin et al. 1997; Tareilus et al. 1997; Walker et al. 1998). However, the functional significance of these additional binding sites has not yet been determined. A dimer of G-protein β and γ subunits (G$\beta\gamma$ dimer) has a negative regulatory effect on all Ca_v2 channels, but not on Ca_v1 channels, through direct binding to the I–II linker (Herlitze et al. 1996; Ikeda 1996; De Waard et al. 1997; Zamponi et al. 1997; see review by Dolphin 1998) and C-terminus (Zhang et al. 1996; Qin et al. 1997) or N-terminus (Page et al. 1998; Stephens et al. 1998). It is now believed that these multiple binding sites comprise a G$\beta\gamma$ protein-binding pocket (GPBP) that interacts with a single G$\beta\gamma$ dimer, therefore 1:1 interaction between Ca_v2 α_1 and G$\beta\gamma$ (Zamponi and Snutch 1998; see review by De Waard et al. 2005). Interestingly, as shown in Fig. 13.3a, the fact that these G$\beta\gamma$ binding sites appear to overlap, or are located close to the β subunit binding sites, is consistent with the antagonistic effects of β subunits on G$\beta\gamma$-induced current inhibition (De Waard et al. 1997; Qin et al. 1997; Zamponi et al. 1997; see review by De Waard et al. 2005). Another important protein interaction site is the so-called "synprint motif" in the II–III linker. The synprint site plays a crucial role in neurotransmission through a tight interaction with SNARE proteins such as syntaxin 1A, SNAP-25 and cystein string protein (see review by Jarvis and Zamponi 2001).

13.2.3 Auxiliary β and $\alpha_2\delta$ Subunits

$Ca_v\beta$ subunits consist of 480–630 amino acids and are widely distributed in various tissues. To date, four isoforms (β1–4) have been identified. Each β subunit isoform is subject to alternative splicing to yield additional variants. In contrast to other auxiliary subunits ($\alpha_2\delta$ and γ subunits), β subunits do not contain hydrophobic segments in their amino acid sequence, and therefore β subunits are cytoplasmic proteins without a transmembrane domain (Fig. 13.1). As mentioned above, it has been reported that β subunits affect calcium channel function primarily through the interaction with the AID on the I–II linker of α_1 subunit (Pragnell et al. 1994; De Waard et al. 1995). The expression of different combinations of β subunits was shown in different regions of the brain, suggesting heterogeneity of β subunit composition among different classes of neurons (Tanaka et al. 1995).

$Ca_v\alpha_2\delta$ subunits are extensively glycosylated proteins of ~170 kDa (Leung et al. 1987; Takahashi et al. 1987; De Jongh et al. 1990; Jay et al. 1991; see review by

Hofmann et al. 1999). Complementary DNA cloning revealed that there were 18 potential *N*-glycosylation sites and three hydrophobic domains (Ellis et al. 1988). The $\alpha_2\delta$ subunits are cleaved post-translationally into two disulfide-linked parts, α_2 and δ, with a single transmembrane segment arising from the δ subunit (De Jongh et al. 1990; Jay et al. 1991) and the large glycosylated α_2 anchored to the membrane by the δ subunit (Fig. 13.1). Four isoforms of $Ca_v\alpha_2\delta$ subunits, $\alpha_2\delta$-1 to -4, are known, together with their splice variants (Ellis et al. 1988; Brust et al. 1993; Klugbauer et al. 1999; Qin et al. 2002).

13.2.4   Regulation of Macroscopic Current Amplitude by Auxiliary Subunits

Modulation of HVA calcium channel current amplitude, all four β subunit isoforms and the $\alpha_2\delta$-1 subunits have been shown primarily to increase macroscopic current amplitude.

Coexpression of β subunits enhances the level of channel expression in the plasma membrane (Williams et al. 1992; Brust et al. 1993; Shistik et al. 1995), probably by chaperoning the translocation of α_1 subunits (Chien et al. 1995; Yamaguchi et al. 1998; Gao et al. 1999; Gerster et al. 1999). The $Ca_v\beta$ subunit has been shown to interact with the AID in the cytoplasmic I–II linker of the α_1 subunit (Pragnell et al. 1994; Witcher et al. 1995), and all four β subunits interacted with the AID of $Ca_v2.1$ and $Ca_v2.2$ *in vitro* with high affinity (K_d= ~5–50 nM) (De Waard et al. 1995; Scott et al. 1996; Canti et al. 2001). The interaction between α_1 and β subunits through the AID plays a critical role in channel trafficking from the endoplasmic reticulum (ER) to plasma membrane (Pragnell et al. 1994; De Waard et al. 1995) by antagonising the binding between α_1 and an ER retention protein (Bichet et al. 2000). In contrast, there is a report that β subunits do not alter gating charge movement (Neely et al. 1993), suggesting that calcium channel expression levels in the plasma membrane are not affected by β subunits. As an additional mechanism, β subunits increase single channel open probability or maximal channel open probability reflected by the ratio of maximal ionic conductance to maximal gating charge moved (Neely et al. 1993; Shistik et al. 1995; Kamp et al. 1996; Jones et al. 1998; Qin et al. 1998; Gerster et al. 1999; Wakamori et al. 1999; Hohaus et al. 2000). Given no change in single channel conductance (Neely et al. 1993; Wakamori et al. 1993; Shistik et al. 1995; Jones et al. 1998; Gerster et al. 1999; Wakamori et al. 1999; Hohaus et al. 2000), the increase in the open probability can be attributed to facilitation of intermolecular coupling between the voltage sensor and channel pore opening. It has been reported that the increase in the open probability was specific to the β subunit isoform (Noceti et al. 1996). A hyperpolarising shift of current–voltage (I–V) relationships by β subunits (Neely et al. 1993; Yamaguchi et al. 1998) also contributes to at least a partial increase in macroscopic current amplitude.

Interestingly, the β3 subunit exhibits biphasic effects on N-type calcium channel currents, which are potentiating and inhibitory, depending on the ratio between α_1 and β subunits (Yasuda et al. 2004a).

Coexpression of $\alpha_2\delta$ subunits augments ligand binding (B_{max}) and α_1 subunit protein expression levels in the plasma membrane (Williams et al. 1992; Brust et al. 1993; Shistik et al. 1995; Bangalore et al. 1996) without increasing channel open probability (Bangalore et al. 1996; Jones et al. 1998; Wakamori et al. 1999; cf. Shistik et al. 1995), single channel conductance (Bangalore et al. 1996; Jones et al. 1998; Wakamori et al. 1999) or shifting the voltage-dependent activation curve in a hyperpolarising direction (Qin et al. 1998; Wakamori et al. 1999; Gao et al. 2000). The mechanism underlying $\alpha_2\delta$-induced potentiation of α_1 expression appears to be different from that of β subunits. A domain of $\alpha_2\delta$, which interacts with α_1 subunits, was proposed to be located in an extracellular region (Gurnett et al. 1997), and therefore it is unlikely that $\alpha_2\delta$ subunits antagonise an interaction between an ER retention protein and an intracellular AID of α_1 subunits. Furthermore, lack of evidence for $\alpha_2\delta$-induced α_1 subunit trafficking was shown using immunohistochemistry (Gao et al. 1999). Recently, it has been reported that $\alpha_2\delta$ subunits prevent N-type calcium channel internalisation and subsequent degradation, and therefore exhibit a stabilising effect on membrane-expressed calcium channels (Bernstein and Jones 2007).

Importantly, β and $\alpha_2\delta$ subunits exhibit a synergistic effect on current amplitude (Mori et al. 1991; Stea et al. 1993; De Waard et al. 1995; Shistik et al. 1995; Yasuda et al. 2004b). The current potentiating effect of $\alpha_2\delta$ subunits can be detected only in the presence of β subunits for Ca_v2 channel isoforms (Mori et al. 1991; Stea et al. 1993; De Waard et al. 1995; Parent et al. 1997) in an oocyte expression system. In a mammalian cell expression system, however, the $\alpha_2\delta$-1 subunit potentiates various VGCC currents even without a β subunit (Jones et al. 1998; Yasuda et al. 2004b). The synergistic effect between β and $\alpha_2\delta$ subunits may be due to an augmentation of channel open probability, which is accompanied by an increase in long openings of 2–9 ms duration (Shistik et al. 1995). In contrast, antagonism between β and $\alpha_2\delta$ subunits has been reported for the channel open probability (Qin et al. 1998; Wakamori et al. 1999).

Despite various regulatory effects on HVA calcium channel currents, the roles of β and $\alpha_2\delta$ subunits on LVA calcium channel currents are ill-defined (see reviews by Perez-Reyes 2003, 2006). At least, it is unlikely that β subunits control LVA channel expression or gating as there is no conserved high affinity interaction domain for a β subunit in LVA calcium channels although an unknown interaction domain(s) may exist. On the other hand, data from *in vitro* recombinant expression systems suggest that $\alpha_2\delta$, and also γ, subunits can modify the expression level and channel gating of LVA channels. The physiological and pathological significance of auxiliary subunit-induced LVA channel modulation remains to be elucidated.

13.3   Physiological Roles of Calcium Channels in Neuronal Function

In addition to roles in muscle and endocrine cells, VGCCs control various neuronal events in the central and peripheral nervous systems, including sensory pathways. Specific isoforms of VGCCs are believed to play pivotal roles in each neuronal event for different neurons. For example, in immature neurons, L- and N-type calcium channels appear to function dominantly in neuronal physiology: L-type for gene expression (Morgan and Curran 1986; Sheng and Greenberg 1990; Murphy et al. 1991; Brosenitsch et al. 1998), N-type for neuronal migration and synapse formation (Komuro and Rakic 1992; Basarsky et al. 1994), and N- and L-type calcium channels for neurite outgrowth (Kater and Mills 1991; Doherty et al. 1993; Moorman and Hume 1993; Manivannan and Terakawa 1994). In contrast, all VGCC isoforms are involved in neurotransmission of mature neurons: T-type for neuronal firing (D. Kim et al. 2001), and N-, P/Q-, R- and L-type calcium channels mainly for presynaptic neurotransmitter release (see review by Meir et al. 1999). Consistent with these diverse roles of different VGCC isoforms between immature and mature neurons, the dynamic change in expression of each VGCC isoform during neuronal development has been reported (Tanaka et al. 1995; Jones et al. 1997; Vance et al. 1998). By means of *in situ* hybridisation, temporal and spatial differences in mRNA expression of various isoforms of VGCC α_1 ($Ca_v1.2$, 1.3, 2.1 and 2.2) and β ($\beta1–4$) subunits have been shown in developing and mature brains, suggesting that not only an α_1 isoform, but also a combination pattern of $\alpha_1–\beta$ subunits is important for each neuronal event, e.g. maturation and neurotransmitter release, of different neurons (Tanaka et al. 1995; see review by McEnery et al. 1998). In agreement with this observation, individually regulated expression of $Ca_v2.2$ and various β and $\alpha_2\delta$ subunit proteins has been found during brain ontogeny (Jones et al. 1997; Vance et al. 1998).

Among VGCCs expressed in native neurons, L-type (Ca_v1) channels appear to be less important for neurotransmitter release at nerve terminals (e.g. Takahashi and Momiyama 1993; but see Bonci et al. 1998). It has been shown that these channels exist on the nerve cell soma and dendrites, as well as on parts of the axonal terminals (Hell et al. 1993; Westenbroek et al. 1998), which is consistent with the effects of these channels on gene expression (Morgan and Curran 1986; Sheng and Greenberg 1990; Murphy et al. 1991; Brosenitsch et al. 1998). In addition, L-type channels may have a more important role in controlling release/secretion from soma or dendrites. For example, it has been shown that L-type channels are involved in dynorphin release from granule cell dendrites in the hippocampus (Simmons et al. 1995). In neurons, T-type (Ca_v3) channels are also preferentially expressed in soma and dendrites, and play an important role in the generation of low-threshold Ca^{2+} spikes that are crowned by burst firing, and thereby control synaptic integration (see review by Perez-Reyes 2003). In contrast to L-type and T-type channels, it has been well validated using specific peptidic antagonists that non-L-type (Ca_v2) channels are involved primarily in neurotransmitter release from synaptic terminals of central and peripheral neurons (see reviews by Wu and Saggau

1997; Meir et al. 1999; Waterman 2000; Fisher and Bourque 2001). In particular, N-type (Ca$_v$2.2) calcium channels and the structurally similar P/Q-type (Ca$_v$2.1) channels are known to be major isoforms distributed predominantly in nerve terminals (Westenbroek et al. 1992, 1995, 1998; Wu et al. 1999) and responsible for presynaptic neurotransmitter release (Hirning et al. 1988; Lipscombe et al. 1989; Takahashi and Momiyama, 1993; Wu et al. 1999). Generally, the contribution of R-type (Ca$_v$2.3) channels appears to be relatively minor compared to N- or P/Q-type calcium channels.

Accumulated reports based on gene mutations of VGCC α_1 and auxiliary subunits reveal further specialised roles of VGCCs in neuronal physiology. Mutations in the *CACNA1F* gene encoding Ca$_v$1.4 subunit, which is distributed in the cell bodies and synaptic terminals of photoreceptors in the retina (Morgans et al. 2001), have been shown to be involved in incomplete congenital stationary night blindness in humans (Boycott et al. 2001). In contrast to the Ca$_v$1 channel family, mice deficient in the Ca$_v$2 channel family (P/Q-, N- or R-type calcium channels) exhibit more systemic phenotypes related to central or peripheral nerve defects. Ca$_v$2.1-null mice develop a rapidly progressive ataxia and dystonia before dying ~3–4 weeks after birth, and without a significant decrease in synaptic transmission, which is compensated by N-type and R-type calcium channels in hippocampal slices (Jun et al. 1999). Deletion of the Ca$_v$2.2 gene does not affect lifespan and apparent behaviour, but results in hypertension and lack of the baroreflex due to sympathetic nerve dysfunction (Ino et al. 2001; Mori et al. 2002). It is of particular interest that Ca$_v$2.2-knockout mice have been shown to be resistant to chronic pain (Hatakeyama et al. 2001; C. Kim et al. 2001; Saegusa et al. 2001). Ca$_v$2.3-null mice behave normally except for reduced spontaneous locomotor activities (Saegusa et al. 2000). Similarly, mice deficient in one of the LVA calcium channels, Ca$_v$3.1, show a normal lifespan without significant developmental abnormalities, whereas burst firing of thalamocortical relay neurons is abolished (D. Kim et al. 2001). Ca$_v$3.1-knockout mice exhibit hyperalgesia to visceral pain by suppressing a negative regulatory pathway of ventroposterolateral thalamocortical neurons, which prevents recurring sensory signal input (Kim et al. 2003).

13.4 N-Type Calcium Channel Diversity

It is well known that there is diversity in biophysical and pharmacological properties of native N-type channels. Specific N-type calcium channel antagonists, particularly ω-conotoxins, have been important tools not only in the elucidation of the role of N-type channels but also for determining N-type channel diversity. The diversity of the N-type channel is likely to arise through a combination of several different mechanisms. First, coexpressed auxiliary subunits modulate not only current amplitude, but also channel gating and steady-state inactivation properties. Second, functional diversity can be explained in part by modulation of α_1 subunits by cytosolic proteins, such as G-proteins (see review by Dolphin 1998). More recently, a wide variety of splice variants of calcium channel α_1 subunits has been identified and shown to exhibit functionally distinct channel properties (see comprehensive review by Lipscombe et al. 2002).

13.4.1 N-Type Calcium Channel Splice Variants

The mammalian $Ca_v2.2$ (N-type calcium channel α_1) subunit has been cloned from various species including human, rat, rabbit, mouse and chick. In parallel, as shown in Fig. 13.3b, splice variants of N-type calcium channels have been identified in loop I–II (exon 10), loop II–III (exons 18a, 19, 20, and 21), the IIIS3–IIIS4 linker (exon 24a), the IVS3–IVS4 linker (exon 31a), the C-terminus (exons 37a/b and 46a) and the 3′ untranslated region (Dubel et al. 1992; Williams et al. 1992; Coppola et al. 1994; Stea et al. 1995; Lin et al. 1997; Ghasemzadeh et al. 1999; Lu and Dunlap, 1999; Schorge et al. 1999; Kaneko et al. 2002; Maximov and Bezprozvanny 2002; Bell et al. 2004).

It should be noted here that nomenclature of splice variants is based on the systemic exon-oriented naming proposed by Lipscombe et al. (2002). For example, if the only difference is in splicing at exon 24a of N-type channels, a pair of variants can be described as $Ca_v2.2e[\Delta24a]$ and $Ca_v2.2e[24a]$ without indicating other splicing sites.

The synprint site in the cytoplasmic II–III loop plays an important role in neurotransmitter release by interacting with SNARE proteins (Fig. 13.3a). Interesting examples of N-type calcium channel splice variants are the skipping/inclusion of exon(s) encoding 21, 22 and 382/263 amino acid residues in the II–III loop region that have been reported for rat, mouse and human $Ca_v2.2$, respectively (Coppola et al. 1994; Ghasemzadeh et al. 1999; Kaneko et al. 2002). Given that the 21 and 22 amino acid residues derived from exon 18a in rat and mouse $Ca_v2.2\alpha_1$ are located in the synprint site, and that the human large deletion form lacks more than one-half of the synprint site, these splicing variants may critically affect neurotransmitter release. In a comparison between exon 18a splice variants of rat, the half inactivation voltage, $V_{1/2,\ inact}$ values obtained from steady-state inactivation of $Ca_v2.2e[\Delta18a]$ channels are ~10 mV more negative than that of $Ca_v2.2e[18a]$ channels, whereas there was no difference in the I–V relationships (Pan and Lipscombe 2000). This

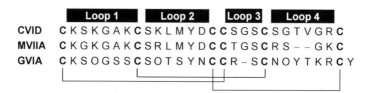

Fig. 13.4 Amino acid sequence alignment of ω-conotoxin CVID (from *Conus catus*), MVIIA (from *Conus magus*), and GVIA (*Conus geographus*). Shown are the positions of the four loops and disulfide connectivity that characterise ω-conotoxins

shift is β subunit isoform-dependent and observed in the presence of either β1b or β4, but not the β2a or β3 subunit (Pan and Lipscombe 2000). In contrast, deletion of 382 amino acids ($Ca_v2.2e[\Delta18a/\Delta19/\Delta20/\Delta21]$) or 263 amino acids (splicing mechanism is not clear) of human $Ca_v2.2\alpha_1$ causes a 25- or 18-mV positive shift in $V_{1/2, inact}$ values (Kaneko et al. 2002). Both deletion forms have unaltered activation kinetics and voltage dependence of channel activation, although the shorter deletion form accelerates inactivation kinetics (Kaneko et al. 2002).

13.4.2 N-Type Calcium Channel Sensitivity to ω-Conotoxins

A pharmacologically distinguishing feature of N-type calcium channels is their sensitivity to block by ω-conotoxins, relatively small (25–27 residues) polypeptides isolated from the venom of the marine snail genus, *Conus* (see reviews by Olivera et al. 1994; Nielsen et al. 2000; Terlau and Olivera 2004; Schroeder et al. 2005). The ω-conotoxins GVIA, MVIIA and CVID, which are isolated from *Conus geographus*, *Conus magus* and *Conus catus*, respectively, have been used extensively as research tools to help define the distribution and physiological roles of specific calcium channels (Adams et al. 1993; Dunlap et al. 1994; Lewis et al. 2000). Structurally, ω-conotoxins are characterised by their high content of basic amino acid residues and a common cysteine scaffold that stabilises the four-loop framework (Fig. 13.4). It has been shown that the positive charges, especially Lys[2], and loop 2, in particular the hydroxyl group on residue Tyr[13], are important for binding to N-type calcium channels (Kim et al. 1994, 1995; Nadasdi et al. 1995; Lew et al. 1997). Futhermore, position 10 (Hyp, Arg and Lys in GVIA, MVIIA and CVID, respectively) was found to be a critical determinant of toxin reversibility (Mould et al. 2004). On the other hand, the interaction site(s) of N-type calcium channels for ω-conotoxins and their mode of action are yet to be fully elucidated. A critical channel motif for binding was found in the extracellular linker between S5 and the P-region in domain III that forms a part of the vestibule of the N-type channel pore (Fig. 13.3a) (Ellinor et al. 1994). This finding suggests a pore-blocking model for ω-conotoxins. Subsequently, residue Gly[1326] in the linker was shown to be a major determinant of GVIA and MVIIA binding, as mutation of this residue to Pro exhibited fully reversible block by these ω-conotoxins (Feng et al. 2001). Thus, single residues on the toxin molecule or the VGCC can have a significant impact on ω-conotoxin dissociation.

There is considerable heterogeneity of N-type calcium channels with regard to toxin sensitivity. GVIA- and MVIIA-resistant but CVID-sensitive transmitter release from preganglionic nerve terminals has been demonstrated in rat parasympathetic ganglia (Adams et al. 2003). GVIA-resistant N-type channel currents have also been reported in frog sympathetic neurons (Elmslie 1997). In PC12 cells, there are two types of N-type channel currents with regard to reversibility of GVIA-induced channel block: reversible and irreversible (Plummer et al. 1989). These N-type calcium channel diversities may be attributed to tissue-specific splice variants or auxiliary subunit compositions of the channels expressed. One of the human N-type calcium channel splice variants, $Ca_v2.2e[\Delta18a/\Delta19/\Delta20/\Delta21]$, exhibits a significantly lower (~15 times) sensitivity to GVIA and MVIIA although the splicing locus is in the cytoplasmic II–III loop region (Fig. 13.3b) (Kaneko et al. 2002). Furthermore, it has been shown that ω-conotoxin binding affinity to N-type channels is modulated by an $\alpha_2\delta$ auxiliary subunit (Brust et al. 1993; Mould et al. 2004; Motin et al. 2007).

Taken together, it is important to consider that there may be a specific variant/auxiliary subunit composition of N-type channels in central and peripheral neurons under particular physiological and pathophysiological conditions.

13.5   N-Type Calcium Channels in Nociception and Neuropathic Pain

Pain is an unpleasant sensory response to tissue damage, and therefore is of primary importance as a warning signal for body protection. Among various sensory neurons, pain sensation is transmitted through small diameter unmyelinated C and myelinated Aδ neurons, so-called nociceptors, whose cell bodies are located in the DRG. The sensory DRG neurons project into superficial laminae of the dorsal horn of the spinal cord and make synapses on secondary sensory neurons, which in turn relay nociceptive signals toward the thalamus. N-type calcium channel currents were first characterised in chick DRG neurons (Nowycky et al. 1985) and the localisation of N-type calcium channels was subsequently observed in sensory nerve terminals in superficial laminae of the spinal dorsal horn using autoradiography (Kerr et al. 1988; Gohil et al. 1994) and immunohistochemistry (Westenbroek et al. 1998; Cizkova et al. 2002; Murakami et al. 2004). Nerve terminals expressing N-type channels were confirmed to be nociceptor terminals as they contain the nociceptive neuropeptide, substance P (Westenbroek et al. 1998). N-type calcium channels are also expressed in the cell soma of DRG neurons (Murakami et al. 2001) and secondary sensory neurons (Westenbroek et al. 1998). Both pre- and post-synaptic N-type calcium channels have been shown to contribute to monosynaptically evoked postsynaptic currents of spinal lamina I neurons (Heinke et al. 2004).

13.5.1   Electrophysiology and a Role for N-Type Calcium Channels in Sensory Neurons

The single channel conductance of native N-type channels of chick DRG neurons is 13 pS (Nowycky et al. 1985; Fox et al. 1987b; Aosaki and Kasai 1989), which is larger than the 7–8 pS reported for T-type, but smaller than the 23–28 pS for L-type calcium channels.

In combination with their pharmacological classification, gating kinetics, especially inactivation kinetics, have been extensively used as an important biophysical property to distinguish between multiple types of VGCCs. Interestingly however, there is a large variation in the inactivation kinetics of native N-type calcium channels in DRG neurons. Fast and almost complete inactivation (>80%) of macroscopic currents within 100 ms for N-type channels was observed in chick DRG neurons, with an inactivation time constant (τ_{inact}) of ~50 ms (Nowycky et al. 1985; Fox et al. 1987a). This was confirmed at the single-channel level (Nowycky et al. 1985; Fox et al. 1987b). In contrast, very slow inactivation over the course of a 150–200 ms test pulse has been reported for N-type channels in chick DRG neurons with a τ_{inact} of ~300 ms in the presence of intracellular 20 mM EGTA (Kasai and Aosaki 1988). The Ca^{2+}–calmodulin binding-dependent channel inactivation has been reported in N-type calcium channels [Ca^{2+} inactivation (CI) region; Fig. 13.3a] as well as in other HVA calcium channels and found to be highly sensitive to Ca^{2+} buffering (Liang et al. 2003). Furthermore, combined inactivation kinetics of the fast and slow components, were observed in chick DRG neurons with τ_{inact} of ~100 ms and > 2 s, respectively with a 1:2 ratio (Cox and Dunlap 1994). Similar two-component inactivation was also reported in rat DRG neurons (Regan et al. 1991). The combined inactivation kinetics of macroscopic currents suggest that at least two kinetically distinct subunit combinations/variants of the N-type calcium channels may exist.

Another important property of N-type calcium channels is the holding potential (HP)-dependent channel inactivation. This was shown by Nowycky et al. (1985) in chick sensory neurons, when they first identified N-type calcium channels. N-type, but not L-type, channel currents were completely inhibited at a HP of −20 mV (Nowycky et al. 1985). Recently, HP-dependent current inhibition was found to occur even in a channel in the closed state with "ultra-slow" kinetics and is regulated largely by the auxiliary β3 subunit, using recombinant channels expressed in *Xenopus* oocytes (Yasuda et al. 2004a). In addition, the inactivation kinetics of whole-cell calcium channel currents are decelerated and often exhibit a non-inactivating current when cells are held at more depolarised HPs in DRG neurons (Fox et al. 1987a, 1987b; Regan et al. 1991; Cox and Dunlap 1994). The non-inactivating current is not due to the residual L-type channels as this observation was confirmed with isolated N-type channel currents from mixed whole-cell currents (Regan et al. 1991).

The population of the N-type channel component in whole-cell calcium channel currents in DRG neurons has been evaluated using ω-conotoxins – specific N-type channel antagonists. The inhibitory effect (% inhibition) of the antagonists, such as GVIA, MVIIC and CVID, on the whole-cell current of DRG neurons varies, ranging from 30 to 70% (Regan et al. 1991; Scroggs and Fox, 1991; Mintz et al. 1992; Piser et al. 1994; Evans et al. 1996; Scott et al. 1997; Hatakeyama et al. 2001; C. Kim et al. 2001; Saegusa et al. 2001; Murakami et al. 2004; Motin et al. 2007). This variation of the ω-conotoxin-sensitive current component may be ascribed, at least in part, to differences in the HPs used as N-type channels are susceptible to HP-dependent inactivation. For instance, a greater inhibition of 60–70% was observed with HPs more negative than −80 mV (Piser et al. 1994; Evans et al. 1996; Scott et al. 1997; Motin et al. 2007), and a reduced inhibition of 30–55% was obtained with HPs more positive than −70 mV (Regan et al. 1991; Scroggs and Fox 1991; Saegusa et al. 2001; Murakami et al. 2004), but there are also exceptions. Additional factors, e.g. animal age, preparation (acute vs culture), and/or culture condition, can modulate the calcium channel population in DRG neurons. Since isolated DRG neurons are comprised mainly of cell soma, the contribution of each VGCC to whole-cell currents obtained does not necessary correlate with contribution to neurotransmitter release at the nerve endings. Nociceptors release neuropeptides, such as substance P, calcitonin-gene-related peptide (CGRP), somatostatin, and excitatory amino acids, such as glutamate, from afferent nerve terminals located in the spinal dorsal horn. It has been demonstrated that N-type calcium channel current blockade leads to significant reduction (50–90%) of the evoked release of nociceptive neuropeptides (substance P and CGRP) from primary afferent nerve terminals (Holz et al. 1988; Maggi et al. 1990; Santicioli et al. 1992; Evans et al. 1996; Harding et al. 1999; Smith et al. 2002). In a co-culture system of DRG and spinal cord neurons, DRG action potentials in response to depolarising current injection evoke excitatory postsynaptic potentials (EPSPs) in spinal cord neurons (Gruner and Silva 1994). The EPSPs are completely inhibited by an N-methyl-D-aspartate (NMDA) receptor antagonist and also by an N-type calcium channel inhibitor, therefore suggesting that the N-type calcium channel is critical for neurotransmission of glutamatergic sensory neurons (Gruner and Silva 1994).

13.5.2 N-Type Calcium Channel Splice Variants in Sensory Neurons

Some of the $Ca_v2.2$ splice variants have been shown to be expressed preferentially/specifically in DRG neurons. For example, $Ca_v2.2e[\Delta24a/31a]$ and $Ca_v2.2e[24a/\Delta31a]$ subunits are expressed predominantly in peripheral (superior cervical ganglion and DRG neurons) and central nerves, respectively (Lin et al. 1997, 1999). $Ca_v2.2e[37a]$ is expressed only in DRG neurons, particularly in capsaicin-responsive neurons (Bell et al. 2004). In chick DRG neurons, a unique 5-bp deletion in the C-terminus leads to a frame shift and a premature stop codon, thereby giving rise to a truncated (>100 amino acid residues) form of $Ca_v2.2$, although a similar truncation has not been identified in mammalian DRG neurons (Lu and Dunlap 1999). It has been shown that $Ca_v2.2$ splice variants expressed in oocytes or human embryonic kidney (HEK) cells exhibit significant differences in channel biophysical and pharmacological properties between each pair of alternative splicing forms. Splicing in the IVS3–S4 linker (exon 31a) affects channel activation but not inactivation properties (Lin et al. 1997, 1999). Inclusion of exon 31a, encoding the two amino acid residues Glu–Thr, causes a 1.5- to 2-fold deceleration of the activation kinetics and a positive shift of approximately 6 mV in the half activation voltage, $V_{1/2,act}$ (Lin et al. 1999). In contrast, the effect of splice variants in the IIIS3–S4 linker (exon 24a) on channel properties is apparently minor (Lin et al. 1999). Toxin sensitivity of N-type channels to ω-conotoxin MVIIC, but not MVIIA, was reduced by half by insertion of four amino acid residues (SFMG) in this site (Meadows and Benham 1999). The recent finding of DRG-specific expression of the isoform of $Ca_v2.2[37a]$, and its strong correlation with nociceptive markers of the capsaicin receptor VR-1 and the TTX-resistant sodium channel $Na_v1.8$, has important implications for pain (Bell et al. 2004). Alternative expression of $Ca_v2.2[37a]$ was identified in 55% of capsaicin-sensitive DRG neurons, but in only 17% of nonsensitive neurons. Moreover, all capsaicin-sensitive neurons that express $Ca_v2.2[37a]$ were found to co-express $Na_v1.8$. Capsaicin-sensitive DRG neurons expressing $Ca_v2.2[37a]$ exhibit significantly larger (~60%) currents than those expressing only $Ca_v2.2[37b]$, without exhibiting a change in activation or inactivation kinetics (Bell et al. 2004). This enhanced macroscopic current with $Ca_v2.2[37a]$ expression is achieved by increased expression of functional N-type channels and prolonged channel open time (Castiglioni et al. 2006). From a therapeutic view point, a unique splice variant may provide a significant target for specific/selective antagonists with minimal side effects.

13.5.3 Pathophysiological Role of N-Type Calcium Channels in Pain – Therapeutic Target for Neuropathic Pain

The pathophysiological role of N-type calcium channels in pain has been demonstrated using peptide toxin antagonists and gene knockout mice (also see recent reviews by Altier and Zamponi 2004; Snutch 2005; McGivern 2006). It has been shown that nociceptive responses are attenuated in mice lacking $Ca_v2.2$. Homozygous mutant ($Ca_v2.2^{-/-}$) mice exhibit no apparent behavioural or morphological abnormalities, survive to adulthood and produce offspring, although one report also showed some lethality (30%) after birth (Ino et al. 2001; Saegusa et al. 2001). In $Ca_v2.2^{-/-}$ mice, N-type VGCC currents were almost completely abolished in DRG neurons as expected, and thereby whole-cell currents were reduced without significant compensation by other HVA calcium channels (Hatakeyama et al. 2001; C. Kim et al. 2001; but see Saegusa et al. 2001). In addition, reduction of N-type channel currents at primary afferent nerve terminals was demonstrated using spinal synaptosomes dissected from $Ca_v2.2^{-/-}$ mice (Hatakeyama et al. 2001). Effects of abolished $Ca_v2.2$ expression have been studied in various acute nociception/pain models, which evaluate spinal reflex response or supraspinal pathway-involved response to noxious mechanical or thermal stimuli. Essentially, it has been suggested that there is no significant difference in acute nociception between wild-type ($Ca_v2.2^{+/+}$) and $Ca_v2.2^{-/-}$ mice, although some results remain controversial (Hatakeyama et al. 2001; C. Kim et al. 2001; Saegusa et al. 2001). In contrast, a clear inhibitory effect of $Ca_v2.2$ deletion has been reported for inflammatory pain (Hatakeyama et al. 2001; C. Kim et al. 2001; Saegusa et al. 2001). Similarly, β3 subunit-deficient (β3$^{-/-}$) mice, where L- and N-type channel currents are diminished in DRG neurons (Namkung et al. 1998), also exhibit an anti-nociceptive phenotype for inflammatory pain (Murakami et al. 2002). A most striking finding is that $Ca_v2.2^{-/-}$ mice exhibit marked reduction of symptoms of mechanical allodynia and thermal hyperalgesia induced by spinal nerve ligation as a neuropathic pain model (Saegusa et al. 2001). Given that allodynia and hyperalgesia are critical clinical symptoms of patients with neuropathic pain, this result could facilitate the development of clinical drugs modulating N-type calcium channel activity.

Consistent with the results obtained in $Ca_v2.2^{-/-}$ mice, sub-nanomolar bolus or continuous intrathecal (spinal) doses of the ω-conotoxins GVIA, MVIIA and CVID have been shown to preferentially block allodynia or hyperalgesia in neuropathic pain models and nociception in inflammatory pain models, but to exhibit controversial anti-nociception in acute pain models in rats (Chaplan et al. 1994; Malmberg and Yaksh 1995; Bowersox et al. 1996; Diaz and Dickenson 1997; Wang et al. 2000a, 2000b; Scott et al. 2002; Smith et al. 2002). It was also confirmed that CVID blocks evoked substance P release in the rat spinal cord (Smith et al. 2002). Inhibition of neurally evoked dorsal horn neuronal activities by GVIA was greater in rats with neuropathic pain than in control rats, suggesting an increased role for N-type calcium channels in neuropathy (Matthews and Dickenson 2001). In support of this finding, an upregulation of $Ca_v2.2$ subunits in dorsal horn lamina II, where nerve terminals of C-fibers are located, was observed by immunohistochemistry after chronic sciatic nerve injury (Cizkova et al. 2002), whereas no significant change in the expression levels of mRNA and protein of $Ca_v2.2$ in DRG and the spinal cord has been reported (Luo et al. 2001). The enhanced immunoreactivity of $Ca_v2.2$ in lamina II may reflect synaptic rearrangement caused by sciatic nerve injury (see review by Bridges et al. 2001). Several ω-conotoxins and small molecule N-type channel antagonists are now in clinical trials. ω-Conotoxin MVIIA (SNX-111/Ziconotide/Prialt; Elan, San Francisco, CA) has been used in clinical trials for pain treatment and was approved for the treatment of intractable pain in 2004 despite concerns regarding dose-limiting side effects (Atanassoff et al. 2000; Jain 2000). Another ω-conotoxin, CVID (AM366; CNSBio, Clayton, Australia) appears to have a wider therapeutic window compared to MVIIA (Scott et al. 2002; Smith et al. 2002), and a Phase II clinical trial has been completed. In addition, a small molecule N-type channel blocker, MK-6721 (NMED-160; Merck/Neuromed, Darmstadt, Germany), is in a Phase II clinical trial in 2006.

Gabapentin (Neurotin; Pfizer, Tadworth, UK) and pregabalin (Lyrica; Pfizer) are unique anti-nociceptive drugs used clinically for the treatment of postherpetic neuralgia, diabetic neuropathy, fibromyalgia and various types of neuropathic pain (see reviews by Taylor et al. 1998; Cheng and Chiou 2006). Compared with gabapentin, pregabalin exhibits similar pharmacological characteristics but is more potent and has improved bioavailability. Gabapentin and pregabalin have been proven to ameliorate allodynia and hyperalgesia in various neuropathy models and inflammatory pain, but not acute pain (Field et al. 1997a, 1997b, 2000; Hunter et al. 1997; Christensen et al. 2001; Feng et al. 2003). The analgesic effect of gabapentin is likely caused by inhibition of release of the nociceptive neurotransmitters, such as glutamate, substance P and CGRP, in the spinal cord (Patel et al. 2000; Fehrenbacher et al. 2003; Feng et al. 2003; Bayer et al. 2004). However, the suppression of neurotransmitter release by gabapentin was observed only under neuropathic and inflammatory pain conditions (Patel et al. 2000; Fehrenbacher et al. 2003; Feng et al. 2003). Although gabapentin and pregabalin are gamma -aminobutyric acid (GABA) analogues, in many cases they do not bind to GABA receptors (for more details of gabapentin action, see Taylor et al. 1998; Cheng and Chiou 2006). According to the accumulated data, it is most likely that gabapentin and pregabalin exhibit their analgesic activity by binding to $\alpha_2\delta$ subunits. Gabapentin was found to bind primarily to the $\alpha_2\delta$-1 subunit (Gee et al. 1996) but also to $\alpha_2\delta$-2 with lower affinity (Marais et al. 2001). It has been shown that a single amino acid residue, Arg[217], in the α_2 component is critical for the high affinity binding, while both the α_2 and δ chains are required (Brown and Gee 1998; Wang et al. 1999). Expression of the $\alpha_2\delta$-1 subunit at both the mRNA and protein level is upregulated remarkably in DRG neurons and to a lesser extent in the spinal cord, according to the development of neuropathy after nerve injury (Luo et al. 2001; Newton et al. 2001; Luo et al. 2002). Critical roles of the $\alpha_2\delta$-1 subunit in neuropathic pain and the analgesic effect of gabapentin have been demonstrated using antisense oligonucleotides and mutant mice, respectively. Antisense oligonucleotides to $\alpha_2\delta$-1, introduced intrathecally, partially diminished both the protein expression of $\alpha_2\delta$-1 subunits and the tactile allodynia caused by peripheral nerve injury (Li et al. 2004). Furthermore, the analgesic effect of gabapentin and pregabalin was abolished in mutant mice containing a single point mutation of Arg[217] to Ala, which is critical for gabapentin binding, within the $\alpha_2\delta$-1 gene (Field et al. 2006). Based on these findings, it is reasonable to speculate that gabapentin modulates VGCC – mainly N-type channel (see Sutton et al. 2002) – currents upon binding to $\alpha_2\delta$-1 subunits. However, the effect of gabapentin and pregabalin on VGCC currents was vague (0–30% inhibition) in earlier studies using isolated neurons and recombinant VGCCs (e.g. Stefani et al. 1998; Kang et al. 2002; Sutton et al. 2002). In cultured DRG neurons, the degree of inhibition was changed by altered auxiliary subunit compositions that are dependent upon culture conditions (Martin et al. 2002). Gabapentin sensitivity has been reported to increase in a neuropathic pain model (Sarantopoulos et al. 2002). This was reinforced by an observation made using transgenic mice that constitutively overexpress the $\alpha_2\delta$-1 subunit in neurons. The transgenic mice reproduced a pathological condition of gabapentin-sensitive tactile allodynia (Li et al. 2006). In the transgenic mice, VGCC currents in DRG neurons are enhanced compared with those of wild type mice and are significantly inhibited (by 40%) by gabapentin, whereas only 5% inhibition was seen for wild type (Li et al. 2006). Taken together, these findings strongly suggest that gabapentin and pregabalin bind to $\alpha_2\delta$-1 subunits, inhibit VGCC (preferentially N-type) currents, suppress neurotransmission at primary afferent nerve endings, and decrease nociception. Importantly, this mechanism is predominant under neuropathic pain conditions.

In future, given the existence of tissue-specific splice variants and auxiliary subunit composition of N-type calcium channels, novel types of N-type calcium channel antagonists or modulators that are specific for certain tissues and diseases should be developed.

13.5.4   Endogenous Modulation of N-Type
Calcium Channel-Mediated Nociception

As described above (Sect. 13.2.2), N-type calcium channels as well as other members of the Ca_v2 channel family are negatively regulated by Gβγ subunits. It is well established that opioids exert their analgesic effect through binding to specific G protein-coupled opioid receptors located pre- and post-synaptically in the spinal cord. Presynaptically, opioids block N-type calcium channel currents via $Ca_v2.2$ and Gβγ subunit interactions, resulting in inhibition of neurotransmitter release and pain relief. The opioid receptor-like (ORL1) receptor was cloned as an opioid receptor analogue but exhibits no binding affinity for opioid ligands (see review by Meunier 1997). Subsequently, nociceptin, also known as orphanin FQ, was identified as a ligand for the ORL1 receptor (Meunier et al. 1995; Reinscheid et al. 1995). The ORL1 receptor is the G protein-coupled receptor and is expressed in the dorsal and ventral horns of the spinal cord and DRG neurons, as well as in various regions of the brain (Bunzow et al. 1994; Wick et al. 1994; Le Cudennec et al. 2002). Similarly, the nociceptin precursor peptide is localised widely in the spinal dorsal and ventral horns and in the brain (Lai et al. 1997; Neal et al. 1999). Nociceptin exhibits both pro- and anti-nociceptive effects, probably depending on the site of application, dose and animal stress conditions (see reviews by Meunier 1997; Calo et al. 2000). It has been shown that intrathecal application of nociceptin inhibits acute nociception and ameliorates inflammatory and neuropathic pain symptoms (Xu et al. 1996; Yamamoto et al. 1997a, 1997b). Consistent with this observation, nociceptin reduces EPSPs mediated by glutamatergic neurotransmission in the spinal dorsal horn (Lai et al. 1997; Liebel et al. 1997; Ahmadi et al. 2001) and inhibits HVA calcium channel currents in afferent sensory neurons, especially small size nociceptors (Borgland et al. 2001). Recent findings have revealed a unique cross-talk between N-type calcium channels and ORL1 receptors (Beedle et al. 2004; Altier et al. 2006). ORL1 receptors can directly interact with $Ca_v2.2$ subunits and inhibit selectively N-type calcium channel currents in DRG neurons and recombinant expression systems. Like other G protein-coupled receptors, there are two distinct mechanisms under this $Ca_v2.2$–ORL1 signaling complex-mediated N-type channel inhibition: voltage-dependent (Gβγ-mediated) and voltage-independent. Notably, however, $Ca_v2.2$–ORL1 can exhibit voltage-dependent inhibition without agonist stimulation when ORL1 receptors are expressed at high densities (Beedle et al. 2004). It has been suggested that ORL1 receptors have low amounts of constitutive receptor activity providing Gβγ to proximal $Ca_v2.2$ interaction sites (GPBP). On the other hand, the ORL1 agonist, nociceptin, can cause both voltage-dependent and -independent N-type channel inhibition. Importantly, prolonged exposure to nociceptin was shown to induce internalisation of the $Ca_v2.2$–ORL1 signaling complex from plasma membrane, and therefore profound inhibition of N-type channel currents (Altier et al. 2006). Together with observations of upregulation of ORL1 receptors in neuropathic and inflammatory pain situations (Jia et al. 1998; Briscini et al. 2002), temporal alterations of ORL1-mediated endogenous analgesic responses make chronic pain mechanisms more complicated. Similar, but much faster, N-type calcium channel internalisation was also reported with $GABA_B$ receptor signaling (Tombler et al. 2006).

13.6   Conclusion

VGCCs are important components of the regulatory pathway for Ca^{2+} ion entry in neurons controlling neurotransmitter release and Ca^{2+}-dependent membrane responses that contribute to the characteristic firing patterns of most neurons. In this chapter, we described the significant role of N-type calcium channels in nociception, particularly pain transmission in chronic pain. The clinical success of the N-type calcium channel inhibitor (ω-conotoxin MVIIA) and $\alpha_2\delta$ subunit ligands (gabapentin and pregabalin) has confirmed the significant role of N-type calcium channels in nociception, as well as validating the effectiveness, without severe side effects, of these types of drugs in pain treatment in humans. In addition, $Ca_v2.2$–ORL1 signaling complex-mediated N-type channel inhibition provides a fresh insight into not only the mechanism of endogenous pain regulation, but also future therapeutic approaches. Recent studies using isoform-specific gene knockdown and knockout mice strongly suggest that T-type calcium channels ($Ca_v3.2$ and $Ca_v3.1$ channels) modulate nociception; pro- and anti-nociceptive effects via peripheral and central mechanisms, respectively (Kim et al. 2003; Bourinet et al. 2005; see reviews by Jevtovic-Todorovic and Todorovic 2006; McGivern 2006). Therefore, T-type calcium channels, $Ca_v3.2$ (peripheral neurons) and $Ca_v3.1$ (central neurons), are also likely to be important targets for analgesic therapeutic agents. Current and future efforts are focused largely on small molecule and isoform (or possibly splice variant)-specific antagonists/modulators of N- and T-type calcium channels that exploit unique aspects of their function under chronic pain conditions to enhance analgesic efficacy and specificity (see Table 13.1).

References

Adams DJ, Smith AB, Schroeder CI, Yasuda T, Lewis RJ (2003) ω-Conotoxin CVID inhibits a pharmacologically distinct voltage-sensitive calcium channel associated with transmitter release from preganglionic nerve terminals. J Biol Chem 278:4057–4062

Adams ME, Myers RA, Imperial JS, Olivera BM (1993) Toxityping rat brain calcium channels with ω-toxins from spider and cone snail venoms. Biochemistry 32:12566–12570

Ahmadi S, Liebel JT, Zeilhofer HU (2001) The role of the ORL1 receptor in the modulation of spinal neurotransmission by nociceptin/orphanin FQ and nocistatin. Eur J Pharmacol 412:39–44

Altier C, Zamponi GW (2004) Targeting Ca^{2+} channels to treat pain: T-type versus N-type. Trends Pharmacol Sci 25:465–470

Altier C, Khosravani H, Evans RM, Hameed S, Peloquin JB, Vartian BA, Chen L, Beedle AM, Ferguson SS, Mezghrani A, Dubel SJ, Bourinet E, McRory JE, Zamponi GW (2006) ORL1 receptor-mediated internalization of N-type calcium channels. Nat Neurosci 9:31–40

Aosaki T, Kasai H (1989) Characterization of two kinds of high-voltage-activated Ca-channel currents in chick sensory neurons. Differential sensitivity to dihydropyridines and omega-conotoxin GVIA. Pfluegers Arch 414:150–156

Atanassoff PG, Hartmannsgruber MW, Thrasher J, Wermeling D, Longton W, Gaeta R, Singh T, Mayo M, McGuire D, Luther RR (2000) Ziconotide, a new N-type calcium channel blocker, administered intrathecally for acute postoperative pain. Reg Anesth Pain Med . 25:274–278

Bangalore R, Mehrke G, Gingrich K, Hofmann F, Kass RS (1996) Influence of L-type Ca channel α_2/δ-subunit on ionic and gating current in transiently transfected HEK 293 cells. Am J Physiol 270:H1521–H1528

Basarsky T, Parpura V, Haydon P (1994) Hippocampal synaptogenesis in cell culture: developmental time course of synapse formation, calcium influx, and synaptic protein distribution. J Neurosci 14:6402–6411

Bayer K, Ahmadi S, Zeilhofer HU (2004) Gabapentin may inhibit synaptic transmission in the mouse spinal cord dorsal horn through a preferential block of P/Q-type Ca^{2+} channels. Neuropharmacology 46:743–749

Bean BP (1989) Classes of calcium channels in vertebrate cells. Annu Rev Physiol 51:367–384

Beedle AM, McRory JE, Poirot O, Doering CJ, Altier C, Barrere C, Hamid J, Nargeot J, Bourinet E, Zamponi GW (2004) Agonist-independent modulation of N-type calcium channels by ORL1 receptors. Nat Neurosci 7:118–125

Bell TJ, Thaler C, Castiglioni AJ, Helton TD, Lipscombe D (2004) Cell-specific alternative splicing increases calcium channel current density in the pain pathway. Neuron 41:127–138

Bernstein GM, Jones OT (2007) Kinetics of internalization and degradation of N-type voltage-gated calcium channels: role of the α_2/δ subunit. Cell Calcium 41:27–40

Bichet D, Cornet V, Geib S, Carlier E, Volsen S, Hoshi T, Mori Y, De Waard M (2000) The I–II loop of the Ca^{2+} channel α_1 subunit contains an endoplasmic reticulum retention signal antagonized by the β subunit. Neuron 25:177–190

Bonci A, Grillner P, Mercuri NB, Bernardi G (1998) L-Type calcium channels mediate a slow excitatory synaptic transmission in rat midbrain dopaminergic neurons. J Neurosci 18:6693–6703

Borgland SL, Connor M, Christie MJ (2001) Nociceptin inhibits calcium channel currents in a subpopulation of small nociceptive trigeminal ganglion neurons in mouse. J Physiol 536:35–47

Bourinet E, Alloui A, Monteil A, Barrere C, Couette B, Poirot O, Pages A, McRory J, Snutch TP, Eschalier A, Nargeot J (2005) Silencing of the $Ca_v3.2$ T-type calcium channel gene in sensory neurons demonstrates its major role in nociception. EMBO J 24:315–324

Bowersox S, Gadbois T, Singh T, Pettus M, Wang Y, Luther R (1996) Selective N-type neuronal voltage-sensitive calcium channel blocker, SNX-111, produces spinal antinociception in rat models of acute, persistent and neuropathic pain. J Pharmacol Exp Ther 279:1243–1249

Boycott KM, Maybaum TA, Naylor MJ, Weleber RG, Robitaille J, Miyake Y, Bergen AA, Pierpont ME, Pearce WG, Bech-Hansen NT (2001) A summary of 20 *CACNA1F* mutations identified in 36 families with incomplete X-linked congenital stationary night blindness, and characterization of splice variants. Hum Genet 108:91–97

Bridges D, Thompson SWN, Rice ASC (2001) Mechanisms of neuropathic pain. Br J Anaesth 87:12–26

Briscini L, Corradini L, Ongini E, Bertorelli R (2002) Up-regulation of ORL-1 receptors in spinal tissue of allodynic rats after sciatic nerve injury. Eur J Pharmacol 447:59–65

Brosenitsch TA, Salgado-Commissariat D, Kunze DL, Katz DM (1998) A role for L-type calcium channels in developmental regulation of transmitter phenotype in primary sensory neurons. J Neurosci 18:1047–1055

Brown JP, Gee NS (1998) Cloning and deletion mutagenesis of the $\alpha_2\delta$ calcium channel subunit from porcine cerebral cortex. Expression of a soluble form of the protein that retains [³H]gabapentine binding activity. J Biol Chem 273:25458–25465

Brust PF, Simerson S, McCue AF, Deal CR, Schoonmaker S, Williams ME, Velicelebi G, Johnson EC, Harpold MM, Ellis SB (1993) Human neuronal voltage-dependent calcium channels: studies on subunit structure and role in channel assembly. Neuropharmacology 32:1089–1102

Bunzow JR, Saez C, Mortrud M, Bouvier C, Williams JT, Low M, Grandy DK (1994) Molecular cloning and tissue distribution of a putative member of the rat opioid receptor gene family that is not a μ, δ or κ opioid receptor type. FEBS Lett 347:284–288

Calo G, Guerrini R, Rizzi A, Salvadori S, Regoli D (2000) Pharmacology of nociceptin and its receptor: a novel therapeutic target. Br J Pharmacol 129:1261–1283

Canti C, Davies A, Berrow NS, Butcher AJ, Page KM, Dolphin AC (2001) Evidence for two concentration-dependent processes for β-subunit effects on α1B calcium channels. Biophys J 81:1439–1451

Castiglioni AJ, Raingo J, Lipscombe D (2006) Alternative splicing in the C-terminus of Ca$_V$2.2 controls expression and gating of N-type calcium channels. J Physiol 576:119–134

Catterall WA (2000) Structure and regulation of voltage-gated Ca^{2+} channels. Annu Rev Cell Dev Biol 16:521–555

Chaplan S, Pogrel J, Yaksh T (1994) Role of voltage-dependent calcium channel subtypes in experimental tactile allodynia. J Pharmacol Exp Ther 269:1117–1123

Cheng JK, Chiou LC (2006) Mechanisms of the antinociceptive action of gabapentin. J Pharmacol Sci 100:471–486

Chien AJ, Zhao X, Shirokov RE, Puri TS, Chang CF, Sun D, Rios E, Hosey MM (1995) Roles of a membrane-localized β subunit in the formation and targeting of functional L-type Ca^{2+} channels. J Biol Chem 270:30036–30044

Christensen D, Gautron M, Guilbaud G, Kayser V (2001) Effect of gabapentin and lamotrigine on mechanical allodynia-like behaviour in a rat model of trigeminal neuropathic pain. Pain 93:147–153

Cizkova D, Marsala J, Lukacova N, Marsala M, Jergova S, Orendacova J, Yaksh TL (2002) Localization of N-type Ca^{2+} channels in the rat spinal cord following chronic constrictive nerve injury. Exp Brain Res 147:456–463

Coppola T, Waldmann R, Borsotto M, Heurteaux C, Romey G, Mattei MG, Lazdunski M (1994) Molecular cloning of a murine N-type calcium channel α1 subunit. Evidence for isoforms, brain distribution, and chromosomal localization. FEBS Lett 338:1–5

Cox DH, Dunlap K (1994) Inactivation of N-type calcium current in chick sensory neurons: calcium and voltage dependence. J Gen Physiol 104:311–336

Curtis BM, Catterall WA (1984) Purification of the calcium antagonist receptor of the voltage-sensitive calcium channel from skeletal muscle transverse tubules. Biochemistry 23:2113–2118

De Jongh KS, Warner C, Catterall WA (1990) Subunits of purified calcium channels. α2 and δ are encoded by the same gene. J Biol Chem 265:14738–14741

De Waard M, Witcher DR, Pragnell M, Liu H, Campbell KP (1995) Properties of the α_1-β anchoring site in voltage-dependent Ca^{2+} channels. J Biol Chem 270:12056–12064

De Waard M, Liu H, Walker D, Scott VE, Gurnett CA, Campbell KP (1997) Direct binding of G-protein $\beta\gamma$ complex to voltage-dependent calcium channels. Nature 385:446–450

De Waard M, Hering J, Weiss N, Feltz A (2005) How do G proteins directly control neuronal Ca^{2+} channel function? Trends Pharmacol Sci 26:427–436

Diaz A, Dickenson AH (1997) Blockade of spinal N- and P-type, but not L-type, calcium channels inhibits the excitability of rat dorsal horn neurones produced by subcutaneous formalin inflammation. Pain 69:93–100

Doherty P, Singh A, Rimon G, Bolsover SR, Walsh FS (1993) Thy-1 antibody–triggered neurite outgrowth requires an influx of calcium into neurons via N- and L-type calcium channels. J Cell Biol 122:181–189

Dolphin AC (1998) Mechanisms of modulation of voltage-dependent calcium channels by G proteins. J Physiol 506:3–11

Dubel SJ, Starr TV, Hell J, Ahlijanian MK, Enyeart JJ, Catterall WA, Snutch TP (1992) Molecular cloning of the α-1 subunit of an ω-conotoxin-sensitive calcium channel. Proc Natl Acad Sci USA 89:5058–5062

Dunlap K, Luebke JI, Turner TJ (1994) Identification of calcium channels that control neurosecretion. Science 266:828–831

Ellinor PT, Zhang JF, Horne WA, Tsien RW (1994) Structural determinants of the blockade of N-type calcium channels by a peptide neurotoxin. Nature 372:272–275

Ellis SB, Williams ME, Ways NR, Brenner R, Sharp AH, Leung AT, Campbell KP, McKenna E, Koch WJ, Hui A, et al (1988) Sequence and expression of mRNAs encoding the α_1 and α_2 subunits of a DHP-sensitive calcium channel. Science 241:1661–1664

Elmslie KS (1997) Identification of the single channels that underlie the N-type and L-type calcium currents in bullfrog sympathetic neurons. J Neurosci 17:2658–2668

Ertel EA, Campbell KP, Harpold MM, Hofmann F, Mori Y, Perez-Reyes E, Schwartz A, Snutch TP, Tanabe T, Birnbaumer L, Tsien RW, Catterall WA (2000) Nomenclature of voltage-gated calcium channels. Neuron 25:533–535

Evans AR, Nicol GD, Vasko MR (1996) Differential regulation of evoked peptide release by voltage-sensitive calcium channels in rat sensory neurons. Brain Res 712:265–273

Fehrenbacher JC, Taylor CP, Vasko MR (2003) Pregabalin and gabapentin reduce release of substance P and CGRP from rat spinal tissues only after inflammation or activation of protein kinase C. Pain 105:133–141

Feng Y, Cui M, Willis WD (2003) Gabapentin markedly reduces acetic acid-induced visceral nociception. Anesthesiology 98:729–733

Feng ZP, Hamid J, Doering C, Bosey GM, Snutch TP, Zamponi GW (2001) Residue Gly[1326] of the N-type calcium channel α_{1B} subunit controls reversibility of ω-conotoxin GVIA and MVIIA block. J Biol Chem 276:15728–15735

Field MJ, Holloman EF, McCleary S, Hughes J, Singh L (1997a) Evaluation of gabapentin and S-(+)-3-isobutylgaba in a rat model of postoperative pain. J Pharmacol Exp Ther 282:1242–1246

Field MJ, Oles RJ, Lewis AS, McCleary S, Hughes J, Singh L (1997b) Gabapentin (neurontin) and S-(+)-3-isobutylgaba represent a novel class of selective antihyperalgesic agents. Br J Pharmacol 121:1513–1522

Field MJ, Hughes J, Singh L (2000) Further evidence for the role of the $\alpha_2\delta$ subunit of voltage dependent calcium channels in models of neuropathic pain. Br J Pharmacol 131:282–286

Field MJ, Cox PJ, Stott E, Melrose H, Offord J, Su T-Z, Bramwell S, Corradini L, England S, Winks J, Kinloch RA, Hendrich J, Dolphin AC, Webb T, Williams D (2006) Identification of the α_2-δ-1 subunit of voltage-dependent calcium channels as a molecular target for pain mediating the analgesic actions of pregabalin. Proc Natl Acad Sci USA 103:17537–17542

Fisher TE, Bourque CW (2001) The function of Ca^{2+} channel subtypes in exocytotic secretion: new perspectives from synaptic and non-synaptic release. Prog Biophys Mol Biol 77:269–303

Fox AP, Nowycky MC, Tsien RW (1987a) Kinetic and pharmacological properties distinguishing three types of calcium currents in chick sensory neurones. J Physiol 394:149–172

Fox AP, Nowycky MC, Tsien RW (1987b) Single-channel recordings of three types of calcium channels in chick sensory neurones. J Physiol 394:173–200

Gao B, Sekido Y, Maximov A, Saad M, Forgacs E, Latif F, Wei MH, Lerman M, Lee JH, Perez-Reyes E, Bezprozvanny I, Minna JD (2000) Functional properties of a new voltage-dependent calcium channel $\alpha_2\delta$ auxiliary subunit gene (CACNA2D2). J Biol Chem 275:12237–12242

Gao T, Chien AJ, Hosey MM (1999) Complexes of the α_{1C} and β subunits generate the necessary signal for membrane targeting of class C L-type calcium channels. J Biol Chem 274:2137–2144

Gee NS, Brown JP, Dissanayake VUK, Offord J, Thurlow R, Woodruff GN (1996) The novel anticonvulsant drug, gabapentin (Neurontin), binds to the $\alpha_2\delta$ subunit of a calcium channel. J Biol Chem 271:5768–5776

Gerster U, Neuhuber B, Groschner K, Striessnig J, Flucher BE (1999) Current modulation and membrane targeting of the calcium channel α_{1C} subunit are independent functions of the β subunit. J Physiol 517:353–368

Ghasemzadeh MB, Pierce RC, Kalivas PW (1999) The monoamine neurons of the rat brain preferentially express a splice variant of α_{1B} subunit of the N-type calcium channel. J Neurochem 73:1718–1723

Gohil K, Bell JR, Ramachandran J, Miljanich GP (1994) Neuroanatomical distribution of receptors for a novel voltage-sensitive calcium-channel antagonist, SNX-230 (ω-conopeptide MVIIC). Brain Res 653:258–266

Gruner W, Silva LR (1994) ω-Conotoxin sensitivity and presynaptic inhibition of glutamatergic sensory neurotransmission in vitro. J Neurosci 14:2800–2808

Gurnett CA, Felix R, Campbell KP (1997) Extracellular interaction of the voltage-dependent Ca^{2+} channel $\alpha_2\delta$ and α_1 subunits. J Biol Chem 272:18508–18512

Harding LM, Beadle DJ, Isabel B (1999) Voltage-dependent calcium channel subtypes controlling somatic substance P release in the peripheral nervous system. Prog Neuropsychopharmacol Biol Psychiatry 23:1103–1112

Hatakeyama S, Wakamori M, Ino M, Miyamoto N, Takahashi E, Yoshinaga T, Sawada K, Imoto K, Tanaka I, Yoshizawa T, Nishizawa Y, Mori Y, Niidome T, Shoji S (2001) Differential nociceptive responses in mice lacking the α_{1B} subunit of N-type Ca^{2+} channels. Neuroreport 12:2423–2427

Heinke B, Balzer E, Sandkuhler J (2004) Pre- and postsynaptic contributions of voltage-dependent Ca^{2+} channels to nociceptive transmission in rat spinal lamina I neurons. Eur J Neurosci 19:103–111

Hell JW, Westenbroek RE, Warner C, Ahlijanian MK, Prystay W, Gilbert MM, Snutch TP, Catterall WA (1993) Identification and differential subcellular localization of the neuronal class C and class D L-type calcium channel $\alpha1$ subunits. J Cell Biol 123:949–962

Herlitze S, Garcia DE, Mackie K, Hille B, Scheuer T, Catterall WA (1996) Modulation of Ca^{2+} channels by G-protein $\beta\gamma$ subunits. Nature 380:258–262

Hille B (2004) Ion channels of excitable membranes, 4th edn. Sinauer, Sunderland, MA

Hirning LD, Fox AP, McCleskey EW, Olivera BM, Thayer SA, Miller RJ, Tsien RW (1988) Dominant role of N-type Ca^{2+} channels in evoked release of norepinephrine from sympathetic neurons. Science 239:57–61

Hofmann F, Lacinova L, Klugbauer N (1999) Voltage-dependent calcium channels: from structure to function. Rev Physiol Biochem Pharmacol 139:33–87

Hohaus A, Poteser M, Romanin C, Klugbauer N, Hofmann F, Morano I, Haase H, Groschner K (2000) Modulation of the smooth-muscle L-type Ca^{2+} channel $\alpha1$ subunit ($\alpha1$C-b) by the $\beta2a$ subunit: a peptide which inhibits binding of β to the I–II linker of $\alpha1$ induces functional uncoupling. Biochem J 348:657–665

Holz G 4th, Dunlap K, Kream R (1988) Characterization of the electrically evoked release of substance P from dorsal root ganglion neurons: methods and dihydropyridine sensitivity. J Neurosci 8:463–471

Hunter JC, Gogas KR, Hedley LR, Jacobson LO, Kassotakis L, Thompson J, Fontana DJ (1997) The effect of novel anti-epileptic drugs in rat experimental models of acute and chronic pain. Eur J Pharmacol 324:153–160

Ikeda SR (1996) Voltage-dependent modulation of N-type calcium channels by G-protein $\beta\gamma$ subunits. Nature 380:255–258

Ino M, Yoshinaga T, Wakamori M, Miyamoto N, Takahashi E, Sonoda J, Kagaya T, Oki T, Nagasu T, Nishizawa Y, Tanaka I, Imoto K, Aizawa S, Koch S, Schwartz A, Niidome T, Sawada K, Mori Y (2001) Functional disorders of the sympathetic nervous system in mice lacking the α_{1B} subunit ($Ca_v2.2$) of N-type calcium channels. Proc Natl Acad Sci USA 98:5323–5328

Jain KK (2000) An evaluation of intrathecal ziconotide for the treatment of chronic pain. Expert Opin Investig Drugs 9:2403–2410

Jarvis SE, Zamponi GW (2001) Interactions between presynaptic Ca^{2+} channels, cytoplasmic messengers and proteins of the synaptic vesicle release complex. Trends Pharmacol Sci 22:519–525

Jay SD, Sharp AH, Kahl SD, Vedvick TS, Harpold MM, Campbell KP (1991) Structural characterization of the dihydropyridine-sensitive calcium channel α_2-subunit and the associated δ peptides. J Biol Chem 266:3287–3293

Jevtovic-Todorovic V, Todorovic SM (2006) The role of peripheral T-type calcium channels in pain transmission. Cell Calcium 40:197–203

Jia Y, Linden DR, Serie JR, Seybold VS (1998) Nociceptin/orphanin FQ binding increases in superficial laminae of the rat spinal cord during persistent peripheral inflammation. Neurosci Lett 250:21–24

Jones LP, Wei SK, Yue DT (1998) Mechanism of auxiliary subunit modulation of neuronal α_{1E} calcium channels. J Gen Physiol 112:125–143

Jones OT, Bernstein GM, Jones EJ, Jugloff DG, Law M, Wong W, Mills LR (1997) N-Type calcium channels in the developing rat hippocampus: subunit, complex, and regional expression. J Neurosci 17:6152–6164

Jun K, Piedras-Renteria ES, Smith SM, Wheeler DB, Lee SB, Lee TG, Chin H, Adams ME, Scheller RH, Tsien RW, Shin H-S (1999) Ablation of P/Q-type Ca^{2+} channel currents, altered synaptic transmission, and progressive ataxia in mice lacking the α_{1A}-subunit. Proc Natl Acad Sci USA 96:15245–15250

Kamp TJ, Perez-Garcia MT, Marban E (1996) Enhancement of ionic current and charge movement by coexpression of calcium channel β_{1A} subunit with α_{1C} subunit in a human embryonic kidney cell line. J Physiol 492:89–96

Kaneko S, Cooper CB, Nishioka N, Yamasaki H, Suzuki A, Jarvis SE, Akaike A, Satoh M, Zamponi GW (2002) Identification and characterization of novel human $Ca_v2.2$ (α_{1B}) calcium channel variants lacking the synaptic protein interaction site. J Neurosci 22:82–92

Kang MG, Felix R, Campbell KP (2002) Long-term regulation of voltage-gated Ca^{2+} channels by gabapentin. FEBS Lett 528:177–182

Kasai H, Aosaki T (1988) Divalent cation dependent inactivation of the high-voltage-activated Ca-channel current in chick sensory neurons. Pfluegers Arch 411:695–697

Kater SB, Mills LR (1991) Regulation of growth cone behavior by calcium. J Neurosci 11:891–899

Kerr LM, Filloux F, Olivera BM, Jackson H, Wamsley JK (1988) Autoradiographic localization of calcium channels with [^{125}I] ω-conotoxin in rat brain. Eur J Pharmacol 146:181–183

Kim C, Jun K, Lee T, Kim SS, McEnery MW, Chin H, Kim HL, Park JM, Kim DK, Jung SJ, Kim J, Shin HS (2001) Altered nociceptive response in mice deficient in the α_{1B} subunit of the voltage-dependent calcium channel. Mol Cell Neurosci 18:235–245

Kim D, Song I, Keum S, Lee T, Jeong MJ, Kim SS, McEnery MW, Shin HS (2001) Lack of the burst firing of thalamocortical relay neurons and resistance to absence seizures in mice lacking α_{1G} T-type Ca^{2+} channels. Neuron 31:35–45

Kim D, Park D, Choi S, Lee S, Sun M, Kim C, Shin HS (2003) Thalamic control of visceral nociception mediated by T-type Ca^{2+} channels. Science 302:117–119

Kim JI, Takahashi M, Ogura A, Kohno T, Kudo Y, Sato K (1994) Hydroxyl group of Tyr^{13} is essential for the activity of ω-conotoxin GVIA, a peptide toxin for N-type calcium channel. J Biol Chem 269:23876–23878

Kim JI, Takahashi M, Ohtake A, Wakamiya A, Sato K (1995) Tyr^{13} is essential for the activity of ω-conotoxin MVIIA and GVIA, specific N-type calcium channel blockers. Biochem Biophys Res Commun 206:449–454

Klugbauer N, Lacinova L, Marais E, Hobom M, Hofmann F (1999) Molecular diversity of the calcium channel $\alpha_2\delta$ subunit. J Neurosci 19:684–691

Komuro H, Rakic P (1992) Selective role of N-type calcium channels in neuronal migration. Science 257:806–809

Lai CC, Wu SY, Dun SL, Dun NJ (1997) Nociceptin-like immunoreactivity in the rat dorsal horn and inhibition of substantia gelatinosa neurons. Neuroscience 81:887–891

Le Cudennec C, Suaudeau C, Costentin J (2002) Evidence for a localization of [^{3}H]nociceptin binding sites on medullar primary afferent fibers. J Neurosci Res 68:496–500

Leung AT, Imagawa T, Campbell KP (1987) Structural characterization of the 1,4-dihydropyridine receptor of the voltage-dependent Ca^{2+} channel from rabbit skeletal muscle. Evidence for two distinct high molecular weight subunits. J Biol Chem 262:7943–7946

Lew MJ, Flinn JP, Pallaghy PK, Murphy R, Whorlow SL, Wright CE, Norton RS, Angus JA (1997) Structure-function relationships of ω-conotoxin GVIA. Synthesis, structure, calcium channel binding, and functional assay of alanine-substituted analogues. J Biol Chem 272:12014–12023

Lewis RJ, Nielsen KJ, Craik DJ, Loughnan ML, Adams DA, Sharpe IA, Luchian T, Adams DJ, Bond T, Thomas L, Jones A, Matheson JL, Drinkwater R, Andrews PR, Alewood PF (2000) Novel ω-conotoxins from Conus catus discriminate among neuronal calcium channel subtypes. J Biol Chem 275:35335–35344

Li C-Y, Song Y-H, Higuera ES, Luo ZD (2004) Spinal dorsal horn calcium channel $\alpha_2\delta$-1 subunit upregulation contributes to peripheral nerve injury-induced tactile allodynia. J Neurosci 24:8494–8499

Li C-Y, Zhang X-L, Matthews EA, Li K-W, Kurwa A, Boroujerdi A, Gross J, Gold MS, Dickenson AH, Feng G, Luo ZD (2006) Calcium channel $\alpha_2\delta_1$ subunit mediates spinal hyperexcitability in pain modulation. Pain 125:20–34

Liang H, DeMaria CD, Erickson MG, Mori MX, Alseikhan BA, Yue DT (2003) Unified mechanisms of Ca^{2+} regulation across the Ca^{2+} channel family. Neuron 39:951–960

Liebel JT, Swandulla D, Zeilhofer HU (1997) Modulation of excitatory synaptic transmission by nociceptin in superficial dorsal horn neurones of the neonatal rat spinal cord. Br J Pharmacol 121:425–432

Lin Z, Haus S, Edgerton J, Lipscombe D (1997) Identification of functionally distinct isoforms of the N-type Ca^{2+} channel in rat sympathetic ganglia and brain. Neuron 18:153–166

Lin Z, Lin Y, Schorge S, Pan JQ, Beierlein M, Lipscombe D (1999) Alternative splicing of a short cassette exon in α_{1B} generates functionally distinct N-type calcium channels in central and peripheral neurons. J Neurosci 19:5322–5331

Lipscombe D, Kongsamut S, Tsien RW (1989) α-Adrenergic inhibition of sympathetic neurotransmitter release mediated by modulation of N-type calcium-channel gating. Nature 340:639–642

Lipscombe D, Pan JQ, Gray AC (2002) Functional diversity in neuronal voltage-gated calcium channels by alternative splicing of $Ca_V\alpha_1$. Mol Neurobiol 26:21–44

Lu Q, Dunlap K (1999) Cloning and functional expression of novel N-type Ca^{2+} channel variants. J Biol Chem 274:34566–34575

Luo ZD, Chaplan SR, Higuera ES, Sorkin LS, Stauderman KA, Williams ME, Yaksh TL (2001) Upregulation of dorsal root ganglion $\alpha_2\delta$ calcium channel subunit and its correlation with allodynia in spinal nerve-injured rats. J Neurosci 21:1868–1875

Luo ZD, Calcutt NA, Higuera ES, Valder CR, Song YH, Svensson CI, Myers RR (2002) Injury type-specific calcium channel $\alpha_2\delta$-1 subunit up-regulation in rat neuropathic pain models correlates with antiallodynic effects of gabapentin. J Pharmacol Exp Ther 303:1199–1205

Maggi CA, Giuliani S, Santicioli IP, Tramontana M, Meli A (1990) Effect of omega conotoxin on reflex responses mediated by activation of capsaicin-sensitive nerves of the rat urinary bladder and peptide release from the rat spinal cord. Neuroscience 34:243–250

Malmberg AB, Yaksh TL (1995) Effect of continuous intrathecal infusion of ω-conopeptides, N-type calcium-channel blockers, on behavior and antinociception in the formalin and hotplate tests in rats. Pain 60:83–90

Manivannan S, Terakawa S (1994) Rapid sprouting of filopodia in nerve terminals of chromaffin cells, PC12 cells, and dorsal root neurons induced by electrical stimulation. J Neurosci 14:5917–5928

Mannuzzu LM, Moronne MM, Isacoff EY (1996) Direct physical measure of conformational rearrangement underlying potassium channel gating. Science 271:213–216

Marais E, Klugbauer N, Hofmann F (2001) Calcium channel $\alpha_2\delta$ subunits – structure and gabapentin binding. Mol Pharmacol 59:1243–1248

Martin DJ, McClelland D, Herd MB, Sutton KG, Hall MD, Lee K, Pinnock RD, Scott RH (2002) Gabapentin-mediated inhibition of voltage-activated Ca^{2+} channel currents in cultured sensory neurones is dependent on culture conditions and channel subunit expression. Neuropharmacology 42:353–366

Matthews EA, Dickenson AH (2001) Effects of spinally delivered N- and P-type voltage-dependent calcium channel antagonists on dorsal horn neuronal responses in a rat model of neuropathy. Pain 92:235–246

Maximov A, Bezprozvanny I (2002) Synaptic targeting of N-type calcium channels in hippocampal neurons. J Neurosci 22:6939–6952

McEnery MW, Vance CL, Begg CM, Lee WL, Choi Y, Dubel SJ (1998) Differential expression and association of calcium channel subunits in development and disease. J Bioenerg Biomembr 30:409–418

McGivern JG (2006) Targeting N-type and T-type calcium channels for the treatment of pain. Drug Discov Today 11:245–253

Meadows HJ, Benham CD (1999) Sensitivity to conotoxin block of splice variants of rat α_{1B} (rbBII) subunit of the N-type calcium channel coexpressed with different β subunits in *Xenopus* oocytes. Ann N Y Acad Sci 868:224–227

Meir A, Ginsburg S, Butkevich A, Kachalsky SG, Kaiserman I, Ahdut R, Demirgoren S, Rahamimoff R (1999) Ion channels in presynaptic nerve terminals and control of transmitter release. Physiol Rev 79:1019–1088

Meunier JC (1997) Nociceptin/orphanin FQ and the opioid receptor-like ORL1 receptor. Eur J Pharmacol 340:1–15

Meunier JC, Mollereau C, Toll L, Suaudeau C, Moisand C, Alvinerie P, Butour JL, Guillemot JC, Ferrara P, Monsarrat B, Mazarguil H, Vassart G, Parmentier M, Costentin J (1995) Isolation and structure of the endogenous agonist of opioid receptor-like ORL1 receptor. Nature 377:532–535

Mintz IM, Adams ME, Bean BP (1992) P-type calcium channels in rat central and peripheral neurons. Neuron 9:85–95

Moorman SJ, Hume RI (1993) ω-Conotoxin prevents myelin-evoked growth cone collapse in neonatal rat locus coeruleus neurons in vitro. J Neurosci 13:4727–4736

Morgan JI, Curran T (1986) Role of ion flux in the control of c-fos expression. Nature 322:552–555

Morgans CW, Gaughwin P, Maleszka R (2001) Expression of the α_{1F} calcium channel subunit by photoreceptors in the rat retina. Mol Vis 7:202–209

Mori Y, Friedrich T, Kim MS, Mikami A, Nakai J, Ruth P, Bosse E, Hofmann F, Flockerzi V, Furuichi T, Mikoshiba K, Imoto K, Tanabe T, Numa S (1991) Primary structure and functional expression from complementary DNA of a brain calcium channel. Nature 350:398–402

Mori Y, Nishida M, Shimizu S, Ishii M, Yoshinaga T, Ino M, Sawada K, Niidome T (2002) Ca^{2+} channel α_{1B} subunit ($Ca_v 2.2$) knockout mouse reveals a predominant role of N-type channels in the sympathetic regulation of the circulatory system. Trends Cardiovasc Med 12:270–275

Motin L, Yasuda T, Schroeder CI, Lewis RJ, Adams DJ (2007) ω-Conotoxin CVIB differentially inhibits native and recombinant N- and P/Q-type calcium channels. Eur J Neurosci 25:435–444

Mould J, Yasuda T, Schroeder CI, Beedle AM, Doering CJ, Zamponi GW, Adams DJ, Lewis RJ (2004) The $\alpha_2\delta$ auxiliary subunit reduces affinity of ω-conotoxins for recombinant N-type ($Ca_v 2.2$) calcium channels. J Biol Chem 279:34705–34714

Murakami M, Suzuki T, Nakagawasai O, Murakami H, Murakami S, Esashi A, Taniguchi R, Yanagisawa T, Tan-No K, Miyoshi I, Sasano H, Tadano T (2001) Distribution of various calcium channel α_1 subunits in murine DRG neurons and antinociceptive effect of ω-conotoxin SVIB in mice. Brain Res 903:231–236

Murakami M, Fleischmann B, De Felipe C, Freichel M, Trost C, Ludwig A, Wissenbach U, Schwegler H, Hofmann F, Hescheler J, Flockerzi V, Cavalie A (2002) Pain perception in mice lacking the β3 subunit of voltage-activated calcium channels. J Biol Chem 277:40342–40351

Murakami M, Nakagawasai O, Suzuki T, Mobarakeh, II, Sakurada Y, Murata A, Yamadera F, Miyoshi I, Yanai K, Tan-No K, Sasano H, Tadano T, Iijima T (2004) Antinociceptive effect of different types of calcium channel inhibitors and the distribution of various calcium channel α_1 subunits in the dorsal horn of spinal cord in mice. Brain Res 1024:122–129

Murphy TH, Worley PF, Baraban JM (1991) L-Type voltage-sensitive calcium channels mediate synaptic activation of immediate early genes. Neuron 7:625–635

Nadasdi L, Yamashiro D, Chung D, Tarczy-Hornoch K, Adriaenssens P, Ramachandran J (1995) Structure–activity analysis of a *Conus* peptide blocker of N-type neuronal calcium channels. Biochemistry 34:8076–8081

Namkung Y, Smith SM, Lee SB, Skrypnyk NV, Kim HL, Chin H, Scheller RH, Tsien RW, Shin HS (1998) Targeted disruption of the Ca^{2+} channel β_3 subunit reduces N- and L-type Ca^{2+} channel activity and alters the voltage-dependent activation of P/Q-type Ca^{2+} channels in neurons. Proc Natl Acad Sci USA 95:12010–12015

Neal CRJ, Mansour A, Reinscheid R, Nothacker H-P, Civelli O, J. WSJ (1999) Localization of orphanin FQ (nociceptin) peptide and messenger RNA in the central nervous system of the rat. J Comp Neurol 406:503–547

Neely A, Wei X, Olcese R, Birnbaumer L, Stefani E (1993) Potentiation by the β subunit of the ratio of the ionic current to the charge movement in the cardiac calcium channel. Science 262:575–578

Newton RA, Bingham S, Case PC, Sanger GJ, Lawson SN (2001) Dorsal root ganglion neurons show increased expression of the calcium channel $\alpha_2\delta$-1 subunit following partial sciatic nerve injury. Brain Res Mol Brain Res 95:1–8

Nielsen KJ, Schroeder T, Lewis R (2000) Structure-activity relationships of ω-conotoxins at N-type voltage-sensitive calcium channels. J Mol Recognit 13:55–70

Noceti F, Baldelli P, Wei X, Qin N, Toro L, Birnbaumer L, Stefani E (1996) Effective gating charges per channel in voltage-dependent K^+ and Ca^{2+} channels. J Gen Physiol 108:143–155

Nowycky MC, Fox AP, Tsien RW (1985) Three types of neuronal calcium channel with different calcium agonist sensitivity. Nature 316:440–443

Olivera BM, Miljanich GP, Ramachandran J, Adams ME (1994) Calcium channel diversity and neurotransmitter release: the ω-conotoxins and ω-agatoxins. Annu Rev Biochem 63:823–867

Page KM, Canti C, Stephens GJ, Berrow NS, Dolphin AC (1998) Identification of the amino terminus of neuronal Ca^{2+} channel α1 subunits α1B and α1E as an essential determinant of G-protein modulation. J Neurosci 18:4815–4824

Pan JQ, Lipscombe D (2000) Alternative splicing in the cytoplasmic II–III loop of the N-type Ca channel α_{1B} subunit: functional differences are β subunit-specific. J Neurosci 20:4769–4775

Parent L, Schneider T, Moore CP, Talwar D (1997) Subunit regulation of the human brain α_{1E} calcium channel. J Membr Biol 160:127–140

Patel MK, Gonzalez MI, Bramwell S, Pinnock RD, Lee K (2000) Gabapentin inhibits excitatory synaptic transmission in the hyperalgesic spinal cord. Br J Pharmacol 130:1731–1734

Perez-Reyes E (2003) Molecular physiology of low-voltage-activated T-type calcium channels. Physiol Rev 83:117–161

Perez-Reyes E (2006) Molecular characterization of T-type calcium channels. Cell Calcium 40:89–96

Piser TM, Lampe RA, Keith RA, Thayer SA (1994) ω-Grammotoxin blocks action-potential-induced Ca^{2+} influx and whole-cell Ca^{2+} current in rat dorsal-root ganglion neurons. Pfluegers Arch 426:214–220

Plummer MR, Logothetis DE, Hess P (1989) Elementary properties and pharmacological sensitivities of calcium channels in mammalian peripheral neurons. Neuron 2:1453–1463

Pragnell M, De Waard M, Mori Y, Tanabe T, Snutch TP, Campbell KP (1994) Calcium channel β-subunit binds to a conserved motif in the I–II cytoplasmic linker of the α_1-subunit. Nature 368:67–70

Qin N, Platano D, Olcese R, Stefani E, Birnbaumer L (1997) Direct interaction of Gβγ with a C-terminal Gβγ-binding domain of the Ca^{2+} channel α_1 subunit is responsible for channel inhibition by G protein-coupled receptors. Proc Natl Acad Sci USA 94:8866–8871

Qin N, Olcese R, Stefani E, Birnbaumer L (1998) Modulation of human neuronal α_{1E}-type calcium channel by $\alpha_2\delta$-subunit. Am J Physiol Cell Physiol 274:C1324–1331

Qin N, Yagel S, Momplaisir ML, Codd EE, D'Andrea MR (2002) Molecular cloning and characterization of the human voltage-gated calcium channel $\alpha_2\delta$-4 subunit. Mol Pharmacol 62:485–496

Regan LJ, Sah DW, Bean BP (1991) Ca^{2+} channels in rat central and peripheral neurons: high-threshold current resistant to dihydropyridine blockers and ω-conotoxin. Neuron 6:269–280

Reinscheid RK, Nothacker HP, Bourson A, Ardati A, Henningsen RA, Bunzow JR, Grandy DK, Langen H, Monsma FJ Jr, Civelli O (1995) Orphanin FQ: a neuropeptide that activates an opioidlike G protein-coupled receptor. Science 270:792–794

Saegusa H, Kurihara T, Zong S, Minowa O, Kazuno A, Han W, Matsuda Y, Yamanaka H, Osanai M, Noda T, Tanabe T (2000) Altered pain responses in mice lacking α_{1E} subunit of the voltage-dependent Ca^{2+} channel. Proc Natl Acad Sci USA 97:6132–6137

Saegusa H, Kurihara T, Zong S, Kazuno A, Matsuda Y, Nonaka T, Han W, Toriyama H, Tanabe T (2001) Suppression of inflammatory and neuropathic pain symptoms in mice lacking the N-type Ca^{2+} channel. EMBO J 20:2349–2356

Santicioli P, Bianco ED, Tramontana M, Geppetti P, Maggi CA (1992) Release of calcitonin gene-related peptide like-immunoreactivity induced by electrical field stimulation from rat spinal afferents is mediated by conotoxin-sensitive calcium channels. Neurosci Lett 136:161–164

Sarantopoulos C, McCallum B, Kwok W-M, Hogan Q (2002) Gabapentin decreases membrane calcium currents in injured as well as in control mammalian primary afferent neurons. Reg Anesth Pain Med 27:47–57

Sather WA, McCleskey EW (2003) Permeation and selectivity in calcium channels. Annu Rev Physiol 65:133–159

Schorge S, Gupta S, Lin Z, McEnery MW, Lipscombe D (1999) Calcium channel activation stabilizes a neuronal calcium channel mRNA. Nat Neurosci 2:785–790

Schroeder CI, Lewis RJ, Adams DJ (2005) Block of voltage-gated calcium channels by peptide toxins. In: Zamponi GW (ed) Voltage-gated calcium channels. www.eurekah.com, pp 294–308

Scott DA, Wright CE, Angus JA (2002) Actions of intrathecal ω-conotoxins CVID, GVIA, MVIIA, and morphine in acute and neuropathic pain in the rat. Eur J Pharmacol 451:279–286

Scott RH, Gorton VJ, Harding L, Patel D, Pacey S, Kellenberger C, Hietter H, Bermudez I (1997) Inhibition of neuronal high voltage-activated calcium channels by insect peptides: a comparison with the actions of ω-conotoxin GVIA. Neuropharmacology 36:195–208

Scott VE, De Waard M, Liu H, Gurnett CA, Venzke DP, Lennon VA, Campbell KP (1996) β subunit heterogeneity in N-type Ca^{2+} channels. J Biol Chem 271:3207–3212

Scroggs R, Fox A (1991) Distribution of dihydropyridine and omega-conotoxin-sensitive calcium currents in acutely isolated rat and frog sensory neuron somata: diameter-dependent L channel expression in frog. J Neurosci 11:1334–1346

Sheng M, Greenberg ME (1990) The regulation and function of c-fos and other immediate early genes in the nervous system. Neuron 4:477–485

Shistik E, Ivanina T, Puri T, Hosey M, Dascal N (1995) Ca^{2+} current enhancement by α2/δ and β subunits in *Xenopus* oocytes: contribution of changes in channel gating and α1 protein level. J Physiol 489:55–62

Simmons ML, Terman GW, Gibbs SM, Chavkin C (1995) L-Type calcium channels mediate dynorphin neuropeptide release from dendrites but not axons of hippocampal granule cells. Neuron 14:1265–1272

Smith MT, Cabot PJ, Ross FB, Robertson AD, Lewis RJ (2002) The novel N-type calcium channel blocker, AM336, produces potent dose-dependent antinociception after intrathecal dosing in rats and inhibits substance P release in rat spinal cord slices. Pain 96:119–127

Snutch TP (2005) Targeting chronic and neuropathic pain: the N-type calcium channel comes of age. NeuroRx 2:662–670

Stea A, Dubel SJ, Pragnell M, Leonard JP, Campbell KP, Snutch TP (1993) A β-subunit normalizes the electrophysiological properties of a cloned N-type Ca^{2+} channel α_1-subunit. Neuropharmacology 32:1103–1116

Stea A, Soong TW, Snutch TP (1995) Determinants of PKC-dependent modulation of a family of neuronal calcium channels. Neuron 15:929–940

Stefani A, Spadoni F, Bernardi G (1998) Gabapentin inhibits calcium currents in isolated rat brain neurons. Neuropharmacology 37:83–91

Stephens GJ, Canti C, Page KM, Dolphin AC (1998) Role of domain I of neuronal Ca^{2+} channel α1 subunits in G protein modulation. J Physiol 509:163–169

Sutton KG, Martin DJ, Pinnock RD, Lee K, Scott RH (2002) Gabapentin inhibits high-threshold calcium channel currents in cultured rat dorsal root ganglion neurones. Br J Pharmacol 135:257–265

Takahashi M, Seagar MJ, Jones JF, Reber BF, Catterall WA (1987) Subunit structure of dihydropyridine-sensitive calcium channels from skeletal muscle. Proc Natl Acad Sci USA 84:5478–5482

Takahashi T, Momiyama A (1993) Different types of calcium channels mediate central synaptic transmission. Nature 366:156–158

Tanaka O, Sakagami H, Kondo H (1995) Localization of mRNAs of voltage-dependent Ca^{2+}-channels: four subtypes of α1- and β-subunits in developing and mature rat brain. Brain Res Mol Brain Res 30:1–16

Tareilus E, Roux M, Qin N, Olcese R, Zhou J, Stefani E, Birnbaumer L (1997) A *Xenopus* oocyte β subunit: evidence for a role in the assembly/expression of voltage-gated calcium channels that is separate from its role as a regulatory subunit. Proc Natl Acad Sci USA 94:1703–1708

Taylor CP, Gee NS, Su T-Z, Kocsis JD, Welty DF, Brown JP, Dooley DJ, Boden P, Singh L (1998) A summary of mechanistic hypotheses of gabapentin pharmacology. Epilepsy Res 29:231–246

Terlau H, Olivera BM (2004) Conus venoms: a rich source of novel ion channel-targeted peptides. Physiol Rev 84:41–68

Tombler E, Cabanilla NJ, Carman P, Permaul N, Hall JJ, Richman RW, Lee J, Rodriguez J, Felsenfeld DP, Hennigan RF, Diverse-Pierluissi MA (2006) G protein-induced trafficking of voltage-dependent calcium channels. J Biol Chem 281:1827–1839

Vance CL, Begg CM, Lee WL, Haase H, Copeland TD, McEnery MW (1998) Differential expression and association of calcium channel α_{1B} and β subunits during rat brain ontogeny. J Biol Chem 273:14495–14502

Wakamori M, Mikala G, Schwartz A, Yatani A (1993) Single-channel analysis of a cloned human heart L-type Ca^{2+} channel α_1 subunit and the effects of a cardiac β subunit. Biochem Biophys Res Commun 196:1170–1176

Wakamori M, Mikala G, Mori Y (1999) Auxiliary subunits operate as a molecular switch in determining gating behaviour of the unitary N-type Ca^{2+} channel current in *Xenopus* oocytes. J Physiol 517:659–672

Walker D, Bichet D, Campbell KP, De Waard M (1998) A β_4 isoform-specific interaction site in the carboxyl-terminal region of the voltage-dependent Ca^{2+} channel α_{1A} subunit. J Biol Chem 273:2361–2367

Walker D, Bichet D, Geib S, Mori E, Cornet V, Snutch TP, Mori Y, De Waard M (1999) A new β subtype-specific interaction in α_{1A} subunit controls P/Q-type Ca^{2+} channel activation. J Biol Chem 274:12383–12390

Wang M, Offord J, Oxender DL, Su TZ (1999) Structural requirement of the calcium-channel subunit $\alpha_2\delta$ for gabapentin binding. Biochem J 342:313–320

Wang Y-X, Pettus M, Gao D, Phillips C, Scott Bowersox S (2000a) Effects of intrathecal administration of ziconotide, a selective neuronal N-type calcium channel blocker, on mechanical allodynia and heat hyperalgesia in a rat model of postoperative pain. Pain 84:151–158

Wang Y-X, Gao D, Pettus M, Phillips C, Bowersox S (2000b) Interactions of intrathecally administered ziconotide, a selective blocker of neuronal N-type voltage-sensitive calcium channels, with morphine on nociception in rats. Pain 84:271–281

Waterman SA (2000) Voltage-gated calcium channels in autonomic neuroeffector transmission. Prog Neurobiol 60:181–210

Westenbroek RE, Hoskins L, Catterall WA (1998) Localization of Ca^{2+} channel subtypes on rat spinal motor neurons, interneurons, and nerve terminals. J Neurosci 18:6319–6330

Westenbroek RE, Sakurai T, Elliott EM, Hell JW, Starr TV, Snutch TP, Catterall WA (1995) Immunochemical identification and subcellular distribution of the α_{1A} subunits of brain calcium channels. J Neurosci 15:6403–6418

Westenbroek RE, Hell JW, Warner C, Dubel SJ, Snutch TP, Catterall WA (1992) Biochemical properties and subcellular distribution of an N-type calcium channel $\alpha 1$ subunit. Neuron 9:1099–1115

Wick MJ, Minnerath SR, Lin X, Elde R, Law PY, Loh HH (1994) Isolation of a novel cDNA encoding a putative membrane receptor with high homology to the cloned μ, δ, and κ opioid receptors. Brain Res Mol Brain Res 27:37–44

Williams ME, Brust PF, Feldman DH, Patthi S, Simerson S, Maroufi A, McCue AF, Velicelebi G, Ellis SB, Harpold MM (1992) Structure and functional expression of an ω-conotoxin-sensitive human N-type calcium channel. Science 257:389–395

Witcher DR, De Waard M, Liu H, Pragnell M, Campbell KP (1995) Association of native Ca^{2+} channel β subunits with the α_1 subunit interaction domain. J Biol Chem 270:18088–18093

Wu LG, Saggau P (1997) Presynaptic inhibition of elicited neurotransmitter release. Trends Neurosci 20:204–212

Wu LG, Westenbroek RE, Borst JG, Catterall WA, Sakmann B (1999) Calcium channel types with distinct presynaptic localization couple differentially to transmitter release in single calyx-type synapses. J Neurosci 19:726–736

Xu XJ, Hao JX, Wiesenfeld-Hallin Z (1996) Nociceptin or antinociceptin: potent spinal antinociceptive effect of orphanin FQ/nociceptin in the rat. Neuroreport 7:2092–2094

Yamaguchi H, Hara M, Strobeck M, Fukasawa K, Schwartz A, Varadi G (1998) Multiple modulation pathways of calcium channel activity by a β subunit. Direct evidence of β subunit participation in membrane trafficking of the α_{1C} subunit. J Biol Chem 273:19348–19356

Yamamoto T, Nozaki-Taguchi N, Kimura S (1997a) Analgesic effect of intrathecally administered nociceptin, an opioid receptor-like1 receptor agonist, in the rat formalin test. Neuroscience 81:249–254

Yamamoto T, Nozaki-Taguchi N, Kimura S (1997b) Effects of intrathecally administered nociceptin, an opioid receptor-like1 (ORL1) receptor agonist, on the thermal hyperalgesia induced by unilateral constriction injury to the sciatic nerve in the rat. Neurosci Lett 224:107–110

Yang N, Horn R (1995) Evidence for voltage-dependent S4 movement in sodium channels. Neuron 15:213–218

Yasuda T, Lewis RJ, Adams DJ (2004a) Overexpressed $Ca_v\beta3$ inhibits N-type ($Ca_v2.2$) calcium channel currents through a hyperpolarizing shift of "ultra-slow" and "closed-state" inactivation. J Gen Physiol 123:401–416

Yasuda T, Chen L, Barr W, McRory JE, Lewis RJ, Adams DJ, Zamponi GW (2004b) Auxiliary subunit regulation of high-voltage activated calcium channels expressed in mammalian cells. Eur J Neurosci 20:1–13

Zamponi GW, Snutch TP (1998) Decay of prepulse facilitation of N type calcium channels during G protein inhibition is consistent with binding of a single $G\beta$ subunit. Proc Natl Acad Sci USA 95:4035–4039

Zamponi GW, Bourinet E, Nelson D, Nargeot J, Snutch TP (1997) Crosstalk between G proteins and protein kinase C mediated by the calcium channel α_1 subunit. Nature 385:442–446

Zhang JF, Ellinor PT, Aldrich RW, Tsien RW (1996) Multiple structural elements in voltage-dependent Ca^{2+} channels support their inhibition by G proteins. Neuron 17:991–1003

Index